PRAISE FOR *LANGUAGE AT THE SPEED OF SIGHT*

"In *Language at the Speed of Sight*, [Seidenberg] develops a careful argument, backed by decades of research, to show that the only responsible way to teach children to read well is to build up their abilities to connect reading with speech and then to amplify these connections through practice, developing skillful behavioral patterns hand in hand with the neurological networks that undergird them. . . . Every teacher of young children as well as those who train them should read this book." **—*Wall Street Journal***

"An important and alarming new book. . . . Seidenberg makes a strong case for how brain science can help the teaching profession" **—*New York Times***

"Seidenberg . . . unravels the science of reading with great flair. He is the ideal guide—and it turns out that we need a guide to reading, even though we've been doing it most of our lives." **—*Washington Post***

"Seidenberg reviews the latest science on reading and makes an impassioned plea for putting this knowledge to use." **—*Scientific American***

"Cognitive neuroscientist Seidenberg digs deep into the science of reading to reveal the ways human beings learn how to read and process language. . . . Seidenberg's analysis is backed up by numerous studies and table[s] of data. His approach is pragmatic, myth-destroying, and rooted in science—and his writing makes for powerful reading." **—*Publishers Weekly***

"The neuroscience underlying [Seidenberg's] findings is complex, of course, but [he] does not often fall into thickets of technicality . . . his discussions are clear and accessible. . . . A worthy primer on the science of comprehending language." **—*Kirkus Reviews***

"No technologically advanced society exists without reading. This is the remarkable story of why and how it all works. From David Letterman's irony to posited Sumerian patent trolls, the writing is lively, informative, and supremely entertaining."

—Daniel J. Levitin, bestselling author of *This Is Your Brain on Music* and *The Organized Mind*

"Have you picked up the idea that reading is something that kids 'just pick up' and shouldn't be rushed into, or that learning to read is something different from 'comprehension,' or that a whole book about reading would be dull? *Language at the Speed of Sight* will disabuse you of all three notions and more—pick it up and marvel at how hard it will be to put it down."

—John McWhorter, author of *Word on the Move* and *Talking Back, Talking Black*

"Few works of science ever achieve Italo Calvino's six qualities of our best writing: Lightness, exactitude, visibility, quickness, multiplicity, and consistency. Mark Seidenberg's new book achieves just that. If every educator, parent, and policy maker would read and heed the content of this book, the rates of functional illiteracy, with all their destructive sequelae, would be significantly reduced."

—Maryanne Wolf, author of *Proust and the Squid*

"A world-renowned expert explains the science of reading with clarity and wit—anyone who loves to read will be fascinated, and teachers will absolutely devour this book."

—Daniel Willingham, author of *Why Don't Students Like School?*

"*Language at the Speed of Sight* is an incisive tour through the fascinating science of reading. From cuneiform to dyslexia to the future of literacy, Seidenberg is a master guide who—lucky for us—is as gifted a writer as he is a scientist."

—Benjamin Bergen, author of *What the F*

LANGUAGE AT THE SPEED OF SIGHT

How We Read, Why So Many Can't, and What Can Be Done About It

MARK SEIDENBERG

BASIC BOOKS
New York

Basic Books
Hachette Book Group
290 Avenue of the Americas, New York, NY 10104
http://www.basicbooks.com/
@BasicBooks

Printed in the United States of America

First Trade Paperback Edition: March 2018

Published by Basic Books, an imprint of Perseus Books, LLC, a subsidiary of Hachette Book Group, Inc.
The Basic Books name and logo is a trademark of the Hachette Book Group.

Print book interior design by Jeff Williams

The Library of Congress has catalogued the original edition as follows:
Names: Seidenberg, Mark S., author.
Title: Language at the speed of sight : how we read, why so many can't, and what can be done about it / Mark Seidenberg.
Description: New York : Basic Books, an imprint of Perseus Books, a division of PBG Publishing, LLC, a subsidiary of Hachette Book Group, Inc., [2017] | Includes bibliographical references and index.
Identifiers: LCCN 2016046731| ISBN 9780465019328 (hardcover) |ISBN 9780465080656 (ebook)
Subjects: LCSH: Reading (Higher education) | Language experience approach in education. | Cognition disorders. | Psycholinguistics.
Classification: LCC LB2395.3 .S44 2016 | DDC 428.4071/1—dc23
LC record available at https://lccn.loc.gov/2016046731

ISBN: 978-0-465-01932-8 (hardcover); ISBN: 978-0-465-08065-6 (e-book);
ISBN: 978-1-5416-1715-5 (paperback)

LSC-C

10 9 8 7 6 5 4 3

For Maryellen
With deepest [love, admiration, and gratitude].

For Claudia and Ethan
Fun. Food. Family.
And for showing me what I didn't know about reading, and the rest.

Look closely at the letters. Can you see,
entering (stage right), then floating full,
then heading off—so soon—
how like a little kohl-rimmed moon
o plots her course from *b* to *d*

—as *y*, unanswered, knocks at the stage door?
Looked at too long, words fail,
phase out. Ask, now that *body* shines
no longer, by what light you learn these lines
and what the *b* and *d* stood for.
b o d y

—JAMES MERRILL

CONTENTS

READING, WRITING, AND SPEECH

CHAPTER 1

The Problem and the Paradox

I DON'T THINK WE'VE MET, but I know two things about you. One is that you are reading these words. And because you're reading these words, I also know that you are an expert at what this book is about. Which is reading.

A friend once gave me a scholarly 380-page book about the history of the pencil. I like pencils. I'm good at using them. I like a fresh pink eraser tip. I just do not need to read about how they got that way. Perhaps reading is for you like the history of pencils is for me. Why read about reading?

Reading is one of the few activities you do every day whether you want to or not. Street signs, menus, e-mails, Facebook posts, novels, ingredients in Chex Mix. You read for work, for school, for pleasure; because you have to, because you want to, because you can't help it. That is a lot of practice over a long period. If it takes thousands of hours to become an expert at something like chess, we readers are in grandmaster territory.

One result is that everyone has an opinion about how they read. When you are introduced at the dinner table as someone who studies reading for a living, you hear all the stories. People tell me whether they read word by word or in big chunks. Whether they read visually or hear the sounds of words in their heads while they read. How they learned to read and how their kids learned. Many confessions from people who say they are slow readers, more often than not from accomplished people in fields that seemingly demand a high level of reading competence—lawyers or school superintendents, say. Most of you think that other people read faster than you. It's the opposite of Lake Wobegon: reading is the land where the folk are all below average.

But here's the rub: people manage to be good at reading without knowing much about how they do it. Most of what goes on in reading is subconscious:

we are aware of the result of having read something—that we understood it, that we found it funny, that it conveyed a fact, idea, or feeling—not the mental and neural operations that produced that outcome. People are unreliable narrators of their own cognitive lives. Trying to understand reading by observing our own reading is hopeless, like trying to understand how a television works by watching *Game of Thrones*. Being an expert reader doesn't make you an expert about reading. That is why there is a science of reading: to understand this complex skill at levels that intuition cannot easily penetrate.

I am a psychologist/psycholinguist/cognitive neuroscientist who has been studying reading since the disco era. I'm not alone: a huge community of scientists studies reading around the world. Many people are surprised to learn there is a science of reading. Really, what is there to study? Words on page; eyes scan words; words are comprehended. It's just like listening, only visual. Book ends here. Beneath this seemingly simple behavior, however, a vast, coordinated network of activities is occurring. A snapshot of a person's brain activity while reading shows that most of the brain is involved: areas involved in vision and language, of course, but also neural systems that control action, emotion, and decision making; several memory systems; and much of the rest. Forget the myth that people only use 10 percent of their brains: we use more than that merely reading. The relationship between the experience of reading and its underlying neurocognitive mechanisms is about as opaque as the relationship between behavior and its psychodynamic causes in Freudian theory. Fortunately we have better methods than Freud for exploring the subconscious basis of behavior. We'll have you lie down and talk to us, but in the barrel of a magnetic resonance imaging machine, not on a couch.

We've learned quite a lot actually.

We understand the basic mechanisms that support skilled reading, how reading skill is acquired, and the main causes of reading impairments.

We know which behaviors of three- and four-year-olds predict later reading ability.

We know how children become readers during the first years of schooling and the obstacles that many encounter.

We know what distinguishes good from poor readers, younger from older skilled readers, and typical readers from those who are atypical because of constitutional factors (such as a hearing or learning impairment) or environmental ones (such as inadequate instruction or poverty).

We understand what is universal about reading (things that all readers do the same way because their brains are essentially alike) and what is not (because writing systems differ).

We have identified the main neural circuits involved in reading and some of the anomalous ways they develop in children with reading impairments.

We even have computational models of learning to read, skilled reading, dyslexic reading, and the loss of reading ability due to brain injury. It takes a deep understanding of these phenomena to develop models that reproduce them.

This vast research base has led to the development of methods that can reliably help many children who struggle to read. Researchers disagree about many details—it's science, not the Ten Commandments—but there is remarkable consensus about the basic theory of how reading works and the causes of reading successes and failures.

Sputnik Lands on USA

The successes of reading science create a paradox: if we know so much about reading, why are literacy levels in the United States so low?

In 2011 America had what our president called a Sputnik moment, occasioned by the release of the latest results from the Programme for International Student Assessment (PISA), a massive appraisal of the reading abilities of fifteen-year-olds in seventy-four countries and municipalities from Kyrgyzstan to Canada. As in previous rounds and again in 2012, US performance was close to the average for the thirty-four member countries of the Organisation for Economic Co-operation and Development (OECD), which conducts the PISA exercise. However, the 2009 round was the first to include data from Shanghai and Singapore, which scored higher than the US, as did Asian neighbors Korea, Hong Kong, and Japan. The Shanghai students lapped everyone else, scoring highest in reading, math, and science by wide margins. These results received far more attention than the fact that the US always scores lower than countries like Australia, Canada, Finland, and New Zealand in such exercises. Government officials and commentators treated the results as a wake-up call about the uncompetitive state of American education, said to pose a threat to the country's future akin to that represented by Sputnik in 1957. The Soviet Union's stunning success spurred a rapid, comprehensive governmental response that seems as much a relic of a bygone era as the satellite itself. Within two years, legislators had created the National Aeronautics and Space Administration and the precursor of the Defense Advanced Research Projects Agency (which funded the development of breakthrough technologies including the Internet), tripled the National Science Foundation budget, and allocated hundreds of millions of dollars for student loans and scholarships under the National Defense Education Act. But 2011 was not 1957, and the second Sputnik moment passed quickly, rapidly dropping out of public discourse (Figure 1.1).

Although the PISA results made the news, there is plenty of in-house data about America's literacy issues. The country is a chronic underachiever. A 2003

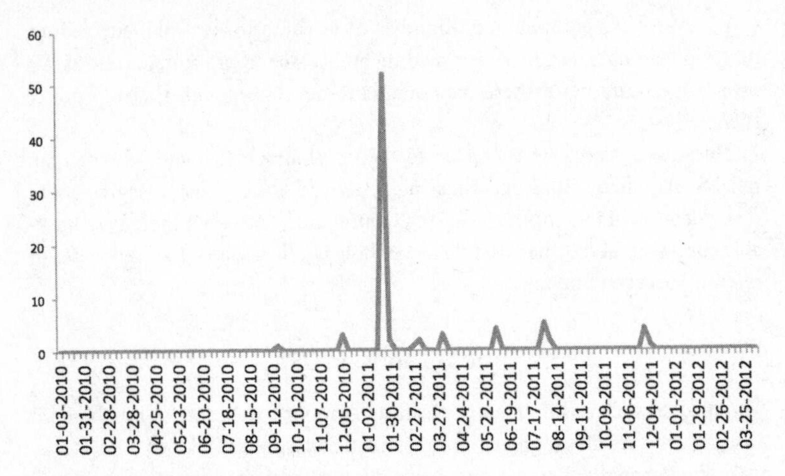

FIGURE 1.1. The number of times the phrase "Sputnik moment" was uttered on CNN, the American news network, over a two-year period. The large spike followed the release of the 2009 PISA results and coincided with the president's State of the Union address.

study found that about 93 million adults read at basic or below basic levels. At those levels, a person might be able to follow the instructions for mixing a batch of cake mix but not understand a fact sheet about high blood pressure. The emergence of our literacy problem is visible in the performance of fourth and eighth graders on the National Assessment of Educational Progress (NAEP), "the Nation's Report Card," an assessment administered by the US Department of Education. Over half the children have scored at basic or below basic levels *every time* it has been administered. At the upper end, we are turning out fewer highly proficient readers than expected given our economic resources. Like everything else about education in the US, the results of assessments like the PISA and NAEP have generated controversy. Are the tests too hard? Are they poorly matched to what American children are taught? If the scores are so low, why does the US continue to have the biggest national economy in the world? The American polity has been in a test-happy phase, and much other data about who can read and how well paints a consistent picture: large numbers of individuals in the US read poorly, as has been true for many years. Although I focus here on the US because it is home, the situation is similar in many other countries with advanced economies.

The consequences of marginal literacy for the affected individuals and for society are vast, as we all know. Reading is fundamental—so the slogan goes—and children's failures to acquire reading skills have rapidly cascading effects on learning other subjects. Even math is implicated, given the heavy emphasis on language in math curricula and instruction (word problems, explaining your work, the practice of embedding math problems in real-world contexts). Having less reading experience makes it harder to learn how to learn and how to think critically and analytically. Reading failures place children at high risk of falling out of the educational system. As adults, poor readers cannot participate fully in the workforce, adequately manage their own health care, or do much to advance their own children's education. If that doesn't convey sufficient urgency, a 2012 report from the Council on Foreign Relations declared, "Large, undereducated swaths of the population damage the ability of the United States to physically defend itself, protect its secure information, conduct diplomacy, and grow its economy."

This information plagues me, especially because the situation is not new. Scores on standardized reading tests have been nearly flat for decades. Hopes that the *Harry Potter* craze would increase children's involvement in reading gave way to deep disappointment when it didn't. We hear the suspicion voiced everywhere—from Ivy League colleges to local middle schools—that important reading skills and habits not measured by standardized tests, such as the ability to engage in close reading of challenging texts, are in active decline. According to Philip Roth, that game is already over. "The evidence is everywhere that the literary era has come to an end. . . . The evidence is the culture, the evidence is the society, the evidence is the screen, the progression from the movie screen to the television screen to the computer. . . . Literature takes a habit of mind that has disappeared."

Anxiety about reading achievement underlies endless debates about how reading should be taught. Parents know it is essential that their children learn to read well, but lacking confidence that schools can do the job, they are driven to seek additional help from tutors or commercial learning centers, if they can afford it. Our knowledge about reading has grown enormously, but, as ever, many people cannot read, or can read only poorly, or are able to read but avoid it. And so I have asked myself whether our science has anything to contribute to improving literacy outcomes in this country and elsewhere.

It might not. Reading failures could simply be collateral damage caused by other, deeper problems, such as poverty. About 15 percent of the US population lives in poverty, as the US Census Bureau defines it, including over 16 million children. Poverty is associated with higher infant mortality, higher risk of atypical neurodevelopment, shorter life span, worse health and health care, higher crime and incarceration rates, lower educational achievement, higher dropout rates, poorer schools with weaker teachers, and, at the bottom of a much longer list, poor reading. Surely addressing poverty would have a bigger impact on literacy than anything inspired by our research.

Poverty is a huge factor in America, as in other countries. Reading achievement is higher in Australia, Canada, and New Zealand than in the US, but not among their aboriginal peoples, for whom poverty rates are high. No discussion of what to do about reading can ignore the noxious conditions that govern many people's lives. However, 93 million is a lot of marginal readers, and they are not confined to the lowest economic stratum. There's Rob Gronkowski, a colorful professional football player, who said he hadn't read a book since *Mockingbird to Remember* in ninth grade. But look around: they walk among us. We could be talking about an airport security agent, an X-ray technician, police officers. High school graduates who cannot handle the literacy demands of community college. "Student athletes" who play big-time college football. "Low-information voters" whose primary news sources are cable news and social media. Or perhaps some readers of this book, highly motivated yet struggling through it. People with only basic reading skills can hold demanding jobs that affect health care, personal safety, law enforcement, business, politics, and every other element of society.

How much of the problem is due to cultural characteristics that conspire against our becoming readers, such as the profusion of screen-based activities: 24/7 movies and television, gaming, apps, cat and music videos, PewDiePie (45 million YouTube subscribers, 2,900 videos)? Are reading and writing gradually being reduced to tools in the service of reading-ish activities such as texting, tweeting, and Facebooking? What about overscheduling: the relentless demands of ballet, piano, violin, drama, tae kwon do, 6 a.m. slots at the hockey rink, soccer practice, matches, and travel? Middle schoolers cannot read while running dribbling and passing drills. Perhaps reading is simply less important than it used to be. Writing was the first information technology, but now there are others. We have screens, they have pictures, sound, video. Text carries less of the communicative load than before. I greatly prefer to read a recipe, but a video demonstration conveys additional information, for people who have the patience to sit through it. You can read about the phenomena I describe in this book but also see some of them in videos or try them out, which I highly recommend (and provide links to allow).

Something is happening here, but we don't know what it is, yet. If you position your ear close to the Internet, however, you can hear the steady thrum of reports and pronouncements about the state of reading: the good (easier access to a larger quantity and greater variety of texts than ever, integration of text with sound and image), the bad (easier access to a larger quantity and greater variety of nontextual media and undemanding text), and the ugly (a long-term decline in the linguistic complexity of K–12 schoolbooks). The National Endowment for the Arts issues grim warnings about the decline of reading, while educators view reading as one of several "literacies" for the modern era.

The results of this real-world experiment in how information is transmitted will not be known for generations. I think it is safe to assume that reading is not going away any time soon, if ever, and that people who are skilled readers will continue to have advantages over people who are not. With this in mind, I focus on a factor that has direct, lasting effects on who reads and how well: education. With some exceptions, children learn to read in school. The only certain way to obviate low literacy is prevention: successfully teaching children to read in the first place. Would more people be better readers if they had been taught differently? How much does schooling affect how well children read and, with it, their engagement in reading?

Here we encounter a problem. There is a profound disconnection between the science of reading and educational practice. Very little of what we've learned about reading as scientists has had any impact on what happens in schools because the cultures of science and education are so different. These cross-cultural differences, like many others, are difficult to bridge.

The gulf between science and education has been harmful. A look at the science reveals that the methods commonly used to teach children are inconsistent with basic facts about human cognition and development and so make learning to read more difficult than it should be. They inadvertently place many children at risk for reading failure. They discriminate against poorer children. They discourage children who could have become more successful readers. Many children who manage to learn to read under these conditions wind up disinterested in the activity. In short, what happens in classrooms isn't adequate for many children, and this shows in the quality of this country's literacy achievement. Reading is under pressure for other reasons, but educational theories and practices may accelerate its marginalization.

Reading science cannot address all of the issues that affect reading achievement—all those sequelae of poverty, for example—but it does speak to serious issues about how reading is approached in schools and at home. A look at the basic science suggests specific ways to promote reading success. These do not require more testing or new federal laws; they do not require vast infusions of money; they are not based on classroom computers that treat learning like a video game or other faddish uses of technology. What they require is changing the culture of education from one based on beliefs to one based on facts.

The next several chapters provide an insider's guide to some beautiful science that looks at reading at levels ranging from the movement of one's eyes across the page to the brain circuits that support reading to genes that affect how those circuits develop. Some of it is paper-clip-and-string research using the simplest of methods, such as recording the errors that children make when they read aloud or having the subjects in an experiment decide if SUTE is a word (it isn't, but it sounds like one, which creates momentary uncertainty even when it is read silently). Some of the research involves exotic methods such as transcranial magnetic stimulation, which briefly disrupts neural processing in parts of the brain relevant to reading. We will look at children who struggle to read and children who can read words but comprehend very little. Of course there is the terrible fascination of patients whose brain injuries impair reading in unusual ways. The late Oliver Sacks's famous agnosic patient mistook his wife for a hat; a patient with an analogous reading impairment, called deep dyslexia, read the word SYMPATHY as "orchestra." Along the way I cheerfully destroy a few myths about reading, including ones that have supported years of commercial schemes said to improve your child's reading in a few short weeks or to make you a super-mega-reader who can whip through books a page at a glance. We will look at educational neuroscience, the study of the brain bases of skills such as reading and math and how they are acquired. Fact: in neuroimaging studies, poor readers show atypically low activation in a part of the brain that processes the spellings of words. Can such findings inform how reading is taught, or do they merely provide neural window dressing for something teachers already know, that poor readers don't read or spell words very well?

With an understanding of the basic science of reading in hand, I take a close look at what happens in classrooms in America. If you are curious about the state of reading, these findings will, at the least, surprise you. If you are (or were) a parent of a school-age child, you might feel the shock of recognition and the dropping of scales from your eyes.

Understanding how the science relates to educational practices is an important step, but it is not enough to generate change. Educators are deeply immersed in their own worldview and well defended against incursions from outside. The education side focuses on "literacy" (literacy practices, cultural attitudes toward literacy, multiple types of literacy including ones that do not involve print), not reading. The scientific perspective is seen as sterile and reductive, incapable of capturing the ineffable character of the learning moment or the chemistry of a successful classroom. That people who enter the field of education do not gain exposure to modern research in cognition, child development, and cognitive neuroscience deprives them of the ability to evaluate what is studied, how it is studied, what is found, and what it means. Instead, they are exposed to the views of a few authorities, the most influential being Lev Vygotsky, who lived in the Soviet Union, wrote in Russian, died in 1934, and never saw an American classroom or a television, calculator, computer, video game, or smartphone. We scientists naively take the importance of our findings and their implications for classroom practice to be obvious, overestimating how much people on the education side know about or value them.

Education as a discipline has placed much higher value on observation and hands-on experience, which brings us back to page one (of this book). Intuitions about reading do not penetrate very far, and observations are influenced by prior beliefs, including what one has been taught about how children learn and how reading works. Educators are people too, subject to the same cognitive limitations and biases as everyone else. The lack of scientific literacy, combined with deep faith in the validity of personal observation, creates vulnerability to claims that are intuitively appealing but unproved or untrue.

People complain about the state of education about as much as they complain about the government. There's an endless, hopeful, desperate search for a game-changing innovation that will make more people smarter, more quickly, with less effort and expense. Some people focus on eliminating poverty itself, which would obviously have enormous impact on education and much else. We can focus at the same time on how to modify more tractable conditions to produce better outcomes. Learning more about the values and beliefs of the educators who teach teachers, design curricula, and create instructional practices could be a powerful impetus to change. In the closing chapters I spell out these concerns as they have arisen in reading education. This analysis clarifies what needs to change, and I discuss some ways we could move forward.

Since change can't be achieved overnight, I've also written this book to give readers—a parent interacting with a teacher, a school principal, or the bureaucrats who run the local school system; a voter whose representatives have the power to affect educational goals and practices, how public education gets funded, and how tax dollars are spent—information they can use now. These are the tools you need: an explanation of what we know about skilled reading, learning to read, the causes of reading difficulties, and the brain bases of normal and impaired reading; a critical analysis of the ways that reading is taught; an account of the controversies in reading education; and an understanding of educational ideology and how it evolved. Like health care, education is a multi-billion-dollar industry involving multiple stakeholders—government, business, educators, parents, children, taxpayers, unions, interest groups, philanthropy—whose perspectives and interests often conflict. In education, as in health care, end users benefit if they are informed, proactive participants rather than passive recipients.

Of course plenty of information about reading is available online, but on the Internet nobody knows your website is a dog. Websites do not come labeled with independent scientific seals of approval; doctrinal biases are not labeled as such; a layperson has no way to evaluate contradictory claims and assertions. Everyone—ranging from the National Council of Teachers of English (not much science there) to the Scientologists (whose teaching method L. Ron Hubbard is said to have devised himself)—has an opinion about reading. School districts post jargon-laden statements of teaching philosophies and curricular goals on their websites. It's a jungle out there, and a person could use some help.

A Science of Reading

Reading is one of the oldest topics in experimental psychology—the first American professor of psychology, James McKeen Cattell, studied it in the 1890s—and yet it is also a favorite topic of researchers in cognitive neuroscience, a field that emerged about a hundred years later. Why this enduring interest in such a familiar activity?

Reading is unique. Reading is among the highest expressions of human intelligence. Although spoken language is usually taken as the capacity that distinguishes people from other species, researchers have debated the degree to which other species's communication systems resemble language. No other species has a linguistic capacity equal to ours, but animal communication systems share some properties with human language. The late African Grey parrot Alex clearly had communicative interactions with his longtime trainer Irene Pepperberg. Was his use of words very much like human speech or an oddly evocative simulation, the result of thousands of hours of intense training? Whatever the answer to that question, we know that no other species has an ability remotely like reading. Indeed, *Homo sapiens* didn't either until the invention of writing about five thousand years ago. Understanding this complex skill means understanding something essential about being human.

Reading is important. Human culture has evolved to the point where this skill is critical to our ability to thrive. For most of human history people were illiterate and yet functioned well enough. That is not the case in modern societies. In a 2007 article bemoaning Minnesota children's low levels of reading achievement, Garrison Keillor observed, "Teaching children to read is a fundamental moral obligation of the society." Given its importance in modern culture, could anyone disagree? Unfortunately, reading is not always an easy skill to acquire; for most children it requires instruction, and for all of us it entails years on task. How children are taught matters a great deal: it can affect whether they become readers or not, their level of reading skill, and the extent to which they enjoy and seek out the experience. In order to teach children effectively and make this essential skill available to as many people as possible, we need to know how reading works. As a scientist I mainly study reading in order to understand the parts and how they fit together; it's a puzzle that interests me deeply. However, my own moral obligation is to pursue how the research can serve to help more children become better readers. This book is a part of that effort.

Reading is a tool for understanding human cognition. The capacity to use language evolved in humans over many thousands of years, the end result being that children acquire it easily and rapidly through interactions with other speakers. Reading is different: it is a technology, like radio, that came into existence because a person—or possibly several—had the insight to invent it. The advent of reading occurred relatively recently in human history, well after humans had evolved capacities to speak, think, perceive, reason, learn, and act. Reading was a new tool created out of existing parts. The fortuitous by-product of this history is that we can use reading to investigate all these capacities. A person doesn't have to be a reading scientist to study reading; they might study vision or memory, for example, using experimental methods that happen to involve having people read words and sentences. This bonus has resulted in the creation of a research literature of exceptional depth and quality.

In short, reading is interesting. It's complex, it's essential, and there is an urgent need to reduce the number of people who read little or not at all and to ensure that future generations will be sufficiently literate to thrive in the world they will inhabit. Out of such elements emerged a science of reading and this book.

CHAPTER 2

Visible Language

WE READ WITH OUR EYES, but the starting point for reading is speech. Informally we think of reading and speech as different ways of representing the same thing. Spoken language, visual language: one is a transposition of the other, just as numbers can be written in decimal or binary. In fact writing and speech are not interchangeable, but they are closely intertwined, each deeply affecting the other, like a couple of linguistic codependents with serious boundary issues.

Reading is secondary to spoken language: there are no natural languages (the term is used to distinguish human language from bee language, body language, markup language, and every other metaphoric extension of the term) that have a written form but not a spoken one. The capacity for spoken language evolved in humans well before writing was invented. For most of human history spoken languages existed without written forms, and many contemporary languages still do. Children's development recapitulates this pattern: their use of spoken language invariably precedes reading and writing.

For linguists who study the formal properties of languages, reading is about as interesting as Morse code. The creation of writing piggybacked on speech, which is where the core, defining characteristics of language are found, they say. Though a brilliant technological innovation of enormous practical importance, writing did not fundamentally change the nature of language. This view is simplistic, however. The origins of reading are one thing, but how it functions in modern literate cultures and in the brains of modern literate individuals is another. It is not merely that written language is supremely important, although of course it is—modern societies exist because of it and could not survive without it—but also that the technology afforded new ways to use language. It matters little whether reading "fundamentally" changed the nature of language or was merely the most important aftermarket add-on in history. Any account of language that excludes reading ignores what language has become because of it.

The intimate but complex relationship between written and spoken language underlies many of the important scientific questions and educational controversies about reading and so will emerge as one of the major themes of this book. Consider:

- The beginning reader's initial task is to learn how the spoken language they know relates to the written code they are learning. This step can be difficult for many children precisely because the two codes are not simple transpositions of each other.
- Because writing systems represent spoken languages, many educators concluded that learning to read should proceed just like learning one's first language. In fact, no one learns to read the way they learned to talk, thus occasioning the question, What happens when several generations of children are taught to read under this mistaken assumption?
- Debates about how to teach reading have focused on whether to promote or discourage the use of sound-based information in silent reading. Phonics methods are controversial because they emphasize connections between print and speech.
- Dyslexia, the developmental reading disorder, involves impairments in relating the written and spoken forms of language.
- Children's progress in learning to read is greatly affected by their experience with spoken language: the quantity and variety of the speech to which they are exposed and how the spoken language or dialect used in the home and community compares to the one used in school.

Whereas learning to relate print to speech is the beginning reader's first challenge, describing how the two are related is mine.

Each of These Things Is Not Like the Other

Yes, spoken and written language are alternative ways to represent the same thing. This observation would be too obvious to mention were it not for the cases where it does not hold. A prominent example is hearing-impaired people born to hearing-impaired parents who learn to sign rather than speak. Sign languages such as American Sign Language (ASL) exhibit the same types of grammatical structures and communicative functions as spoken languages, proving that the capacity for language can be realized using another modality. However, the grammar of ASL is radically different from the grammar of English. Because texts are not written in ASL, deaf individuals who learn ASL as their first, "spoken" language must learn a second language, English, for reading and writing. Using one language for conversation and a second language for reading? Not easy.

The other way to decouple spoken and written language is neuropathology, such as a stroke or other brain injury. In the absence of pathology, a person who can read and understand a sentence such as JOHN GAVE THE BOOK TO MARY can do the same when it is spoken. Under some deeply unfortunate conditions, comprehension is disrupted in one modality while leaving it relatively intact in the other. Some patients can understand words such as BOOK or GAVE when they are spoken but not written, and others show the opposite pattern. These individuals are of great interest to researchers precisely because such disruptions in the marriage of speech and reading are exceptional: it takes *brain damage* to cause them.

Spoken and written language share several other basic characteristics, two of which are of particular interest because they are true but didn't have to be. First, both print and speech are comprehended rapidly and automatically. You, reader, are understanding this sentence right here as you move along, and the same would be true if I were saying it to you. Language comprehension in both modalities is relentless and almost unstoppable, short of shutting one's eyes while reading or saying "lalalalala" while listening. It is not that comprehension works without error or that the result is the same for everyone. Rather, comprehension waits for no one, closely trailing the flow of words whether written or spoken. It is impressive that our brains work this rapidly and surprising that the two modalities behave so much alike, given their many intrinsic differences. Sign languages are also understood in this immediate, online manner.

The other important similarity is that people's reading and spoken-language skills tend to be very closely matched. For adults, the positive correlation between them is about 0.9 (the maximum is 1.0). Dyslexics are exceptional because their reading is much poorer than their spoken language. Some texts are tough going (I'm thinking of you, Henry James and Stephen Hawking), but the audiobooks aren't *Winnie the Pooh*.

There are plenty of books about the nature of spoken language, and so if reading and speech were merely very similar, this one could end right here. Once we look beyond these broad commonalities, however, they differ at every turn.

Speech evolved in the species.

Reading is a cultural artifact, like money.

Speech is universal: in the absence of pathology, everyone learns to talk.

Reading is like Wi-Fi: only some people have it.

Children learn a spoken language through interactions with other language users.

Reading is taught, beginning with alphabet songs and bedtime stories and continuing through several years of schooling.

Speech is fast fading: the signal is gone once it is produced.

Writing systems were created as a way to transcend the impermanence of speech. This text is not disappearing as you read it.

Speech is messy. Producing a coherent utterance is a complex action: deciding what to say, picking the words and grammatical structures that express the intended meaning, loading a program to articulate the sequence of words, and then running the program, all done on the fly. The process is so demanding that the end product is littered with dysfluencies and errors.

Writing is cleaner than speech because it isn't produced under the same time constraints and can be edited. Speech is more like a rough draft, text like the corrected version.

That one has some obvious exceptions. Some formal modes of speaking are grammatically immaculate, and informal writing can be messy. A television performer such as Stephen Colbert is a professional talker skilled at producing fluent, well-formed utterances most of the time. Conversely, there is Wikipedia, the Great Wall of Words, a system for creating texts that are full of grammatical, spelling, and usage errors.

Finally, speech and reading are opposites with respect to who controls the rate at which comprehension can occur. Readers decide how rapidly to proceed. Listeners are at the mercy of speakers performing the difficult task of producing coherent, mostly fluent utterances. In exchange for ceding control to the speaker, the listener gains one advantage: speech affords more opportunities for clarification. Circumstances permitting, speakers can be asked about what they have said, whereas readers cannot query authors. (We have book tours and Reddit Ask Me Anythings for that.)

The net result is that properties of spoken and written language diverge. The distributions of words in speech and text differ; many words occur more often in one than the other. For example, which of the following words occur more often in speech and which more often in text?

```
yeah such right actually obviously thus sh*t which
like diverge*
```

* The examples alternate between more common in speech (`yeah`) or in writing (`such`).

Texts for children and for adults also include more complex sentence structures than occur in their own speech. There are other differences, but let's forgo the exhaustive list and settle for the main point: because of the ways they are perceived and produced, spoken and written language are not simple variants of each other.

Mashup!

These dissimilarities arise from properties of the auditory and visual modalities and the corresponding sensory and motor systems. What happens if properties that are associated with one modality are ported to the other? Could a hybrid combine the best features of both? Or would it be a hideous mutant? The modern world has done a few such experiments for us. One is texting. Texting is a mashup of written and spoken language, created in response to the challenge of using language under the novel conditions created by smartphones. The result is a form of written language with several characteristics of speech:

- It's fast(er) fading. I type a text message; it's read by the other party, hopefully the intended one. The text remains on the screen for a while but then drifts off. For the user, text messages are more transient than traditional texts but more persistent than speech. Of course text messages are far more permanent than they appear, having left the screen but not necessarily the Internet's colossal long-term memory.
- As with speech, control over the rate at which information becomes available rests with the producer. Typing a text message is more like formulating a spoken utterance; if the text is dictated using your phone's amazing speech-recognition software, then it is based on actual speaking. Text messages are less amenable to editing than ordinary texts. I said that speech was more like a first draft and a standard text more like the final version? Texting is a rough draft with limited vocabulary and syntax, spelling errors, neologisms, abbreviations, and, for the exuberant, emoticons, emojis, and other sprinkles. These characteristics largely result from the constraints imposed by smartphone technology, combined with software limitations such as imperfect autocorrection. We use the shortcuts that are available in this medium to approximate a real-time conversation even though we're tapping virtual keyboards.

Texting is a case in which merging some characteristics of text and speech is successful, creating a novel hybrid that fills a new, if limited, communicative niche. Early fears that texting's idiosyncrasies would corrupt grammar and spelling have turned out to be unfounded, although they have created new genres of humor. Internet amusements aside, the evidence to date suggests that people—even tweens who text a few thousand times a month—do not find it difficult to maintain a distinction between texting and more formal types of writing.

Other hybrids are disasters because the properties of the two codes are mismatched. Some kinds of texts can be understood when read but not when heard. For many years academic philosophers made a point of giving talks that consisted of reading a characteristically abstruse paper aloud. The text was written knowing that readers would be able to proceed at their own pace, rereading, pausing, and consulting other sources as necessary. The spoken form eliminates these options, making the same text much harder to understand—often unintelligible, I'm told.

The education of many generations of hearing-impaired people has been affected by the misuse of hybrid forms of English. Like other spoken languages, English was shaped by our hearing and speaking capacities. Signed languages such as ASL are shaped by the characteristics of the visual-gestural modality. Signed English is a code that educators of the hearing-impaired devised so that English grammar could be used in both signing and reading, a logical goal. However, taking a grammar adapted to the properties of one modality (speech/hearing) and porting it to a different modality (vision/gesture) creates an epic mismatch. Signed English can be easily comprehended and produced by hearing people but is tedious and awkward for the hearing impaired, the intended beneficiaries.

Hear My Words

Reading and speech share a common linguistic core, but the modality differences mean that some kinds of information are easily communicated in one but little or not at all in the other. This means that there is more to language than afforded by either code itself. That the codes are complementary rather than wholly redundant contributes to the symbiosis between them and to my rejection of the view that reading is merely the handmaiden of spoken language.

Some types of information are represented in the written code but not in speech. Spaces indicate the boundaries between words in alphabetic writing systems; there are no "spaces" between words in fluent speech. Spoken languages have homophones, words that are pronounced the same but are otherwise unrelated. They are ambiguous when spoken but often spelled differently (e.g., BLUE/BLEW, PEAR/PARE/PAIR). Typographical conventions such as using capitalization to indicate proper nouns do not have spoken equivalents in English.

That is small stuff compared to what's left out. Writing systems represent spoken languages, but they are not close representations of speech itself. The speech signal contains clues to everything from the speaker's sex, age, race, and education to whether they are happy or anxious, alert or distracted, telling the truth or dissembling. We do not rely on writing systems to represent this information, even though it is often essential to comprehending a text. Consider the following monologue:

> Are you talking to me?
> Are you talking to me?
> Are you talking to me?
> Are you talking to me?
> Are you talking to me?
> Are you talk? . . .
> Are you talking to me?
> Are you talking to me?
> Are you talking to me?
> Well, you must be, because I'm the only one here.

A person might experience feelings of dread merely reading the transcript of Robert De Niro's scabrous monologue from the 1976 movie *Taxi Driver*. But the quote is not from that movie. It's from *Wise Guys*, a 1986 gangster comedy in which Harvey Keitel does a send-up of his friend De Niro. The example illustrates the fact that text conveys limited information about how an utterance is spoken, data that can signal the difference between psychopathy and parody. And speaking of humor, here is John Cleese's favorite joke:

> A grasshopper hops into a bar and onto a stool. The bartender says, "We've got a drink named after you." The grasshopper says, "What, Norman?"

The joke only succeeds if the reader assigns the appropriate intonation—a property of speech—to the punchline "What, Norman?" (roughly the same as in "Who, me?"). Punctuation provides helpful clues but does not fully specify the humor-intensive intonation. In written form the joke relies on the reader's ability to mentally supply this information. Thus a joke can be funny even if it is read silently rather than told by John Cleese. Successfully "hearing" the relevant intonation is part of the pleasure.

These examples illustrate an intrinsic property of writing: properties of speech that are relevant to comprehension are not systematically represented. Reading is the use of a written code (orthography) to represent language. As in speech, the meaning of a text is greatly affected by phonology—the sound patterns of language. Writing systems vary in how much sound-based information they represent, but all of them omit a great deal. Properties such as pitch (Elijah Woods's is higher than Seth Rogan's), timing (speech rate, the length and placement of pauses), and loudness convey important information but are represented little or not at all. We're left with two questions. How do we manage to avoid regularly misinterpreting what we read? And wouldn't it have been smarter to include this information in the written code? Our keyboards are packed with characters that could be commandeered for the purpose.

The basic answer to both is that reading works because we are able to supply this missing information, most of the time, without its being an explicit, codified part of the writing system. Although it could be fully represented in writing, the costs would greatly outweigh the intermittent benefits. The system is a compromise that works quite well but not perfectly.

The principal mechanism at work is that many kinds of phonological information can be safely omitted because they are predictable from other things we know—about the topic, about the context in which a sentence occurs, about spoken language. Explicit marking would be superfluous. The ability to use what we know to go beyond the information given is a fundamental property of human perception and cognition. We fill in missing information all the time. We see shapes and letters where the parts are only implied, as in visual illusions and "incomplete" type fonts. We "hear" words when we read. Permit me to demonstrate. You just read the word PERMIT in that last sentence, and you unconsciously gave it the iambic (weak-strong) stress pattern: perMIT. But if you read I JUST GOT MY NEW PARKING PERMIT, you give the same spelling pattern the trochaic (strong-weak) stress: PERmit. We "hear" the appropriate stress pattern in our mind's ear. This solution is apparently good enough, and we're smart enough for it to work, but do we have to like it? Wouldn't life be simpler—and reading a little easier—if the writing system deigned to indicate syllabic stress using, say, accent marks?

Actually, no. As it has evolved over many centuries, written English has come to respect a very deep principle about what makes a code (such as a writing system or encryption cipher) efficient. Informally it can be stated as, Formulate your message in a way that makes it likely to be understood but avoid including more information than that requires. Explicit marking of syllabic stress, as in PerMIT ME TO DEMONSTRATE HOW TO USE THE PARKING PERmit, would be overkill because we can determine stress patterns well enough without it. Using capitalization, accent marks, or other typographic elements to explicitly indicate syllabic stress would only gild this particular linguistic lily.

Adding notational elements also comes with costs: more symbols to learn, more to write, more to read, more paper and ink or bytes. These far exceed the costs of occasional errors. Worse, providing too much information makes reading more difficult. Modern Hebrew provides a beautiful example of the trade-offs between explicitness and efficiency. Like English and other alphabetic writing systems, the modern Hebrew alphabet includes symbols for consonants and vowels. Unlike English, Hebrew can be written either with or without the vowels. The pointed version, in which small diacritic marks indicate the vowels, is used for beginning readers. Skilled readers transition to the unpointed version in which the vowels are omitted. What happens, then, when adults are asked to read texts in which the vowels are put back in? They read more slowly. It's orthographic TMI.

For both English stress patterns and Hebrew vowels, the moral is that readers are better off with texts that omit some information because we can reliably fill it in based on our vast experience, which tells us what is likely to be correct. The operative phrase here is "vast experience," which is what it takes to sort out the complex details of systems that have statistical tendencies but not inviolable rules. For example, many noun-verb pairs are of the PERmit/perMIT type, with strong-weak stress for the noun and weak-strong stress for the verb (others are CONSORT, DETAIL, PROTEST). The language allows many exceptions, including ones where both the noun and the verb have the strong-weak pattern (use the ANCHOR to ANCHOR the ship) and ones where both have the weak-strong pattern (RELEASE the press RELEASE). This characteristic is not limited to syllabic stress; it is a fundamental characteristic of language. The term for a system, like language, that exhibits rule-like regularities but also admits instances that deviate from these central tendencies in varying degrees is "quasiregular." Languages are quasiregular, but chess is not, because the pieces are not allowed to deviate intermittently from their designated movements.

We are able to speak and comprehend language with great skill despite its quasiregularity—indeed, because of it. Communication requires shared knowledge, and so languages must be systematic rather than arbitrary. However, the demands of comprehending and producing language require additional flexibility because speakers produce forms that deviate from standard patterns and listeners must be able to comprehend them. Many shortcuts that promote fluent speech eventually enter the language, such as "gonna," "hafta," and "tryna," which partially overlap with the source words. The product of these conflicting pressures is quasiregularity. These patterns can be mastered with extensive practice, which is easy to obtain if you've grown up speaking a language and become a fluent reader. Mastering stress patterns is much harder for people learning English as a second language, who often exhibit "stress deafness."

In short, writing lets readers make use of the knowledge that allows spoken language to work so well, without getting overly specific. This solution does not work perfectly, but aiming for perfection would make it worse.

Just Cheat

And then there's cheating. The degree to which writing underrepresents speech creates room for misinterpretation. The risk can be reduced by taking advantage of ways the basic alphabetic code can be embellished, which range from punctuation (the Oxford comma, which shoots and leaves) to fonts and styles to ornaments such as emoticons and emojis. Users of English take these options for granted, but many writing systems admit far less of this stuff: there is little underlining in Chinese or italic in Hebrew, for example. In the writing systems that allow such embellishments, they are often used to fill phonological gaps, adding missing information on an as-needed basis. You may be constitutionally opposed to emojis and the effusive use of italics and exclamation points, but these devices are weapons in a revanchist struggle against phonological underspecification in the written code.

Clive James, a prolific writer of Australian origin and a witty presence on British television for many years, is possessed of an enviable ability to uncover many of the elegant sentences that are buried in the combinatorial explosion of possibilities afforded by the English language. A man who writes "Everyone has a right to a university degree in America, even if it's in Hamburger Technology": *that* is a humorist. A writer who has translated Dante's *Inferno* has no use for typographical embellishments. The closest James will come to an emoticon is a discussion of John le Carré's Smiley. But what if failing to use typography in a familiar way misrepresents the meaning of an utterance? Here is James reviewing a British reality TV program:

> Nobody should ever watch *Classic Car Rescue* for any reason, but if you did happen to see it you might find one of its unprepossessing male presenters pointing at a car and saying: "That is a car!" The information content of such a statement is zero. . . . The statement "That is a car!" can tell you nothing, unless the car is disguised as, say, a heap of cardboard boxes.

The program sounds dreary, but the utterance "That is a car!" does not lack content. James's print version conveys that the fellow was stating the obvious with mindless enthusiasm. It seems far more likely that he was expressing an opinion about the object, which could be conveyed by simply using italics in the familiar way to indicate heavy stress on the initial word: "*That* is a car." The phrase exemplifies a familiar phrasal template used to identify an entity as a special example of its kind. Of Dr. Martin Luther King Jr.'s "I Have a Dream": *that* was a speech. Of a special dish: *that* is food. Of an automobile exhibiting qualities that distinguish an object of automotive envy from a Pontiac Aztek: *that* is a car. "Nobody should ever watch *Classic Car Rescue* for any reason" is hyperbole, but "the information content of 'that is a car' is zero" is fallacious because of James's disdain for using a trivial typographical convention to accurately represent the form and meaning of the man's utterance. *That* was disingenuous.

It may not please the National Council of Teachers of English (or concern Clive James in the slightest), but typographical variation is a fine way to represent heavy word stress, a property that can greatly change meaning:

> *Fred* didn't take the test yesterday. (Somebody else did.)
>
> Fred *didn't* take the test yesterday. (Although he could have.)
>
> Fred didn't *take* the test yesterday. (He posted it on the Internet.)
>
> Fred didn't take *the* test yesterday. (He took a different—lesser?—one.)
>
> Fred didn't take the *test* yesterday. (He took something else.)
>
> Fred didn't take the test *yesterday*. (He took it some other day.)

These differences in emphasis are a property of spoken language that is easy to represent in writing using italics, boldface, caps, or color.

These options are fine as far as they go, which unfortunately isn't very. Typography works pretty well for individual words but not for prosody—phonological patterns that extend over sequences of words. Prosody can make "Are you talking to me?" sound threatening and "What, Norman?" funny. Like other writing systems, written English represents very little prosodic information despite its impact on meaning. A few simple patterns can be represented fairly reliably, such as uptalk, in which a statement is spoken with question intonation—as in "And so, one time? I was at band camp? And we weren't supposed to have pillow fights? But we had a pillow fight! And it was so much fun!" This one is easy because the question marks successfully evoke question intonation even when the sentence is an assertion. The technique is not a precision instrument, however, and it helps if you've seen the movie.

Mischief can ensue. Figurative language is a potential weapon of self-destruction in public settings such as politics and popular culture. Verbal irony works by creating a contradiction between the intended meaning of an utterance and its literal meaning. The speaker's ironic intent is often conveyed by prosody, which is not represented in the written code. The transcription of an ironic remark will better represent the unintended meaning, a formula for spectacular miscommunication. The utterance "The Beatles are more popular than Jesus" could be said in a manner that conveyed astonishment, regret, or glee, distinctions that are lost in a literal transcription. Journalists are ostensibly required to report what was said, not what they judge the speaker to have meant, but quoting only the exact words is not neutral or "objective." By omitting information about how the utterance was said, a quotation can convey a different meaning than the intended one.

It is a short step from The Beatles to Ronald Reagan. His Russian bombing remark ("My fellow Americans, I'm pleased to tell you today that I've signed legislation that will outlaw Russia forever. We begin bombing in five minutes") was audacious humor as spoken by a president of the United States, but the utterance can read quite differently, as it did to the Soviets ("The USSR condemns this unprecedented and hostile attack by the US president"). Nonliteral expressions are high-risk behavior for public figures because of their susceptibility to misinterpretation, especially in print, which is why an ironist like David Letterman did not grow up to be president.

Typographic conventions can reliably convey some kinds of useful information but are utterly ineffective with others. They are objects of prejudice because what they do convey can usually be expressed more clearly using language itself and because their genuine communicative utility is undermined by other factors: their slackish use in place of language, their decorative functions, their appropriation by middle school children. The net result is that they act as useful adjuncts to a writing system rather than being systematically incorporated. We want these gimmicks around because they are helpful and entertaining, but they are not sufficiently well mannered to gain full membership in the orthographic club.

What Is It Like to Be the Word "Bat"?

Writing does not specify features of speech that carry important information, which readers can usually fill in. Making a habit of this could do things to a person's brain. For literate individuals, what we know about the written code deeply penetrates our knowledge of spoken language, changing how we think about speech, how we understand it, and how it is represented in the brain.

Say the word BAT out loud. What is that sound made of? We think of "bat" as comprising three sounds, called phonemes, which I'll write as /b/, /a/, and /t/, spoken in that order. TAB consists of the same phonemes spoken in the opposite order. Everyone knows this. Right?

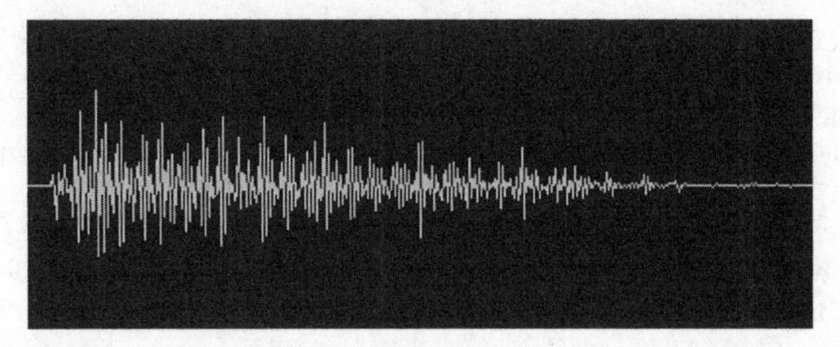

FIGURE 2.1. Waveform for the spoken word "bat."

Figure 2.1 shows the sound wave produced when I recorded the word "bat" on my laptop. Time runs from left to right, and the height of the waves indicates their intensities (roughly, loudness). Where are the three phonemes in this waveform? The boundaries between the letters in BAT are clear, but the boundaries between the sounds in "bat" are not. (The same is true if the spoken word is viewed as a spectrogram, a more detailed image that also represents sound frequency). It isn't possible to select parts of the waveform that correspond to each of the phonemes because the word is pronounced as a continuous articulatory gesture, not three discrete ones. Parts corresponding to the /b/, for example, also include parts of sounds that come later. The letters of a word are like beads on a string, but the sounds are more like a cascading waterfall.

Why do we have the overpowering sense that "bat" consists of three phonemes? Because we read and spell. English is written with an alphabet. We learn to treat the spoken word *as if* it consisted of three discrete sounds because it is written with three discrete letters. This abstraction allows units in the written code (graphemes, which are single letters or combinations, such as SH) to correspond to units in the spoken code (phonemes). A beginning reader can then learn systematic mappings between the two.

Although we think of speech as consisting of discrete phonemes, it is easy to demonstrate that it doesn't. Think of the common activity in which a Muppet (or person) models sounding out a simple word. Letters appear on the screen and the Muppet says the sounds associated with them, one at a time, "b" . . . "a" . . . "t" . . . , gradually decreasing the pauses between them. Sometimes the letters are displayed far apart on the screen and gradually brought closer together as a visual cue. The sounds do not meld into "bat" no matter how rapidly in succession they are spoken because it does not consist of three discrete segments. A discontinuity always occurs at the very end when the rapidly but discretely enunciated phonemes are followed by the word pronounced as a whole. How to get from one to the other, the Muppet does not say. The activity is useful because the child learns about letters and their sounds. It encourages the fiction that words consist of discrete segments even as it demonstrates that they do not.

Learning to treat spoken language as if it were composed of phonemes is an important step in learning to read an alphabetic writing system. Spoken-language games such as rhyming take readers part of the way there. Hearing that "cat" and "hat" have the same ending "at" makes it easier to isolate the differing initial sounds. However, it takes exposure to the spellings of words—learning to read and write—to complete the full phonemic illusion.

If the sense that spoken words consist of a series of letter-like sounds arises from learning an alphabet, do people who cannot read perceive speech differently than people who can? Definitely. It is trivial for you, a literate speaker of English, to decide if two spoken words end with the same sound (e.g., "bat," "sit") or not ("bat," "sick"). Preschool children who are ready for reading can too. In a classic study, illiterate Portuguese adults (tested in their language, of course) found the task extremely difficult, and their accuracy was poor. A person who does not read treats a word like "bat" as a pattern without discrete parts, more like the speech wave pictured above. Learning the alphabetic code reveals components hidden within spoken words.

This fact is crucial. Using spoken language does not require knowledge of phonemes: the Portuguese adults could speak the language; they just couldn't read. Learning to read changes the representation of speech, promoting the emergence of an abstract unit, the phoneme. Representing spoken words this way makes it easier to read the alphabetic code, which in turn solidifies representing speech as phonemes. This feedback loop is a critical mechanism in the development of reading skill. Learning to read is more difficult when the development of phonemic representations is impaired or discouraged by educational practices.

There are easier ways to observe the impact of alphabetic knowledge on spoken language than studying Portuguese illiterates. We'll do an experiment. On every trial the subject (a reader like you or me) will hear a pair of words and decide if they rhyme. Some pairs will be rhymes, such as "bank"/"sank," and some will be nonrhymes, such as "beer"/"sank." The task is easy. The question is, What happens if the rhyming words are spelled similarly (e.g., GOAL/COAL) or dissimilarly (BOWL/COAL)? Keep in mind that the experiment only involves rhyming: the subjects hear the words, with no reading or writing involved. The results come out very clearly: it is harder to decide if two words rhyme if they are spelled differently. Subjects do not make many mistakes—they know that "bowl" and "coal" rhyme—but it takes measurably longer to decide. Apparently the spellings of words affect how we hear them. The effect is so strong you can replicate it with an obliging friend and a stopwatch app.

Spelling's impact on the brain's representation of speech can also be seen using neuroimaging and brain-stimulation methods. For people who can read, there are no pure representations of the sounds of words in the brain because they've been contaminated by spelling. Language is a virus, as the musician-artist Laurie Anderson said, but orthography is the virus that infiltrates language, as observed by the noted British cognitive neuroscientist Uta Frith: "Learning an alphabetic code is like acquiring a virus [that] infects all speech processing, as now whole word sounds are automatically broken up into sound constituents. Language is never the same again."

Reading is powerful. The human capacity to use spoken language has existed for eons, yet it is readily modified by a few years of exposure to print. That is a very good thing, because bringing print and speech into alignment makes reading feasible.

Reading and speech are different because of the modalities they involve, the ways they are acquired, the information they convey, and the conditions under which they are used. Different, yes, but closely bound and mutually dependent, each one changed by knowledge of the other, from behavior to brain.

CHAPTER 3

Writing: It's All Mesopotamian Cuneiform to Me

WHAT IS THE MOST IMPORTANT invention in history? In the era of the Internet and the listicle (and who invented that?), the question is easy to ask and has been many times. The answers are boringly predictable: wheel, printing press, lightbulb, computer, Internet, and so on. The bar code and laser scanner are huge for people in retailing; according to one poll, the invention that MIT undergraduates could least do without was the toothbrush (followed by car, computer, mobile phone, microwave oven). These lists are more interesting for what they consistently omit: writing.

The first thing that happened to reading is writing. For most of our history, humans have been able to speak but not read. Writing is a human creation, the first information technology, as much an invention as the telephone or computer. Had the US Patent and Trademark Office been there, writing could have been patented—as "a system for inscribing messages on semipermanent media" perhaps—whereupon a Sumerian patent troll could have sued the Phoenicians, Egyptians, and Chinese for infringement and demanded licensing fees. People probably don't think of writing as an invention because they do not clearly distinguish writing from speech. Writing's origins are also obscure: if there was an ancient Thomas Edison, that person neglected to use their astonishing invention to record their own name. And yet the creation of writing was one of the greatest achievements in human history. The development of modern civilization could not have occurred without the massive increase in the creation, retention, and transmission of information that writing afforded. Without writing there would *be* no printing press, lightbulb, computer, or Internet.

Writing systems are fascinating but also exhausting, like the Egyptian rooms in the British Museum. The sheer number of writing systems and range of elements they have been fashioned from are a vivid testament to human creativity. Writing seems to come in about as many styles as beer. Orthography mavens are like birders, tracking exotic species identified by visual features, sound, and habitat. They have Pinterest boards and websites. Projects on cuneiform and Egyptian hieroglyphic are middle school standards, occasions for retelling stories about the ingenious clay tablets and inscrutable pictographs. In some cultures the young find it pleasing to have symbols from exotic writing systems inscribed on their bodies, often to comic effect though they may not know it.

Writing systems are immensely interesting in their own right (or, with a nod to John Lennon, in their own write), but they are not just for hobbyists or bad tattoos. The properties of writing systems are important because of their influence on how we read. Their apparent diversity raises a fundamental question: How many ways *are* there to read? Does reading work in essentially one way—because human brains and other relevant body parts are alike—or do the many writing systems indicate that reading is accomplished in multiple ways, a case of different strokes (of the pen) for different folks?

The surprising finding is that what is most salient about writing systems—their visual heterogeneity—camouflages what is most important about them: that they are essentially alike. The great insight was that spoken language could be represented in the medium we call writing. Several thousand years of experimentation with such systems ensued. The development of writing systems was shaped by human capacities, cultural factors, and some accidents of geography and history, eventually resulting in a high degree of convergence in how they work. They are similar because we are similar, as are the functions and uses of writing and the ways we read.

Interestingly, these same considerations militate against there ever being a single, universal writing system. The pairings of writing systems and languages are not random. Like writing systems, spoken languages are also "essentially alike," with some variation in how they are organized. Some types of writing systems work better with some types of spoken languages. Sufficient congruence between the properties of writing systems and the spoken languages they represent is critical to their functionality. The misalignments that periodically occurred during that long period of experimentation were fatal for some writing systems but also the source of breakthroughs in the advancement of this technology, the one we truly cannot live without.

The primal question about writing is how it originated. How did writing come into being, when, and how many times? Was it devised by a genius inventor? Or is writing, like the computer, a technology that has many sources but no originator? The similarities to that other primal question, about the origin of human life, are obvious. We have a deep-seated urge to understand how not only we but also our greatest artifacts, institutions, and ideas came into existence. Just as our existence is commonly attributed to a supernatural being, writing has been regarded with such awe that it too has been thought to have divine origins. The Babylonians believed that the god Oannes, part fish, part man, emerged from the sea to impart language, science, and writing; in Egypt, it was Thoth, also the god of wisdom, magic, language, and arithmetic.

Writing can also be seen as the result of either intelligent design or evolution. For some scholars, writing was such a profound advance, so unlike what came before and so fully realized from the start, that it could only have been designed by a supreme intellect, an ancient Newton or Einstein. Others interpret the evidence as showing that writing systems resulted from an evolutionary process that proceeded in fits and starts and punctuated equilibria on a very long time scale, shaped by adaptations and mutations that produced intermediate forms that eventually gave way to contemporary ones. The controversies are similar as well, as in the interpretation of gaps in the fossil record for evolution and in the archaeological record for writing. Writing systems are not biological organisms—they don't have genes to pass on to their offspring, for one thing—which is where the close analogies end. Writing systems are the product of cultural evolution, which is loosely related to the biological kind.

Unlike the origin of the human species or the invention of the wheel, tangible evidence about the origins and development of writing exists because it provided a record of itself. The clay tablets turned out to communicate far more than their creators intended and to be more durable too. We do not know whether the inscriptions on the Pharaohs' tombs accomplished their intended eschatological functions, but they succeeded in preserving information about the Pharaohs' lives, the lives of their people, and their culture, including the ingenious hieroglyphic writing. The drama of the discovery and deciphering of these artifacts is another source of writing's fascination, the stuff of symbology and *Indiana Jones* movies: ancient objects carrying faint messages from deep in the human past, the literal beginnings of recorded history. From the troves of writing samples that have been recovered, anthropological archaeologists have managed to establish a remarkable amount about early writing and the peoples who produced it. For example, much is known about the domestication of cattle in Sumeria in 3000 BC because scholars were able to identify signs for cattle and dairy products and ones for quantities signifying the number of cattle in a herd and amounts of milk and cheese. It was a scholarly tour de force to determine that the first person whose name is known to us from a written record is Enmebaragesi of Kish, a Sumerian king who lived around 2600 BC. The impact of this research runs much deeper than the accumulation of such facts, however. Viewing a tablet that recorded some sort of transaction involving sheep and oil in an ancient village whose location can be pinpointed on Google Maps, we experience a connection to those people that is intimate yet farther removed in time than even the most elaborate genealogy or oral history.

What the archaeological record does not offer is definitive answers to the primal question. The available evidence is frustratingly incomplete and about as jumbled as a pathological hoarder's stash. It reveals a great deal about what was created: various writing systems, their properties, how they changed over time. But specific claims about where, when, and by whom they were invented are (literally) educated guesses. Consider cuneiform, often celebrated as the first writing system, said to have originated in Sumeria in the village of Uruk, where the crucial tablets, about 4,000 of them, were found. Every element of this account is uncertain. Thousands of inscribed tablets have been found across a wide swath of the modern Middle East; many predate cuneiform and may have influenced its development, but few have been deciphered, and most are in storage, unexamined. The cuneiform tablets may have been brought to Uruk rather than originating there. The language may not be Sumerian but that of a neighboring people who created the tablets. Some scholars argue that the cuneiform system was invented out of whole clay, so to speak, by a single Sumerian genius, others that it developed gradually over a wide geographical area. Enmebaragesi is the name, but artifacts discovered in the 1980s suggest that this king might have been a queen or a fictional composite.

Although the archaeological record is indeterminate, the primal question is too interesting to forgo, leaving scholars to vigorously debate various scenarios in accordance with Benford's law of controversy: "Passion is inversely proportional to the amount of real information available." For example, most observers have concluded that writing systems were independently invented three times: in Sumeria, China, and Mesoamerica. Working from the same data as everyone else, anthropologist Jared Diamond proposed that the Chinese picked up the concept from their Egyptian trading partners, who adopted it from the Sumerians. Diamond's version adds a novel twist to an old story, it keeps the discussion alive, it could be true—and if more of those unanalyzed clay tablets are deciphered, some day it may be possible to tell.

The incompleteness of the archaeological record is unfortunate because the experiment called the invention of writing is over and cannot be replicated with a better data-collection plan. The combination of deep-seated questions and a rich but incomplete historical record is toxic stuff for scholars, whose essays exhibit the "false and damaging certainty" that writer Janet Malcolm finds in reporting about contemporary events that, like the origins of writing, have emotional resonance but inherent ambiguity, such as hideous crimes committed under murky circumstances. We demand clear narratives in such cases, even if the truth is hopelessly underdetermined by the evidence. The stories told about where and when writing was invented are what I would call honest confabulations: serious attempts to bring coherence to an immense yet inadequate body of evidence, the docudramas of the academic world, "based on a true story."

What the History of Writing Has to Say About Reading

The primal question concerns the activities of individuals and peoples at particular times and locations. Whatever the interest of that question and prospects for answering it, we can also look at writing as a product of the human species. The properties of writing systems, their similarities and differences, and the ways they changed over time are data that can help to address the hardest question about writing systems, which is not who developed them but how *anyone* could have. How did humans manage to progress from drawing *pictures* of horses to writing *about* them? The invention of writing forces us to confront a general puzzle about human creativity: Where do new ideas come from? Writing seems to be a prime example of a cultural artifact that is sui generis—lacks precedent. Writing did not exist for all of human history until it did. Something happened. But where did the very idea of writing come from if it did not already exist? This puzzle also arises in connection with the several great innovative leaps that occurred during writing's long, slow elaboration and refinement. For example, alphabets represent phonemes, but the phonemic abstraction depends on exposure to an alphabet. How could phonemes have been discovered without alphabets that represent phonemes?

These issues are more complex than whether Enmebaragesi was king or queen, but they may be more answerable because the evidence is not limited to the archaeological record. What we now know about human psychology, biology, and culture is relevant to interpreting what happened to people very much like us several thousand years ago. We can also use what has been learned about how we read to identify crucial properties of writing and track how they developed. The answers to these questions do not turn on information at the level of names, dates, and places and so the archaeological gaps are of less concern. There is insight to be gained about the mechanisms by which writing systems developed from looking at the capacities of the people, the properties of the languages they spoke, and the conditions under which they lived and by asking, What are the problems for which writing systems are the solution?

The standard story about the origin of writing goes like this. Humans have been creating representational images for more than 30,000 years, the approximate date of the oldest known cave paintings. The paintings are depictions of things their creators saw, mainly animals, objects, and body parts. Early writing is said to have built on this capacity, using simplified depictions of objects, called pictographs. The use of pictographs limited communication to what can be rendered in this manner. Illustrations similar to Figure 3.1, found on countless websites, show how signs in Sumerian cuneiform, the best documented of the earliest writing systems, changed over the 3,000 years it was in use. The illustration conveys that the writing system mostly began with pictographs, which gradually became more abstract, greatly expanding what could be represented and communicated. On this view, modern forms of writing came about when ancient peoples developed notations that overcame the limits of depiction, analogous to the transition from realism to abstraction in the history of Western art.

This story has great intuitive appeal, but it is the science project version. It is true that overcoming the limits of depiction was an enormous advance that took eons to achieve. The surprising twist, however, is that writing did not originate with pictures that gradually became more abstract. The monumental development was using graphical elements, some of which were pictographic, in a radically new way, to represent language. It was this innovation that took so long to achieve, and how it could have occurred is the great question. The gradual loss of the pictographic element in some signs was a minor development in comparison; rather than leading to writing, it occurred *because* writing was successful and came into wide use.

Of all the events that occurred in the long development of writing, four are crucial to understanding the essential properties of writing systems and their relation to how we read:

> *Moving from depiction to symbol:* Writing systems emerged when pictographs and other graphical elements were used to symbolize language rather than signify things. A picture of a bird, for example, could be used to represent the spoken word for bird. It is then a symbol for a sound pattern. Using a picture for a purpose other than signifying the pictured entity was not merely counterintuitive, it went unintuited for eons. Once the trick of using graphical elements as symbols was discovered, they could be pictographic or abstract, and they could be used to represent

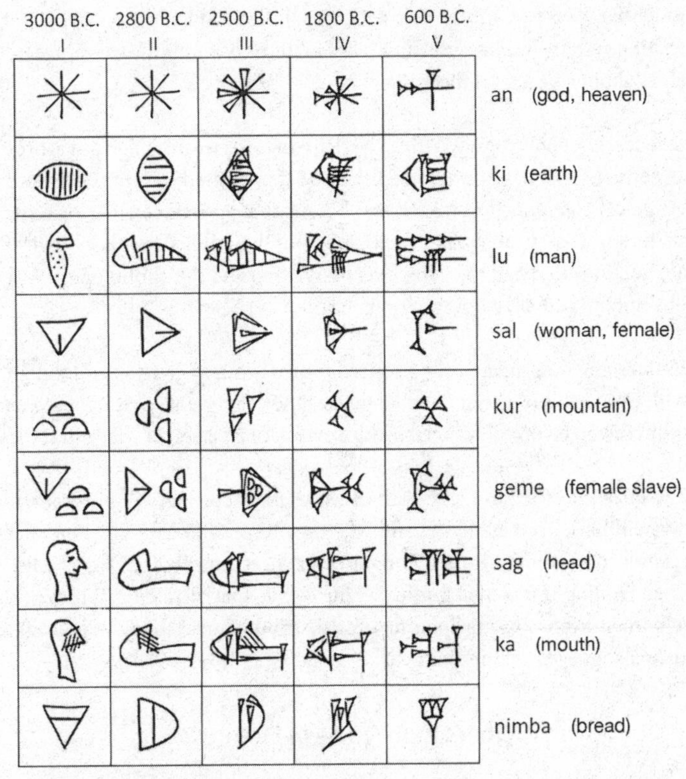

FIGURE 3.1. This illustration (and variants found on numerous websites) conveys that writing began as pictures but evolved into more abstract symbols. The illustrations are based on one in Kramer (1963), popularized by DeFrancis (1989). It isn't quite right because almost all of the earliest cuneiform signs were abstract. From Kramer (1963), 304–305. Copyright © 2015 by University of Chicago Press. Reprinted by permission..

many types of information (e.g., words, initial sounds of words, concepts, categories of objects, grammatical elements).

Representing entire languages: The proto-writing systems only represented some elements of a language, mainly words for objects and quantities. It took another couple of thousand years to develop workable writing systems that fully represented language. The major advance was determining how a relatively small set of symbols could represent a much larger set of words. The general solution, which every successful writing system employs, is using combinations of symbols that represent clues about sound and meaning.

Discovering phonology: Writing systems require treating spoken words as consisting of parts, which can then be represented by a limited set of graphical elements. We take it as obvious that speech consists of units such as words, syllables, and phonemes, but these units are phonological abstractions that had to be discovered. Writing and the phonological way of thinking coevolved over a long period.

Establishing congruence: The properties of writing systems need to align with properties of the spoken languages they represent. Writing systems only converged on this crucial feature over a long period of trial and error.

The questions then are why each of these advances was so important, how they could have been achieved, and why they took so remarkably long. Reading is indeed an "unnatural act" compared to speech, as Philip Gough, a distinguished reading researcher, put it, but the events that were crucial to writing's development were supremely unnatural, distributed over many years, regions, cultures, languages, and individuals.

What Ought to Be in Pictures?

Having developed the necessary capacities and tools, *Homo sapiens* began to produce representational drawings and has done so ever since, a testament to their power, interest, and apparently inexhaustible variety. Prehistoric cave paintings have been discovered at numerous sites around the world, suggesting that creating such images is a very basic human impulse. Why the pictures were created and the functions they served we don't know, of course, but they tell us that the people who produced them possessed the confluence of perceptual, cognitive, and motoric skills that are required to merely depict an object. If the painted image was produced to be communicative rather than decorative or as a way to kill Pleistocene time, other sophisticated capacities would have been involved, such as formulating a communicative intention and understanding how the image would be experienced by someone else. Humans have a "theory of mind" that allows us to make such attributions about the mental states of others.

The cave paintings are remarkably advanced. They are not children's drawings or line drawings. To the modern eye they read as artful representations that required considerable technical skill. Given the age and characteristics of these images, their production is a justly celebrated landmark in human history. We can be awed by the cave paintings, admire their creators, and marvel at their continuity with art of the modern era, but their limitations are most relevant to the development of writing. There was more to know about the horses of the period than can be determined from the paintings at Lascaux. How would the painter have represented the sound of their breathing or how long they slept?

Writing emerged as a way to overcome the limits of representational drawings as the need to communicate messages of greater specificity and variety increased in step with cultural advances, including the development of agriculture, trade, and permanent settlements, and technologies for producing semipermanent records. The astonishing fact is that tens of thousands of years elapsed between depictive drawing and the emergence of writing. What *took* so long? Using graphical elements to represent language rather than the world.

The significance of this advance and why it was so difficult to achieve can been seen by examining how pictures can be used to communicate messages. The obvious way is using a picture to represent the message itself, as when a picture of a horse conveys information via its resemblance to a horse. The message is vague: it might represent a particular horse, horses in general, an event where horses were prominent, or something else. Communication is also limited by constraints on what can be depicted. The expressive range can be expanded somewhat with tricks, such as using images that have strong associations with other things or combining thematically related pictures, as in the international "symbol signs" found in airports. The sign for restaurant, for example, does not depict one. It depicts a fork and knife, which are cues to the concept. Still, the limitations are severe.

Early cuneiform employed several kinds of signs. Some were pictographs, such as the one for barley, a common grain, which was a simplified but clearly recognizable image of a sprig of grain. Like a cave drawing, the pictograph conveyed information via its resemblance to the object itself. The sign is iconic, meaning that the relation between form and meaning is not arbitrary, as indicated by the solid line (Figure 3.2a).

The pictograph differed from a cave drawing because it was also used to represent the object's name, a spoken word (Figure 3.2b). The pictograph now functions as a linguistic symbol for "še" (as in "shepherd"), the Sumerian word for barley. The association between the pictograph and the spoken word is arbitrary (dotted line): the sign would have worked just as well had the word been "ga" or "mark." The association between the spoken word and its meaning is also arbitrary: "še" could have meant cow or head.

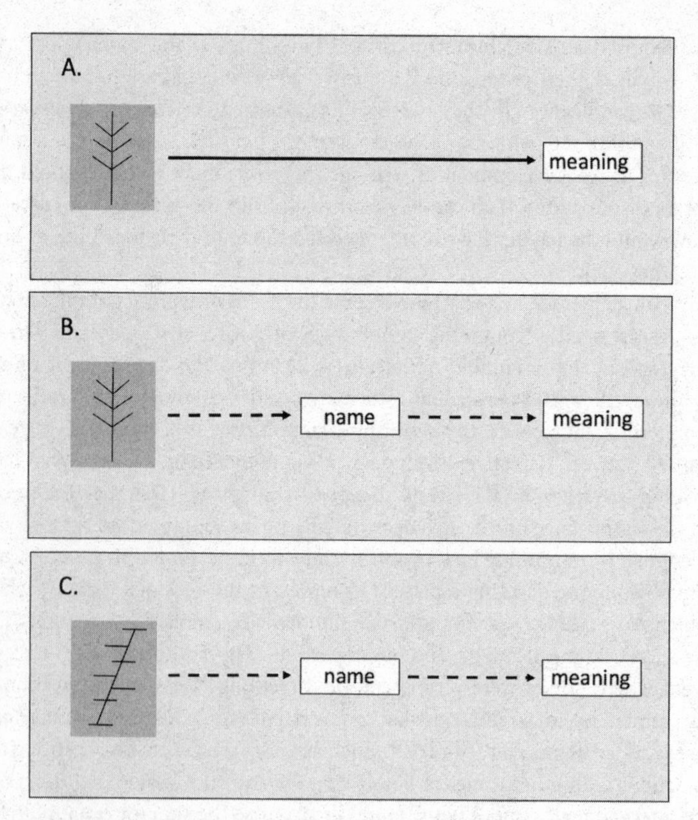

FIGURE 3.2. From pictograph to abstract symbol. In the simplest case, a pictograph (a) conveys meaning via resemblance to the corresponding object (solid line). When the pictograph instead represents the name for the object (b), it functions as a linguistic symbol. Because the visual sign no longer resembles what it represents (dotted lines), it can be abstract (c).

The difference between using a pictograph to indicate barley or the word for barley may seem minor since they work equally well at picking out barley and not fish or bread. However, the change is huge, incorporating several major advances in one swoop.

First, it is a step away from the deep impulse to use depiction to communicate. Want to record a stampede of horses? Draw a picture of it. Want a list of what to get at the Sumerian farmer's market this week? Draw pictures of fish and barley. Using a pictograph as the symbol for a word requires setting aside its most prominent feature, its resemblance to the object. A rose is a rose is a rose, but it could also represent the word for rose.

Second, this usage of pictographs undercuts the need for them. Since the association between an object and its name is arbitrary, the sign does not have to be a picture; it could as well be squiggles (Figure 3.2c). In fact, the earliest Sumerian cuneiform (the precursor to the full writing system) mainly consisted of signs that did not resemble their referents at all, such as the one for sheep: ⊕

Signs that incorporate arbitrary relations between form and meaning are the entrée to building writing systems out of arbitrary symbols such as letters. Signs like barley were transitional objects, a pivot between depiction and writing. Barley had it both ways: it conveyed its meaning via resemblance to the object and its association with the name for the object. The abstract signs demonstrated that the pictographic element was not necessary, even for objects that could be easily drawn. Thus the entries in the later columns in Figure 3.1 are not just objects like barley rendered more abstractly, barley descending a staircase, so to speak. They represent the word, eventually completely divorced from the object's visual characteristics.

Regardless of the exact provenance, Sumerian cuneiform provides clear evidence for the shift from depicting objects to representing language, the great enabling advance in the development of writing systems.

Token → Label → Word

Using graphical elements as symbols to represent language was a great leap forward, initiating several thousand years of activity that resulted in modern writing systems. How could this practice have begun? Perhaps the issue should be put aside because the answer is lost in the sands of time and the Middle East. There is a middle ground between ignorance and certainty that is worth exploring, however, because there are only two serious proposals on the table.

Proto-cuneiform exhibited two puzzling characteristics. First, it was an orderly system of considerable complexity. It apparently did not develop like a child's vocabulary, starting with a single word or two and increasing over time. Second, the signs are mainly abstract rather than iconic; as a consequence, most proto-cuneiform signs have never been deciphered. Such signs are at odds with the view that writing began with pictures.

How could this system have come into being? I've alluded to one possibility, the lone boffin theory: the system was created by a single, unknown individual (or perhaps close collaborators) who had an extraordinary insight, an achievement roughly comparable to that of a Copernicus or Mendel. (Another such person would have to have lived in China and someone else in Mesoamerica, modern Central America.) An individual working on the system in relative isolation for many years, as those scientists did, could explain why it appeared in highly developed form. This theory could be disconfirmed by evidence showing that proto-cuneiform was derived from an earlier system, as some scholars have concluded from inscriptions found on mostly undeciphered artifacts of uncertain provenance that probably predate it.

The other possibility is described in an elegant theory developed by the archaeologist Denise Schmandt-Besserat. The theory holds that a specific, independently established series of incremental cultural developments gave rise to writing as an inadvertent by-product. We are again in the murky territory of Sumeria several thousand years BC, and so naturally the theory is not definitive. However, it accounts for a range of facts and resolves some major puzzles in a principled rather than post hoc way, by connecting two sets of events that would otherwise seem unrelated. And it does not rely on undocumented geniuses or undeciphered writing systems.

Tablets are not the only thing the Sumerians and their neighbors left behind. Thousands of small objects, called tokens, have been found at sites throughout the Middle East. They come in abstract shapes such as cones, disks with markings and holes, and shell-like forms. For many years their functions were unknown. One scholar described some tokens as "mysterious conical clay objects, looking like nothing in the world but suppositories." The tokens appeared around 8000 BC, long before cuneiform. Schmandt-Besserat and others established that they were used in trading, serving the record-keeping functions later assumed by writing. Tokens of a particular shape represented a type of object (such as sheep or oil) or quantity (number of sheep, amount of oil). Crucially, a token was a symbolic object, its shape unrelated to what it designated: the sheep token, for example, did not resemble a sheep.

Like other cultural developments in this era, it took several thousand years, but the Sumerians eventually developed the technology to produce such tokens in large quantities and sealable clay containers to hold them. To keep track of the contents of a container without having to break it open, the owners began to make labels, an impression of the token on the container's exterior. Smart! Because the shapes of tokens were unrelated to what they stood for, the labels based on these shapes inherited this property.

A label, then, was closely associated with both a type of token and what the token symbolized, physical objects that were not visible (think sheep label, sheep token, actual sheep). Since the association was between the label and something in the viewer's head, it could work whether attached to the container or inscribed on a tablet. Labels that migrated to tablets were the source of many early cuneiform signs. The cuneiform for sheep was abstract because the token and its label did not resemble sheep.

The theory explains both of the puzzles. Cuneiform came into the world in a highly developed form because it was derived from the older, established token system. The abstract signs can be traced to the abstract shapes of the tokens. Finally, the theory scores extra points for showing why these properties occurred together—because they had a common source, the token system.

That is a good theory. But, as Steve Jobs used to say in his product introductions, there's one more thing. The theory explains another seemingly incidental fact that turns the idea that writing originated with pictures on its head. In early cuneiform, abstract signs were mainly used for common objects, such as sheep; pictographs were mainly used for less common objects, such as boar. If writing had started with pictures, the pattern should have been the reverse.

It was not that common objects such as sheep were harder to depict than less common ones like boar. Rather, this is an ancient example of the trade-off between explicitness and efficiency. A sign can be more abstract if it is used often enough for people to remember the arbitrary association. More detail is needed if the sign is only used intermittently. The sign for sheep, common animals that were traded, would be used more often than the one for boar, which were rare and not traded. Five thousand years later the same principle explains why a formatting shortcut can be an arbitrary combination of keystrokes if the command is performed frequently but not if it is a rare one such as "insert drop cap."

Schmandt-Besserat's theory yields a surprising twist on the role of pictographs in early writing. Rather than starting with pictographs and gradually abandoning them in order to gain expressive power, cuneiform started with mostly abstract signs. Remembering the arbitrary associations between such signs and the words and referents they represent is manageable as long the signs are few in number and used frequently, as in proto-cuneiform. Growth in the use of the system required creating signs for entities such as boar that were used less frequently. The pictographic element provided a much-needed memory cue for such items, allowing consistent interpretations to be maintained among geographically dispersed users. On this view, pictographs, not abstraction, were the mechanism for expanding the linguistic scope of early writing.

If this chronology is correct, we owe the existence of writing systems to the fact that the Sumerians lacked the technology to produce highly representational trading tokens. Writing would have taken even longer to gestate had there been a Sumerian version of 3-D printing turning out realistic sheep tokens. For the tokens to work, people had to be able to tolerate arbitrary associations between the tokens' forms and what they signified. We know that the people had this capacity because they could speak. Arbitrary mappings are an intrinsic property of language: sound patterns act as symbols for words and other linguistic elements. The sound of the word for sheep is unrelated to its referent; sheep by any other name would smell like sheep. In the token system, the form of the token, one object, was unrelated to that of its referent, another object. Thus language provided one of the key building blocks of the token system, which eventually led to writing, a system for representing language. Perfect!

A Picture Is Worth *How* Many Words?

Using pictographs and abstract signs to represent words was not just game changing, it was life-on-earth-changing because it opened the door to full writing systems. The further problem was to determine how to represent an entire spoken language rather than a few hundred important words. With hindsight, we know what is required. Language is a code that allows an infinite number of messages to be expressed using a finite set of primitive elements: the phonemes, syllables, and morphemes that are the building blocks of words. A writing system based on a suitable set of primitives can represent any word and thus an entire language. This guiding principle had to be discovered, as did how to implement it in a practical way for various types of spoken languages.

Following the initial breakthroughs, the development of proto-cuneiform into such a system proceeded apace—over the next 2,000 years. The system did not expand slowly because geniuses are only born that often. The main factors were cultural: how the system was used, by how many people, using which technologies. Only a limited system known to some individuals was required to manage the books in ancient Mesopotamia. Perhaps because it fulfilled this function well, the far greater potential power of writing was not recognized. Understandable; in our own era, the Internet was in use for many years (on the modern time scale) doing a very good job of allowing early adopters in academia and government to send e-mail, chat, and exchange documents well before the explosive expansion of functions that followed Tim Berners-Lee's invention of the World Wide Web.

Most proto-cuneiform signs, whether pictographic or abstract, were logographs—signs that represent words. Using a different sign for each word becomes infeasible because of limits on how many can be memorized and how efficiently they can be written and read. Natural languages contain far too many words to use a different symbol for each one; thus no writing system is purely logographic. Cuneiform users developed two strategies to overcome this limitation: they assigned additional functions to signs and combined signs to represent additional words. The big breakthrough was using signs to represent the names of objects, but with this successful proof of concept, they could be repurposed in other ways. Too many. Cuneiform eventually developed into a writing system that could represent a language, but it was monstrously complex and superseded by other types of writing that had come along during its long evolution.

Cuneiform took the repurposing strategy to an extreme bordering on chaos. Because signs had multiple functions, it became increasingly difficult to determine how a sign was being used in a given context. Did the barley sign represent barley or one of the other meanings attached to it (e.g., the unit of measurement equal to one-thirtieth of a cubit)? Did it stand alone, or was it part of a multisign combination? Was it there to represent one of the meanings associated with BARLEY or as a cue to other words pronounced "še" or possibly "ši" or "šel"?

The representation of abstract concepts illustrates both the breakthroughs and the complications that ensued. The Sumerians arrived at a strategy for representing abstract concepts, a major step. The words ARROW and LIFE, for example, had the same pronunciation, "ti." An arrow is easy to depict; life is not. Using the arrow pictograph allows abstract words such as LIFE to be represented via the shared pronunciation. The Sumerians had invented homonyms, words like WATCH that have the same sound and spelling but different meanings (a time piece, to look). The LIFE cuneiform is a simple example of the rebus principle, representing a word such as BELIEF by pictures of a bee and a leaf, cues to its sound. The cost is that interpreting even a simple arrow is no longer straightforward: the sign's resemblance to an arrow is relevant to ARROW, but only its sound is relevant to LIFE.

Combining graphical elements to create words is an essential property of modern writing systems, but the cuneiform implementation included triumphs and disasters. Compounding and forming words out of a cue to sound and a cue to meaning were winners, carried forward into modern writing systems. The disasters included representing a word such as FOOD ("gu") by the signs for HEAD ("sag") and BOWL ("sila"). The meanings of the components were relevant but not their pronunciations. Polyphony—using one sign to represent several semantically related words, all pronounced differently—was also a major fail.

This collection of tricks and kludges yielded a full but ultimately unsustainable writing system. If each sign has multiple functions and combines in multiple ways with other such signs, it creates a combinatorial explosion of possible interpretations. Using this system must have been more like puzzle solving than fluent reading, and it died out after many years of service.

Writing Is for Sound and Meaning

Although the Sumerians did not get writing right, they were on the right track. The logic underlying writing systems is simple:

> Spoken languages use sound (phonology) to represent meaning (semantics);
>
> Writing systems represent spoken languages;
>
> therefore,
>
> Writing systems represent sound (phonology) and meaning (semantics).

Writing systems are fundamentally similar because they are all solutions to the problem of representing sound and meaning in visuographic form.

This key idea was hidden in plain sight, in pictographs such as BARLEY (Figure 3.2). The sign conveyed information about both sound (via the association with the spoken word for barley) and meaning (via the pictographic cue). That was the right idea, but such signs did neither job very well. Later writing systems kept the principle but incorporated better ways to accomplish it.

The barley sign's representation of meaning wasn't adequate because it was carried by the pictographic element, which eventually had to go because it was too limiting. If this clue is removed, as in abstract signs such as SHEEP, the form-meaning association is arbitrary and has to be memorized, again only workable when the number of signs is small. The BARLEY sign also represented the sound of the word but via another arbitrary association. Approaches to overcoming these limitations followed two paths. One was separating the sound and meaning cues; the other was finding more systematic and efficient ways to represent the two simultaneously. Both ideas have endured into the modern era.

In the more advanced versions of cuneiform, some words were represented by two signs, a cue to meaning and a cue to sound. The linguist Mark Liberman has termed this the "charade principle," after the age-old party game: if the clues are FRUIT (semantic cue) and LINE (phonological cue), the word must be LIME. Excellent idea; had it been pitched as contemporary Silicon Valley start-up, cuneiform might have made it through several funding rounds before falling into a death spiral because of poor implementation choices, such as using signs as both semantic and phonological cues and allowing too many types of combinations.

Egyptian hieroglyphic writing developed in a remarkably similar way, but with significant enhancements. The system employed pictographs such as an owl, initially used to designate the pictured entity. Like cuneiform, the system evolved during the roughly 3,000 years it was in active use. The owl pictograph was repurposed to represent the spoken word for owl and later the initial sound in the word. Hieroglyphic writing eventually included pictographs for the twenty-four consonants in the language, a major advance, as well as ones for two-and three-consonant combinations. Like cuneiform, the system was complicated because it did not employ separate sets of signs for phonology and semantics. The owl pictograph, for example, was also a semantic determinative, a type of sign indicating a general category, in this case deities (gods, king). Figure 3.3 shows how semantic and phonological cues were used to represent EGYPT. The pictographic information was not essential, and it was not retained in the "everyday" hieratic form of hieroglyphic (sometimes called cursive hieroglyphic).

The strategy of combining phonological and semantic cues was also employed in the character-based Chinese writing system of the second millennium BC. Unlike cuneiform and hieroglyphic, descendants of this system are still in use. Despite its antiquity, there isn't a standard English terminology for describing elements of Chinese writing. Written Chinese is usually inaccurately described as "logographic" for lack of a better term. The system is not logographic because it does not employ a separate sign for every word. It includes some logographs, but they are a small minority (Figure 3.4).

About 80 percent of the characters in modern written Chinese consist of two parts, a cue to meaning (the radical) and a cue to sound (the phonetic). In Figure 3.4, the HORSE logograph is pronounced MǍ.The diacritics over vowels

FIGURE 3.3. Hieroglyphic for "Egypt" illustrates the combination of sound and meaning cues. The component glyphs are *crocodile skin* (left), *owl*, *bread* (top right), and *crossroads* (lower right). The first three are cues to the word's pronunciation, which contained the consonants KMT. *Crossroads* was a semantic determinative for place names.

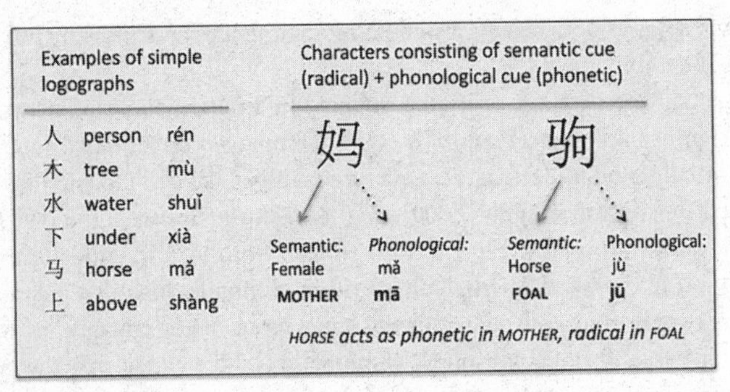

FIGURE 3.4. Chinese writing contains some logographs (left, some examples), but most words are represented by combinations of characters (right). In MOTHER, the HORSE character provides a cue to pronunciation and is positioned on the right. In FOAL, it provides a cue to meaning and is positioned on the left.

indicate tones that signify distinct vowels. MOTHER, pronounced MĀ, consists of the HORSE and FEMALE characters. HORSE (the phonetic) provides the phonological cue, FEMALE (the radical) the semantic cue. Some characters can act as either radical or phonetic; thus, HORSE is a phonetic in MOTHER but a radical in FOAL, pronounced JŪ.

The use of sound + meaning cues in Chinese is effective but has an odd side effect: a word's meaning can be affected by the meaning of the phonological cue. Etymologically, the HORSE character in MOTHER is a phonetic. However, the sound and meaning elements are closely linked. Experiments by Tianlin Wang in our lab found that reading MOTHER mentally activates the semantics of HORSE. This supposedly irrelevant semantic information apparently leaks into the meaning of the word; thus there is a bit of HORSE semantics in MOTHER. In another expression of its power, the way a word is written modifies concepts such as mother.

Japanese writing took the separation of phonological + semantic cues further, using distinct sets of graphical symbols. Japanese was originally written using Chinese characters. The modern writing system retains such characters, termed kanji. Kanji are used for most content words and, owing to their history, have both "Chinese" and "Japanese" pronunciations. A second system, kana, is used to represent the one hundred or so syllables in the spoken language. Kana are used to write semantically light elements, such as grammatical inflections and particles, and *gairaigo*, borrowed words, such as コンピュータ ("konpyuta," computer). The scripts are intermixed in everyday texts.

In written Japanese, then, the principle of conveying meaning through semantic and phonological cues applies to the component scripts: Kanji are more closely associated with semantics and kana with phonology. As ever, the two are not fully separated; the pattern is only quasiregular. A kanji's meaning depends on how it is pronounced; in some cases special-purpose kana (*furigana*) are appended for clarification. Conversely, the kana syllabary has two main subtypes: hiragana, used for native words and grammatical elements for which there are no kanji, and katakana, used for loanwords, scientific terms, and other specialized vocabulary. The two are also distinguished typographically, katakana being more angular than hiragana. Thus, the type of kana that is used to write a word provides some general clues about its semantics.

Intermixing kana and kanji in texts creates a balance that allows the system to be read efficiently. The language can be written entirely in kana, but is very difficult to read that way; it throws off the 和 ("wa," harmony).

Alphabetic writing systems improved on the idea that phonology and semantics could be represented simultaneously using a common set of symbols. Recall that words in Semitic languages such as Hebrew and Arabic are built around roots written with a consonantal alphabet. The words containing a given root are phonologically similar but also tend to be semantically related. KTB, for example, is the root of a family of words related to writing: ***kāṯaḇ****ti* כתבתי "I wrote," ***kāṯaḇ*** כתב "he wrote," ***katteḇeṯ*** כתבת "reporter," and others. The Hebrew root GDL, related to the concept of largeness, occurs in words including the adjective BIG, the verb GROW in various tenses, and the nouns MAGNIFIER, SIZE, and TOWER. This system too is quasiregular: the triconsonantal roots aren't all as closely correlated with semantics as KTB, and at best they only supply partial information about a word's meaning. A word is fully identified by the combination of semantic and phonological cues, plus information from the context in which it occurs.

Finally, alphabetic writing systems such as the one you are reading represent the sounds of words at the level of consonant and vowel phonemes. The correspondences between sound and meaning are largely arbitrary, however. The English word for small domestic feline is "cat" but could as well have been "dog" or "lump." Words that are similar in sound (such as "cat," "cut," "bat") are not similar in meaning; words that are similar in meaning (e.g., "cat," "lion," "panther") differ in sound. Alphabets seem to have taken care of phonology, but where's semantics?

The answer is morphology. Words consist of one or more morphemes, minimal units of meaning analogous to phonemes and graphemes. Words are formed by combining these units in principled ways, as in these examples:

LOOK	one morpheme
LOOKED	two morphemes (the-ED morpheme indicates the past tense of the verb)
LOOKOUT, OUTLOOK	two morphemes (when each is a word, the combination is a compound)
DIET	one morpheme (but two syllables)
DIES	two morphemes (but one syllable)

The association between the sound pattern "cat" and its meaning is arbitrary (as is the spelling-meaning association); however, what is learned about "cat" is relevant to other words in which it occurs: cats, catty, catnip, bobcat, and so forth. Morphemes such as cat make similar phonological and semantic contributions to many words. Because the spelling is alphabetic, CAT represents both phonology and semantics. Thus, alphabetic writing systems also conform to the phonology + semantics principle: they represent the sounds of words as sequences of phonemes, which also correspond to meaning-bearing morphemes.

Finding Phonemo

Everyone seems to have figured out that writing systems must represent sound and meaning, but it took a very long time to develop fully functional systems. The main obstacles were on the phonology side. Spoken words had to be treated as consisting of component parts, which could then be represented by a much smaller number of graphical symbols. The would-be architects of writing systems had to develop something that we now consider an ordinary, teachable aspect of learning to read: phonological awareness. Children demonstrate their understanding that words consist of parts when they can tell if two spoken words rhyme or begin or end with the same sound, or count the number of syllables in a word and identify component phonemes. The question is, How did our ancestors start treating spoken words this way? And how did they converge on the units that were right for their language in accordance with the Goldilocks principle: not too big, not too small, just right. The two developments were closely linked via the successes and failures that occurred during the long evolution of writing.

Although the exact events are unknown, the fact that different types of spoken languages require different types of writing was clearly crucial. The languages spoken in the greater Mesopotamian area where the landmark developments occurred included ones that were typologically unrelated—as dissimilar as English and Japanese. Many times during those eons, the writing system for one language was borrowed for use with an unrelated language. Trying to representing one spoken language in a writing system devised for some other type of language might sensitize a person to differences between them—the sounds they use, how they combine, the patterns that result. I can illustrate why these inadvertent mismatches were such crucial events using Japanese, Hebrew, and English, three typologically, historically, and geographically distinct languages that employ different types of writing.

Japanese is written with the kana syllabary intermixed with kanji. The use of symbols to represent syllables dates from proto-cuneiform. Scholars have noted that the words in the languages for which writing first developed (Sumerian, Old Chinese, Mayan) were mainly monosyllabic. In using signs to represent words, the Sumerians had inadvertently developed a notation for syllables, which in time began to be used for their phonological value. Languages with more complex word structure did not develop writing as quickly.

Accounts of the emergence of writing emphasize cultural and geographic factors rather than linguistic ones. Writing emerged in Sumeria, for example, because the location, a fertile area between two rivers, favored the development of agriculture and animal domestication, giving rise to a sedentary rather than nomadic culture in which trade flourished, creating the demand for writing. However, it's not just that the Sumerians lived in an advantageous place: *they also happened to speak Sumerian.*

The use of a syllabary for Japanese resulted from a classic mismatch. Japanese was initially written with Chinese characters, which did not suffice because the languages are unrelated. It still doesn't: modern Japanese cannot be written wholly in kanji. The fact that major elements of Japanese phonology and grammar could not be represented using Chinese characters created awareness of the need for additional graphical symbols, the seed of the kana system.

Hebrew uses a consonantal alphabet, whose invention is commonly credited to the Phoenicians, living in the area of modern Lebanon and Israel between 1200 and 1000 BC. The Phoenician achievement built on two successive mismatches, a relay from Sumerian to Egyptian to Phoenician.

Egyptian was initially written with cuneiform, which represented syllables in Sumerian, an unrelated language. The phonological gaps were filled by creating symbols for single consonants and combinations of consonants.

Phoenician, in turn, could not be written with either cuneiform or hieroglyphic because it was a Semitic language organized around triconsonantal roots. As we know from Phoenician's modern descendants, Hebrew and Arabic, a consonantal alphabet is sufficient for this type of language. The Egyptians had developed symbols for consonants, but their language could not be written with them alone.

The Phoenician writing system employed representations of single consonants as in the Egyptian system, but not semantic determinatives, because the consonantal roots also provide semantic cues. The Phoenicians' graphical symbols—letters—were wholly abstract rather than pictographs, but so were later cuneiform and hieratic hieroglyphic. The Phoenicians' achievement can then be summarized as this: they codified a nonpictographic, consonantal alphabet that was sufficient to represent a Semitic language with a triconsonantal root system through judicious integration of techniques used in earlier systems.

English is written with an alphabet representing both consonants and vowels (I'll call this the CV alphabet to distinguish it from the consonantal type). Western scholars have traditionally held that writing systems evolved from primitive to advanced forms, the apex being the CV alphabet, one of the extraordinary achievements of Greek culture around 800 BC. CV alphabets aren't just different on this view; they are a superior technology that enabled the development of superior peoples and civilizations. For Jean-Jacques Rousseau, these alphabets marked the transition from primitive to civilized culture: "The depicting of objects is appropriate to a savage people, signs of words and of propositions, to a barbaric people, and the alphabet to civilized peoples." Eric A. Havelock, a distinguished twentieth-century authority on the Greek alphabet, made much the same point more politely in 1982: "The invention of the Greek alphabet, as opposed to all previous systems including the Phoenician, constituted an event in the history of human culture, the importance of which has not as yet been fully grasped. Its appearance divides all pre-Greek civilizations from those that are post-Greek. On this facility were built the foundations of those twin forms of knowledge: literature in the post-Greek sense, and science, also in the post-Greek sense." Or as a linguist of the 1950s wrote, "It is generally accepted on all grounds an alphabetic system is best." High praise.

The development of writing systems that represent spoken words at the level of consonant and vowel phonemes was enormously important, of course, especially for people who speak languages such as Greek and English. The primary question, how the Greeks found the vowels, needs to be separated from the secondary claim, that CV alphabets are superior.

CV alphabets are another mismatch story. Phoenician was a Semitic language; Greek is Indo-European. Looking again at their modern descendants, we see that one type can be written without vowels, and the other cannot. By borrowing the Phoenician system, the Greeks discovered the need to represent vowels. Interestingly, vowels had surfaced earlier, though inadvertently. Signs in cuneiform mainly represented syllables. In Sumerian and many other languages that adopted cuneiform, syllables could consist of a single vowel. For example, Sumerian cuneiform included a sign for the syllable /a/, meaning water. The writing system therefore included symbols for vowels but they were not recognized as such. Symbols for isolated consonants did not develop because syllables cannot consist of a single consonant. Thus, whereas cuneiform had vowels and Phoenician and Egyptian hieroglyphic had consonants, the Greek alphabet incorporated both.

It may be that the full alphabet came together in Greece at that time because the intellectual culture was sufficiently advanced to support it. The succession of innovations in the history of writing suggests that the main factor was that the people happened to speak Greek, a language that required both vowels and consonants to be readable, and to live in Greece, not far from regions where a consonantal alphabet was used. The means to represent both consonants and vowels was evident from other writing systems, and the scholars of the era combined the relevant pieces to create an orthographic tour de force.

Is the full alphabet superior to other types of writing? It is certainly unique in one respect: every language can be written with an alphabet of manageable size because words can always be represented by component phonemes. Alphabets could be seen as superior because they are a general solution to the writing problem. This implies that the world would be a better place if everyone used alphabets and that other types of writing live on only because they are too entrenched to be replaced. These are weak arguments, however. The fact that every language can be written in an alphabet does not mean that doing so will be the most efficient solution. The goal is not to use the writing system with the broadest applicability; rather, it is to use the one that is well-matched to the properties of the spoken language. Alphabetic writing systems owe their existence to the fact that writing systems can be mismatched to spoken languages. Using a full alphabet where another type of writing system is more suitable is also a mismatch.

Consider again Japanese, Hebrew, and English. Japanese has relatively few syllables (around one hundred) and can be written with a syllabary. Writing Japanese with an alphabet yields a poor result because it obscures the syllables that are the foundation of the spoken language by adding unessential phonemic detail.

Writing Hebrew with a full alphabet also provides too much information, making reading more difficult. Writing Hebrew with a syllabary would be an orthographic nightmare because it would obliterate the root structure. For the

KTB words, for example, a different symbol would be required for each syllable containing K—one for KĀ, another for KI, a third for KO, and so forth—and similarly for T and B. The KTB words would no longer have anything in common. It is a problem if a writing system does not respect the basic principle governing the structure of words in the language it represents.

English? It has around 15,000 syllables, depending on who's counting, far too many for a syllabary. It does not have the root structure of languages like Hebrew and cannot be written with a consonantal alphabet. English requires an alphabet representing both vowels and consonants.

These examples suggest that to be effective—and to have survived into the modern era—the properties of a writing system must align with properties of the spoken language it represents. A unit that is well suited to one type of language may be disastrously bad for another. A unit can be too big (a syllabary for English) or too small (a CV alphabet for Japanese). A single, universal writing system isn't feasible because one size does not fit all. The match between units in speech and in writing is not perfect, and some outlier cases have survived. However, the relationship is strong enough that many combinations yield writing systems that range from inefficient to unreadable. In general, spoken languages have ended up with the writing systems they deserve—which aren't always CV alphabets.

The cultural contexts in which writing systems are used can change in ways that affect their continuing viability. The history of writing is littered with systems that were fully adequate in their time and place but were overtaken by new demands. CV alphabets may not be optimal for all languages, but they are very well suited to keyboarding, still the main means of interacting with computers and other screen devices. That is placing pressure on writing systems such as Chinese, which some argue is transitioning to being written with an alphabet. Pinyin, a system for representing Mandarin alphabetically, was developed for use in reading instruction, but also works for keyboarding. Chinese speakers who keyboard extensively with pinyin begin forgetting how to write the Chinese characters, a condition jokingly known as "character amnesia." Whether alphabets will continue to serve as adjuncts to Chinese writing or eventually replace it remains to be seen. Skeptics focus on the mismatches between spoken Chinese and alphabetic representation or favor annotating characters with the corresponding alphabetic code, but others think that alphabets will suffice.

The problem that writing systems solved is how to convey messages of much greater variety and specificity than afforded by pictures, ideographs, and other visual elements. The general solution was to use an efficient graphical code to represent spoken language at a level appropriate for a type of language. The characteristics of writing systems are determined by the ways they represent spoken languages and the ways spoken languages, in turn, represent meaning. Writing systems are alike because they represent phonology and semantics, though the solutions vary in detail. This property has an important implication for how we read: reading is not just about spelling; it is inherently also about phonology and semantics because that is what writing systems represent.

HOW WE READ

CHAPTER 4

The Eyes Have It

READING EXISTS BECAUSE OUR ANCESTORS loosed the shackles of depiction and solved the hard problem of representing spoken language in visual form. The new medium retained the expressive capacity of spoken language but overcame its major limitation: its impermanence. Linguistic records that last longer than spoken utterances are an obvious win, but there was more. Speech is great and powerful, but it has some terrible design features. Utterances can only be produced as a series of words rather than in big chunks. Speakers have to plan what to say next while they finish saying the current bit. Formulating and generating a mostly grammatical utterance that more or less expresses an intended message with a modicum of fluency is hard. We can hear the difficulty in everyday speech: the mispronunciations, grammatical errors, and pauses; how often we start a sentence and change direction in midstream; how much we rely on gestures and facial expressions to help. These phenomena reflect the intrinsic challenges of speech production rather than personal deficiencies.

As listeners we are stuck with what gets served. An utterance unfolds. Words appear sequentially, lasting only as long as it took someone to say them, and then disappear to make room for the next part. Listening would be just like Lucy wrapping chocolates as they came down the conveyor belt if the device stopped, backtracked, and changed speeds at unpredictable intervals. Listeners have to wrap spoken words as they pass by, with no option to stuff a few in their mouth or pocket if they fall behind.

For a capacity that is so deeply ingrained in our species, our abilities to produce and comprehend speech are remarkably mismatched. We can comprehend speech at faster rates than we can fluently produce it.

The reading chair is a lot more comfortable. Behold a text, ready for consumption, nicely plated, the effort that went into preparing it tactfully hidden in the kitchen. We, not the person who produced it, control the pace at which it is ingested. And—best thing ever—text does not disappear as it is read. Language may have originated in the spoken channel, but reading is version 2.0.

Given that the written code didn't inherit the limits of spoken language, did it expand our linguistic capacities? Absolutely. The medium allows writers to create texts that are far more complex than speakers can produce. It's the difference between oral storytelling in preliterate cultures and *The Amazing Adventures of Kavalier and Clay*. Luckily, readers are also able comprehend these more complex constructions. Reading is the prime example of a technological add-on that extended our capacities beyond their natural limits.

Then there is the brute-force, zero-to-sixty increase in comprehension speed that reading affords. This isn't because the speed of light is faster than the speed of sound. It is because of the difference in who controls the rate of transmission. Listeners are obligated to spend as much time on each word as it took the speaker to say it, whereas readers do some optimizing, adjusting the amount of time they spend on words, lingering as long as needed, moving on when they are ready. There is no skipping ahead in listening (it takes a recording for that). The differences in comprehension rate can be substantial. The exact numbers vary, of course: on the reading side, they depend on the skill of the reader, the complexity of the text, whether it is being read for the first time or reread for the *n*th, and how closely the text is read; on the speech side, it depends on the content of the spoken message and how it is said (properties such as rate, accent, loudness). Although no single number can capture the modality difference, its general character can be grasped by looking closely at an example. Picture a good reader (a college freshman perhaps) finally getting around to the first *Harry Potter* book, deeply immersed, reading with care and interest.

That person can read with good comprehension at about four to five words per second. *Harry Potter and the Sorcerer's Stone* has approximately 77,000 words. At five words per second, the book could be read in a little over four hours, theoretically.

The number seems unrealistic because it ignores the many events that affect actual reading times: opening the book (or ebook), finding your place, rereading a little of the text to get reoriented, rereading a sentence, losing your place, turning the page, daydreaming, petting the cat, removing the cat, and the many other little disturbances of everyday reading life. Or pausing to experience what you've read—thinking, feeling, anticipating—or rereading a section for pure pleasure, efficiency be damned. Let's not even consider all the distractions packaged with your e-reading device. If you didn't finish *Sorcerer's Stone* in 4 hours don't feel badly. You might have actually spent about that much time on just the reading part. Think of 4 hours as the EPA mileage estimate, which is better than obtained in the real world but good enough for comparative purposes.

The *Sorcerer's Stone* audiobook, in contrast, is approximately 8.5 hours long. This total includes time unrelated to the text itself (e.g., interstitial music); on the other hand, it is a performance recorded to be easily comprehended in that medium. The audiobook runs about twice as long as our EPA estimate of reading time. Although the numbers are only approximate, the 2:1 ratio between listening and reading times is realistic. Average speaking rates in English are in the 120 to 180 words per minute (wpm) range, and guidelines for recording audiobooks recommend 120 to 150 wpm, which is about half the typical skilled reading rate (4 to 5 words per second = 240 to 300 wpm). Most audiobook and podcast players allow the speed to be cranked up, but playback at 300 words per minute is hard to comprehend except for short stretches. The audio version of the *Harry Potter* book may be perfect for that long car trip with the children, but it is not a speedy way to take the story in.

Do these numbers hold for other types of speech and text? Let's take another shot. Aaron Sorkin is a TV writer/showrunner known for his rapid-fire dialogue. Perhaps you have seen the monologue from his show *Newsroom*, which won an Emmy for Jeff Daniels. Daniels's character is speaking pretty fast, about 489 words in 3.5 minutes, which is about 2.3 words per second. Such well-formed sentences, and so many of them! The speaker could do this, of course, because he's working from a script, doing an actorly simulation of extemporaneous speech. And yet his speaking rate is still considerably slower than a reading rate of four to five words per second.

Things are looking good for reading. With a text that doesn't fade and control over the pace, we can read at speeds that exceed the limits of speech comprehension. If speed is the only consideration, text wins. The question then is how far we can push this advantage. We all have our set-point reading speeds. Could they be improved? People train to improve their running times and laps per minute; why not reading times? Children could prep for reading bees instead of spelling bees, competing to read the fastest with the fewest pronunciation or comprehension errors.

Let's wait on that one. Reading is accomplished by a visual system that evolved to solve other kinds of problems—recognizing faces and objects, perceiving the world as a stable, three-dimensional space, distinguishing the colors, textures, and movements of what is outside over there—not to decipher abstract two-dimensional patterns. Reading's advantages over speech only extend as far as the eye can see.

Eyes Like Yours

Reading is principally a linguistic experience, employing knowledge that is shared with speech. However, the nature and arrangement of the visual symbols on the page, coupled with the fact that they are perceived by eye rather than by ear, give rise to some characteristics that are specific to reading. Using language in this modality has advantages, but we're stuck with properties of the visual system that impose their own limits.

Much is known about vision as it occurs in humans and other species, from the development of the neural substrate to the perception of a three-dimensional world populated by identifiable objects and individuals situated in complex, rapidly changing visual scenes. The most important information about the role of vision in reading comes from looking at the eyes themselves. Technologies that track exactly where the eyes focus and how they move became available in the late 1960s. The first trackers were large, expensive devices, the equivalent of the large, expensive mainframe computers of the era. Now, as with computers, the technology is better, smaller, and cheaper. Lightweight eye trackers can be worn like a hat, collecting data about where people look when they are driving or taking in the *Mona Lisa*. Put one on a baby's head, and we can see what they like to look at. Knowing where people look on web pages or when they are watching TV matters greatly to content producers and television network executives.

In the 1970s, researchers in a few labs decided to use this new tool to study reading. They were gambling because it was not clear whether anything could be learned just from watching people's eyes. Most of the reading process happens in your mind, not your pupils. Eye movements might be useful for studying the lower-level *seeing* part of reading, perceiving the squiggles on the page, but their potential to reveal anything about the higher-level *thinking* part—using knowledge of language and the world to understand the message that the squiggles convey—seemed limited.

All this was obvious until the studies were done. Eye movements proved to be far more revealing than expected. Just the bare facts about how the eyes take in the written code explain a great deal about the nature of reading. Even more surprising, many mental events that occur while we read can be inferred from their effects on eye movements. It takes a properly designed experiment—we are not just looking deeply into a person's eyes to discern what they are thinking or watching someone while they read—but eye movements are a powerful tool for studying reading and thinking.

Eye movements involve two decisions: where to look and when to move. As can be seen by looking at a video (such as this one: http://goo.gl/VnOxuT), the subject in an eye-tracking experiment reads a text displayed on a computer screen. An eye-tracking device directs an imperceptible infrared light at the person's eye. The angle of the beam's reflection off the eye is used to triangulate where the person is looking. As the eye moves from left to right, it makes brief pauses, called fixations. The rapid jumps between fixations are called saccades. The number of fixations, the amount of time spent on each, and the size of the saccades are the major determinants of reading times. (The saccades themselves only add about twenty to thirty milliseconds each.)

Eye movement recordings also show that reading does not just proceed from left to right; the subject makes occasional backward saccades in order to reread part of the text. These "regressive eye movements" are common even for skilled readers. The sweeps that occur when the eye reaches the end of one line and moves to the beginning of the next line are also recorded.

This jumpy, fixation-saccade-fixation pattern characterizes how everyone reads, yet seems inefficient on the face of it. The reader pauses at one spot to read what's there and almost immediately begins planning where to look next, then executes the jump, landing on the next fixation spot. Lather, rinse, repeat, many times per text. What if we could just scan the text in a continuous manner, like a nice steady pan shot? It's not unthinkable. The visual system is capable of another type of eye movement called smooth pursuit. Try it:

Make a "we're number one" fist, index finger pointing up, with your right hand. Position your hand about a foot in front of you on the left side. Fixate on your index finger, then keep your eye on it as you slowly move it from left to right, then right to left. Your eyes should be able to follow these movements without making any jumps. That is smooth pursuit.

Now change one small thing: Make two of those "we're number one" fists. Position your hands a couple of feet apart, comfortably in front of you. Fixate on the index finger of your left hand and make the same slow scan from left to right—but without moving your finger. Your hands indicate where to start and finish, but *you* determine the visual path. Can you still move your eyes smoothly? No. You'll have the same problem if you try to read this text smoothly from left to right. You can try to eliminate the saccades by practicing, but good luck with that. The visual system is only capable of smooth pursuit if the eyes are tracking a moving stimulus (like your index finger), not scanning a static one (like a text). Unless we radically change the way that texts are presented, reading requires fixations and saccades.

The Perceptual Span

The crucial property of the visual system for reading is that the amount of information that can be taken in on a fixation is limited. Say you were reading along and your eyes landed on the A in the word AMOUNT in the previous sentence. How many letters can be seen? How wide is the window, the *perceptual span*?

I will tell you, but as with listening and reading speed, the answer comes with some fine print. We do not perceive exactly the same number of letters on each fixation. The number does not vary much, but it is affected by properties of the display (size of the letters, how densely they are packed) and properties of the text (whether the fixated word or the word to its right is easy or hard). The numbers I'll present represent upper limits obtained under favorable conditions—the best-case scenario.

Say you are reading the sentence

Graphology means personality diagnosis from hand writing.

and your eye has landed on the A in personality (indicated by the underscore). How many letters are seen on this fixation and in how much detail? Here the dotted line indicates a wide window:

```
 ................................................
Graphology means personality diagnosis from hand writing.
```

Can readers take in all of these letters on a fixation? People sometimes think they do. However, the window might be somewhat narrower:

```
      ......................................
Graphology means personality diagnosis from hand writing.
```

or even much narrower:

```
                ..................
Graphology means personality diagnosis from hand writing.
```

How wide is the actual window? The data come from a clever application of the eye-tracking methodology called the moving window technique. A subject's eye movements are monitored as a text is read on a computer

```
The little girl was happy to win the race last weekend.
                      *
-------------------------------------------------------

xxe little girl wxxxxxxxxxxxxxxxxxxxxxxxxxxxxxxxxxxxxxx
      *
xxxxxxxxle girl was hapxxxxxxxxxxxxxxxxxxxxxxxxxxxxxxxx
            *
xxxxxxxxxxxxxxxxxxs happy to win xxxxxxxxxxxxxxxxxxxxxx
                      *
```

FIGURE 4.1. Moving window experiment. *Top line*, normal format; *below*, moving window sequence. Letters are visible within a window around the current fixation (indicated by *); others are replaced by x's. The window follows along as the reader saccades to the next fixation. Three successive fixations while reading a single line are shown. From Bélanger and Rayner (2015), Figure 1, p. 222. Copyright © 2015 by SAGE Publications, Inc. Reprinted by permission.

screen. The trick is that the computer has been programmed to rapidly change the display during the saccade from one fixation to the next. This happens too quickly for subjects to notice. Figure 4.1 illustrates how the display changes on a typical trial as a reader makes a series of saccades across one line of text. A normal display is at the top; the moving window display is below it. The person begins reading from the left, fixating on the first T in LITTLE. During the saccade from there to the I in GIRL, the display changes, crossing out the letters outside the window, and then it changes again during the saccade to the first P in HAPPY. The experimenters determined the size of the perceptual span by varying the size of the window (number of letters to the left and right of fixation that were not replaced by x's).

The predictions are simple. Replacing letters that are too far away from fixation to be perceived—ones that are outside the perceptual span—should have no effect on the reader. Replacing letters that are within the perceptual span should interfere with reading. The interference effect shows up as shorter saccades and longer fixations. Let's call the question: How wide is the window before replacing the letters has an effect?

Answer: there is no effect if the letters are replaced more than about fifteen to sixteen letters to the right and three to four letters to the left of fixation (these counts include spaces between words). Thus the perceptual span has about twenty slots. Replacing letters to make a smaller window disturbs normal reading. The window is asymmetrical to the right because English is read from left to right. As you can probably guess, in Hebrew, which is read from right to left, the span is asymmetrical to the left.

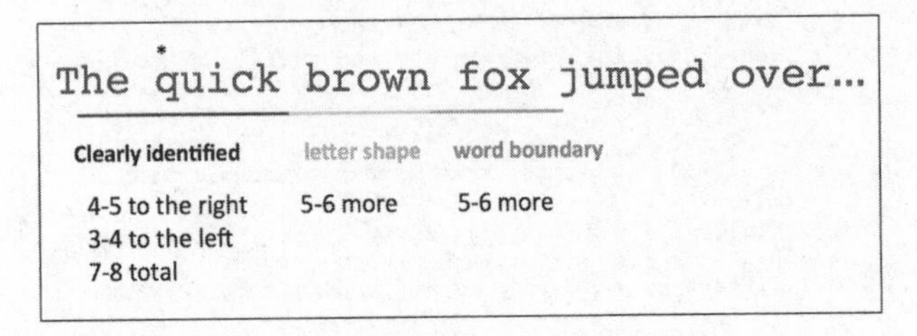

FIGURE 4.2. Perceptual span. Acuity is high around the fixation point (indicated by *) but drops off to the right. The entire span over which any information is obtained is only about 20 slots.

The letters within this twenty-slot window are not seen equally well (Figure 4.2). With the eye fixated on the letter Q, about four to five letters to the right and three to four to the left can be clearly identified. Five to six positions further to the right, only the general shapes of letters can be identified, such as whether they are ascenders or descenders (b, j) or sitting on the line (o, e). Finally, for the five to six slots at the right edge of the window, only the presence or absence of letters can be perceived. This minimal information is useful; the empty slots indicate word boundaries, which helps the reader calibrate where to fixate next. Still, it is surprising how few letters can be seen clearly.

Properties of the retina are to blame. The letters that are close to the fixation point project to the fovea, the area of the retina with the highest acuity (resolution). Moving away from the fovea, acuity drops rapidly. Letters that project to areas farther from the fovea are therefore seen less clearly. This change in acuity is due to the uneven distribution of receptor cells across the retina. The fovea has the densest concentration of cones, the high-acuity receptors. The density of the cones drops off rapidly moving away from the fovea, and with it, sensitivity to detail. Properties of the perceptual span closely track these basic facts about the neurobiology of the retina.

With this information in hand, look again at the pattern of eye movements in a video recording. The fixations are not spaced very far apart. The distances from one fixation to the next vary, but most are about seven or eight letters, with a few that are shorter or longer. The way we space our fixations results in most letters' projecting to the fovea, minimizing the need to guess them from lower resolution images.

The narrowness of the perceptual span seems like a vision problem that should be fixable. Corrective lenses allow people to see letters more clearly in the fovea. What about correcting the poor acuity outside the fovea? Some masters of the eye-tracking methodology have tried it. The technique is

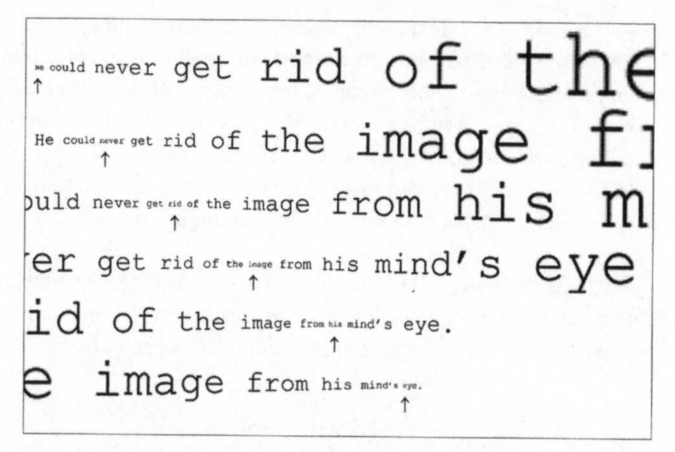

FIGURE 4.3. Parafoveal magnification technique. Each line shows how the display appeared while a single line of text was being read. The arrow indicates where the reader was fixating. Letters around the fixation were presented at a normal size but the surrounding words were artificially increased to compensate for the drop-off in acuity away from fixation. The sizes of the letters changed each time the reader made a saccade. From Miellet, O'Donnell, and Sereno (2009), Figure 1, p. 722. Copyright © 2009 by SAGE Publications, Inc. Reprinted by permission.

called parafoveal magnification (parafovea refers to areas adjacent to the fovea). The trick is to compensate for the decrease in acuity by increasing letter sizes in proportion to distance from the current fixation. The displays they used are illustrated in Figure 4.3. The arrow indicates the current fixation. The lines represent how the display changed on each fixation as one line of text was read. The researchers also varied window size from seven to twenty-one slots using the letter-replacement method from Figure 4.1. The main question was whether the perceptual window would be wider when the letters in the parafovea were larger than normal.

The subjects, who were college students, rapidly adapted to reading with parafoveal magnification. Surprisingly, the manipulation had very little impact, positive or negative. Reading times with magnification closely matched those for normal text, and comprehension was equally good. Again it was window size that had the big effect: with the window reduced to only seven normally displayed letters, reading times were seriously slowed. This finding confirms that information outside the fovea is useful even though it is imprecise.

This experiment shows that the limits on the width of the span are cognitive as well as visual. Under normal conditions, the drop-off in retinal acuity is the major limiting factor. That is the reality of everyday reading. Technology can't compensate much because only a few of those bigger letters fit within the perceptual span. Moreover, the magnified letters have little impact because of limitations on how much information can be processed at a time. Reading is a demanding task. The amount that we can see on a fixation seems pretty well matched to how rapidly we can make sense of it. Magnifying letters in the periphery is ineffective because we're already wrapping chocolates as fast as we can.

Although the logic was solid, compensating for the acuity drop-off is not a promising way to accelerate reading. Performance might improve somewhat with extended practice using the technique, but it isn't a game changer.

We Like to Look

The perceptual span seems pretty small. Only seven or eight letters seen clearly on each fixation? My last name has more letters than that. Here, at the very onset of reading, the eye apparently creates a bottleneck. How bad is the damage, and are there any other ways to mitigate it?

The impact depends on two factors. One is fixation duration, how long we look at the words within the window. The other is how much of a text has to be fixated. How far apart can the fixations be spaced without disrupting comprehension? Is skipping allowed?

Fixation durations depend on the difficulty of a word, the difficulty of the text, how closely it is being read, and the skill of the reader. For skilled readers and a moderately difficult text like this paragraph, the average fixation is in the range of 200 to 250 milliseconds. Representative data from a skilled reader are seen in Figure 4.4. Here the average fixation duration was 222 milliseconds. As in this example, most fixations are between two hundred and three hundred milliseconds. Some can be as short as fifty milliseconds; others, more than ten times that long. These numbers are typical: average fixations are longer for less-skilled readers, and they can be shorter for highly skilled readers. Changes in basic eye movements characteristics with development are shown in Table 4.1.

Which words are fixated? Almost all content words such as nouns and verbs. Some longer ones (like UNIVERSITY or PROPAGANDIZE) can be fixated twice. People also refixate words when they make regressive eye movements. Figure 4.4 illustrates the common finding that 20 to 25 percent of the words are skipped, but skipped does not mean unread, only that they did not get their own fixations. Most are short function words—determiners (A, THE), prepositions, auxiliary verbs, the occasional pronoun—that get a free ride while the

Jerry is usually quite grouchy until he has had his morning coffee and read the paper.

Fixations	*	*	*	*	*	*	*	*	*	*	*	*	*
Order	1	2	3	4	5	7	6	8	9	10	11	13	12
Duration (ms)	311	218	266	202	182	233	178	193	215	227	233	145	288

FIGURE 4.4. An example of eye fixations while reading, measured in milliseconds. Fixation durations typically range from under one hundred milliseconds to longer than a second. Saccade lengths also vary (in this example, from three to fifteen letter spaces). Two of the thirteen saccades are regressions. From Starr and Rayner (2001), Figure 1, p. 158. Copyright © Elsevier. Reprinted by permission.

TABLE 4.1. DEVELOPMENTAL CHANGES IN EYE MOVEMENTS DURING READING

	Grade Level						
	1	2	3	4	5	6	Adult
Fixation duration (ms)	355	306	286	266	255	240	233
Fixations per 100 words	191	151	131	121	117	106	94
Frequency of regressions (%)	28	26	25	26	26	22	14

Note: Average fixation durations, frequency of fixations, and frequency of regressive eye movements all decrease as reading skills increase. In the United States grade 1 children are typically six years old and at the start of reading instruction. *Source:* Starr & Rayner (2001).

eye is fixating an adjacent word. For example, the IS in the Figure 4.4 sentence might be read while the eye was fixating on JERRY.

The fact that people fixate most words isn't an accident or lifestyle choice. It is what happens when the perceptual span collides with the properties of words in English. The words on page 68 are about 5 letters long on average, ranging from 1 to 15, plus one for the space. If only seven or eight letters are seen clearly on a fixation, most content words will be fixated, especially since they are also carrying some of the function words.

Is other skipping allowed? Content words are sometimes skipped if they are highly predictable. In reading

The personable Neil Patrick Harris . . .

if your eyes land on the word PATRICK, you can probably pick up enough information to the right to confirm that the word is indeed HARRIS, allowing you to jump over it. Few content words have this high degree of predictability, however. Function words are more predictable, but there is little benefit to skipping them because they convey less information, and many piggyback on content words anyway. Readers can opt to make larger saccades, but that is like reading a text in which words have been intermittently omitted. We sometimes skip words if the text is familiar or boring, or we decide it doesn't need to be fully comprehended. But let's be clear: skipping results in a loss of information because words that aren't seen can't be read and usually can't be inferred.

Speed Reading and Reading Speed

The late Nora Ephron famously felt badly about her neck, but that's minor compared to how people feel about their reading. We think everyone else reads faster than we do, that we should be able to speed up, and that it would be a huge advantage if we could. To read ten times faster than one's current rate would be to possess a cognitive superpower. You could read as much as a book critic for the *New York Times*. You could finish *Infinite Jest*. You could read *all of Wikipedia*. So, how fast *can* people read?

Reading speed is obviously going to depend on factors such as readers' skills and goals and whether they are reading Richard Feynman's lectures on physics or TMZ.com. But let's just do some cold, hard calculations based on facts about the properties of eyes and texts.

About seven to eight letters are read clearly on each fixation.

Fixation durations average around 200 to 250 milliseconds (four to five per second).

Words in most texts are about five letters long on average.

4 fixations per second = 240 fixations per minute

240 fixations × 7 letters per fixation = 1,680 letters per minute

1,680 letters/6 (five letters per word plus a space) = *280 words per minute*

That's a reasonable number to take as a reference point for normal reading. The number is generous because it assumes, unrealistically, that all seven letters are recognized without error on every fixation, no regressive eye movements occur, and the text is read at a steady rate without pausing. It also leaves out time taken up by saccades, line sweeps, blinking, and other minor events, and it ignores the important fact that many longer words are fixated more than once. The estimate is a bit conservative because readers do take in some information from the parafovea, all the letters in a word don't always have to be recognized, the average length can be a little longer than five letters per word, some people average closer to five fixations per second, and some shorter words are not fixated. It is a good, representative value to use as an anchor.

Given these calculations, how long should it take to read a 309-word text from, say, a book by Maria Montessori, the famous educator and namesake of countless preschools? Since most words are fixated once, some are skipped, and some are fixated twice, the number of fixations will be about 80 to 90 percent of the number of words. A 309-word text would then take around a minute to read. The text is included at the end of this chapter if you'd like to check. Mileage varies: typical reading speeds can be faster or slower than my estimate, and the text can be read more or less closely. The exact number of words per minute is far less important than the fact that this value cannot be greatly increased without seriously compromising comprehension. You can skim the text more quickly, but that's obviously different: you might understand the gist and pick up some specific facts but miss most of the content.

I've described the behavior of a typical skilled reader. How much could performance be improved? Some people claim to know the secret to becoming a superreader and are happy to share it—for a modest fee.

The consuming urge to read more quickly has created a steady market for techniques that promise large increases in reading speeds over the values we've considered. The breakthrough product was Evelyn Wood Reading Dynamics, which caught on in the 1960s with a boost from President John F. Kennedy, said to have been a fan who brought Evelyn Wood instructors to the White House to train his staff. For Evelyn Wood, a longtime school teacher, improving people's reading was a calling. "Why have I dared to believe that reading down the page, at speeds held by many experts to be impossible, can be done? Perhaps it is because my deep religious convictions teach me that 'a man is saved no faster than he gains knowledge;' that 'no man can be saved in ignorance;' that 'the glory of God is intelligence.'" Her precepts about reading were adapted for a succession of technologies (books, cassettes, CDs, DVDs, computer program, a website, YouTube videos). The method has been packaged and repackaged more often than Beatles songs. The Evelyn Wood brand is still around, although it is a minor player in the market it created.

In its 1970s heyday, everyone tried Evelyn Wood Reading Dynamics. Steve Allen, first host of NBC's *The Tonight Show*, pitched the product in the black-and-white era. Marshall McLuhan was an abstruse 1970s media theorist mainly remembered for a hilarious cameo in the movie *Annie Hall* (and for hilarious proclamations such as "Literacy remains even now the base and model of all programs of industrial mechanization; but, at the same time, locks the minds and senses of its users in the mechanical and fragmentary matrix that is so necessary to the maintenance of mechanized society"). McLuhan, a university professor, took up speed reading, boasting to skeptical colleagues that he could read 1,500 words a minute and thirty-five books a week. As an early theorist about the impact of information age technologies, McLuhan did not accord special status to print. Although initially a convert who thought that speed reading revealed deeper patterns not evident in conventional reading, he eventually concluded that it was only useful for tasks like scanning junk mail.

In the modern era, anyone can be a speed-reading expert/entrepreneur. Consider Howard Stephen Berg, author of numerous books with titles like *Super Reading Secrets*, who claims to read 25,000 words per minute (that's 416 words per second). Berg hawked a product called Mega Reading, said to allow people to achieve dramatic increases in reading speed, into the thousands of words per minute, without sacrificing comprehension. His big break was getting into the 1990 *Guinness Book of World Records* as the world's fastest reader. Like purveyors of other miracle self-improvement products, the speed-reading industry relies on demonstrations and on personal testimonials ("After only eight weeks I was able to demonstrate a reading speed of over 30,000 words per minute and comprehension level of 95 percent!"), while avoiding close cross-examination.

The continuing market for these products is a testament to many things: the intensity of the desire to read more quickly; the performance anxiety that reading inspires; the fact that for many people reading is a burdensome but sometimes unavoidable task that they would like to complete as quickly as possible. The potential payoff is so great that people are willing to pay $19.95 plus shipping and handling to give it a shot. As an expert in reading rather than marketing or psychiatry, my concern is whether such systems work. Is it possible to increase reading speed and efficiency using specialized techniques? Reader: save your money. The gap between what is promised and what can be attained is huge, so huge as to have attracted the periodic attention of consumer protection agencies such as the Federal Trade Commission (FTC). What is claimed cannot be true given basic facts about eyes and texts. Unless we redefine reading as rapid page turning, deleting the bit about comprehension, people are as likely to read thousands of words per minute as they are to run faster than the speed of light.

I've already mentioned the one simple, guaranteed way to increase reading speed: skimming. Skimmers sample a text, making fewer fixations and larger saccades. People read for different purposes, and sometimes skimming is all that is required. You're reading a long article on the web about scenarios for economic recovery in southern Europe. You're not interested in the details but would like to get the gist. You skim. You've received the annual multipage Christmas letter from an oversharing childhood friend from Michigan. You skim. There is a trivial sense in which these texts are being read rapidly, but very little is being comprehended. We should call this Quote-Unquote Reading or Sorta Reading rather than speed reading.

Skimming is one of the techniques emphasized in speed-reading courses, although not under that name; the injunction is to read only the information that is important, skipping the rest. The catch, of course, is that one can't be sure which information is important without having read the text. Still, there are times when skimming is very useful. We just wouldn't want to confuse it with what we do when we read a favorite author, a chapter in a biology textbook, or a letter from the IRS.

That's skimming, and it's trivial. The holy grail is increasing reading speed without sacrificing comprehension. The question is whether any of the commercial products do more than teach skimming. As Woody Allen put it in a joke from the Evelyn Wood era, "I took a speed reading course and read 'War and Peace' in twenty minutes. It involves Russia." To avoid the negative connotations that came to be attached to the "speed reading" label, the schemes are marketed as "power reading," "breakthrough rapid reading," "mega-reading," and "reading dynamics for speed, comprehension, and retention." These systems recycle the same methods, changing the wrapping. Speed-reading systems (epitomized by Reading Dynamics) focus on changing how a person reads a standard text. Newer methods use screen-based technologies (computers, pads, smartphones) to change how the text is presented. I'll explain what is supposed to happen and what actually happens.

Change the Reader

The come-on is that the only barrier to reading at warp speed is bad habits. It's a variant of the trope that people only use *n* percent of their brains: we only use a fraction of our reading capacity. Tapping into this unused capacity just requires a bit of retraining. Speed-reading programs focus on modifying readers' behavior in three ways. Remarkably, all of them were laid out in an obscure 1958 book, *Reading Skills*, coauthored by Evelyn Wood and Marjorie Barrows. Written for teachers, the book focused on struggling readers who were still in school. Speed is not emphasized, and the term "speed reading" does not appear. But the methods that Wood and Barrows recommended for helping poor readers became the foundation for speed reading: (1) take in more information at a time, (2) eliminate subvocalization (covertly saying words while reading), and (3) eliminate regressive eye movements. Let's examine each.

Method 1: Take in More Information at a Time

Readers are supposed to learn to taken in bigger chunks of text by training their eyes to process information in the periphery and using specialized techniques for scanning the page. Wood and Barrows introduced the strategy of using a finger to guide the eyes across the page in a zigzag pattern; another method is to move your finger down the center of the page in order to read down, a line at a time, rather than from left to right. In the precomputer era, the ambitious (or anxious) reader could buy gadgets to develop such skills at home. The Book of the Month Club, then a major force in the publishing industry, sold reading timers and contraptions to increase reading speed. The club's interest was obvious: if people read faster, they might buy more books. The problem with such methods should also be obvious: they flagrantly defy constraints imposed by the visual system. The injunction to take in whole lines, paragraphs, or pages cannot be achieved by the human visual system, short of growing additional cells on one's retina. We cannot will ourselves to recognize more letters in the periphery any more than we can will ourselves to hear sounds in the dog-whistle frequency range. Nothing about the structure of language, as displayed on a standard text, makes scanning in a zigzag pattern an efficient way to absorb the content. The important information does not appear at predictable locations or at predictable intervals. It follows, of course, that there couldn't have been any evidence that these methods were effective, but there did not have to be: the companies were selling wish fulfillment, not proven techniques.

Although we can't just will ourselves to taken in more letters at a time, there are some people for whom the perceptual span is larger by a few slots to the right: deaf individuals who use American Sign Language (ASL). Bélanger and colleagues found that for skilled deaf readers, the perceptual span as measured in their experiment was eighteen slots to the right of fixation, compared to fourteen for skilled hearing readers. The effect is unquestionably due to the use of sign language. Signing makes use of several types of visual information, including hand configurations, hand movements, spatial locations, and facial expressions. In conversation, fluent signers look at each other's faces, not their hands. They pick up the important information conveyed by space and movement because they have become exceptionally good at processing information outside the fovea. The study showed that this capacity carries over to reading.

Here is a natural experiment testing a basic tenet of speed-reading programs. Deaf signers become more sensitive to letters in the visual periphery, just as the speed-reading programs recommend. However, this doesn't improve their reading: reading speed and comprehension were nearly identical for the two groups in the study. The additional information that deaf subjects took in was low-quality stuff that was of little use. This capacity is helpful in other contexts but not reading. In comprehending sign language, detecting the approximate location where a gesture originated is linguistically important. Serious video game players are also skillful at detecting information in the visual periphery because a blob might turn out to be an enemy carrying a semiautomatic plasma rifle. The benefit does not transfer to reading because readers need letters, not blobs. Exercises to take in more information at a time are therefore pointless for reading.

Given that people can't actually increase the amount of useful information they take in on a fixation and no magic arises from running a finger down the page, what happens when people use these methods? They skim. Your speed-reading dollars buy a convoluted path back to skimming.

Method 2: Eliminate Subvocalization

Most people have the sense that they are saying words to themselves (or hearing them) as they read. Speed-reading programs appeal to the intuition that this habit slows reading. People who subvocalize can only read at the slower rate at which we produce speech, it is argued. By eliminating subvocalization, reading can proceed on a faster, visual basis. Speed-reading programs exhort people to suppress subvocalization, providing exercises to promote the practice.

I said at the outset of the book that people's intuitions about reading are limited and often misleading. I now present The Subvocalization Fallacy. The sensation that you use information related to the pronunciations of words while you read is not an illusion. However, it is not subvocalization, which is covert speech, as when you repeat something to yourself sotto voce to help to remember it. Skilled readers do something different: they mentally activate the phonological code that allows one to hear the differences between PERmit and perMIT in the mind's ear. The fallacy in the argument against subvocalization is in equating phonology with speech. Using the phonological code doesn't limit the reader to the rate at which speech can be produced because *there's no speaking involved.*

Once this fallacy is exposed, a different set of questions arises: What if using phonological information makes it easier to read, not harder? What if it is skilled readers who rely more on phonological information, not less-skilled readers? What if the inability to use phonological information efficiently is one of the main characteristics of reading impairments? What if skilled readers cannot prevent themselves from activating phonological information because it is so deeply integrated with spelling and meaning in writing systems and in the neural circuits that support reading?

All of these what-ifs are indeed the case, as established by several decades of research. Speed-reading schemes would improve reading by eliminating one of the main sources of reading *skill*. Imagine if the same approach had been applied to vision. People blink a lot. We can't see when we blink, so if we didn't blink we could take in information more rapidly: speed seeing! Treating blinking as a bad habit ignores the functions that it serves and the costs associated with trying to eliminate it. According to speed-reading logic, learning to suppress the activation of phonology, a highly overlearned, automatic, reflexive behavior, will make you a better reader. I wouldn't put it past a Tibetan monk to accomplish this with many years of concentrated practice. I also wouldn't recommend it if your goal is reading well.

I return in later chapters to the role of phonological information in silent reading and learning to read. Here I'll just summarize by saying, *Thairs moar two reeding than meats the I.*

Method 3: Eliminate Regressive Eye Movements

The third leg of the speed-reading stool is no regressive eye movements; read it right the first time. Like phonology, regressive eye movements serve a useful function, and eliminating them makes it harder to read, not easier. Regressive eye movements are an intrinsic element of normal skilled reading. They don't only occur because a text has been misread; they also allow readers to enhance their understanding beyond what could be obtained on the first pass. Some looking back is also inevitable because of the nature of language. Sentences unfold in a linear sequence, but the messages they convey often do not. Consider this sentence:

> If he doesn't know the woman who runs the front office, your brother shouldn't ask her for a favor.

The interpretation of the pronoun HE isn't clear until the later phrase YOUR BROTHER, whereas the interpretation of HER depends on an earlier phrase, THE WOMAN. Languages are filled with temporary ambiguities about what goes with what. The efficient coping strategy—the one that skilled readers discover—incorporates intermittent regressions as one component. We have ways to eliminate them (discussed in the next section), but they won't make you a more efficient reader. Just annoyed.

Now we can restate the major tenets of speed-reading programs more accurately: Expand visual perceptual capacities beyond physical limits. Suppress the use of information that contributes to reading skill, resisting the close integration of spelling and sound developed through years of practice. Eliminate beneficial regressive eye movements. Focus attention and effort on trying to read in a highly unnatural manner, at the expense of concentrating on the text.

If such programs are based on bogus principles, why is there always a market for them? I think that Evelyn Wood has a lot to answer for, or would if she weren't dead. The template is there in the Wood and Barrows book, the Ur-document in the speed-reading enterprise:

- Describe the obvious benefits of reading faster.
- Sow doubts about the person's reading ability.
- Describe methods that seem plausible and appeal to common sense.
- Use personal authority and credentials to provide assurance that they work.

Wood and Barrows could not have had any evidence for their claims because they didn't have the necessary tools (an eye tracker, say) or research skills. They described what seemed true based on their personal experience as teachers, but they were wrong. The eyes of a slow, plodding reader, they wrote, move like this:

The picture is a caricature. No reader's eye movements look like that. The saccades are too short and too evenly spaced for normal-sized text, and poor readers' eye movements are far more erratic. Here is their depiction of regressive eye movements:

Lovely swirls but wrong again. To remind yourself what actual eye movements look like, watch "Eye Tracking Reading Study," a fascinating YouTube video (http://goo.gl/SLsP6L) of eye movements in skilled and unskilled reading produced by one person, a child reading Swedish, her first language, and then

English, her second. Finally, the claims about the pernicious effects of subvocalization were simply asserted as though they were patently obvious, but they failed to distinguish covert speech from mental phonology. These are the "bad habits" that speed-reading programs have encouraged people to eliminate ever since. Properties of the visual system like the perceptual span are not mentioned.

Although Wood and Barrows did not have the tools to establish their claims, neither did anyone else. It took years of research to establish the basic facts about eye movements and subvocalization that disproved those claims. By that time, Evelyn Wood Reading Dynamics was a large, established enterprise that spawned many copycat programs, and myths about speed reading had entered the cultural consciousness. Although we now know the claims are false, they are as difficult to eliminate as bedbugs. The number of people who understand the relevant research—which now includes you—is far smaller than the number of people who would like to read a lot faster. The claims that are made in marketing such products don't have to be true. My point in going back to Wood and Barrows is that they never were.

Evelyn Wood is no longer the icon of speed reading, off in retirement with Joe Camel, Mr. Whipple, and Josephine the Plumber. But now there are scores of Evelyn Woods. The Internet is littered with blogs and websites by reading experts who will help you or your child read faster and jump higher. Most of the proprietors have credentials similar to Evelyn Wood's. They are mainly former teachers intent on sharing the knowledge they acquired during their long professional careers. They confidently list the keys to skilled reading, but they don't all match. They sell repackaged materials freely available elsewhere on the Internet. It's an Internet cottage industry. Some sites are better, some are worse, and there's no way for a parent (a homeschooler, for example) to tell them apart.

And what about Howard Stephen Berg, the world's fastest reader? He reached his apotheosis in an infomercial for Mega Reading produced with Kevin Trudeau (author of *Cures They Don't Want You to Know About* and other such books), which eventually came to the attention of the FTC. Its decision is worth reading, not skimming. Whereas Berg and Trudeau claimed that Mega Reading was successful in teaching anyone, including adults, children, and disabled individuals, to significantly increase their reading speed while substantially comprehending and retaining the material, the FTC decided that "in truth and in fact [it] is not successful in teaching anyone." The FTC settlement (which covered several Trudeau products, including memory and hair-farming systems) barred the defendants from making deceptive claims, which ended the Mega Reading infomercial. Berg has remained free to promote products using the many alternative delivery systems made possible by the Internet: YouTube videos, personal websites, and social media. The protean Berg currently offers seminars on "how to learn anything faster and better, sharing his knowledge of techniques demonstrated to improve reading, memorization, materials organization, math, and writing skills." He sells his several books. His entry in the *Guinness Book of World Records* was removed after a year, however, and Guinness no longer anoints a world's fastest reader.

Change the Text

Reading speeds might increase if there were a way to deliver information to the visual system more efficiently than conventional formats like this one. The ancient Greeks experimented with a method called boustrophedon (literally, ox turning, referring to the ox's reversal of direction at the end of plowing one row to start the next one). Texts were written bidirectionally, left to right on one line, then right to left on the next. This method would seem to allow reading to proceed continuously, uninterrupted by line sweeps. Try it. Start reading the next paragraph normally, left to right, but then at the end of the line, pivot and read right to left. Then continue plowing back and forth.

Here we have a nice normal first line.

.siht ekil nettirw eb dluoc enil txen ehT

Wow that is pretty deeply unpleasant.

.bad Not ?method this about What

No way! These "fixes" make reading harder, not easier!

Bidirectional reading was one of those little experiments during the development of writing that didn't work out. However, modern screen-based technologies afford other possibilities. Children raised on tablets and computers might rapidly adapt to new ways of presenting text, if a more effective method should exist.

We have seen one example of modified text presentation, the parafoveal magnification technique, which is neither effective nor feasible. A method called rapid serial visual presentation (RSVP) seems more promising. A text is presented at a single location on a screen, one word (or sometimes a few) at a time. It was developed for research purposes in the 1960s. When personal computers became common, it was sold as a reading improvement tool; now there are apps. A YouTube video presents Edgar Allan Poe's "The Raven" in this format. The technique is also available on several free websites; be sure to try it. RSVP asks the question, What would reading be like if there were *no* eye movements? The text is delivered at a spot on the screen, like a series of flash cards. The user sets how long each card is displayed. As in the "Raven" video, the usual procedure is to present one word at a time. Readers are liberated from having to decide how much time to spend on each word because that is set in advance, and saccades, regressive eye movements, line sweeps, and page turning have been eliminated. A reader can fully concentrate on comprehending the text as it flashes by.

Was the "Raven" video encouraging? The text is presented at about 278 words per minute, within the skilled reading range, yet requires extra effort to understand. Every word, whether DOOR or MORROW, is displayed for the same amount of time. The reader loses control over the rate of transmission and, with it, the ability to allocate reading time intelligently. Useful parafoveal information and the option to reread have been eliminated. The experience feels like stalking the text rather than reading it.

In laboratory studies college students could read with RSVP at up to seven hundred words per minute with good comprehension, about triple their normal speeds. A quick tryout of the procedure on a website yields a surge of pleasure at the realization that one can indeed understand texts that whiz by so quickly. Such short demo experiences are crucial to sustaining a market for these products, but the acid test is longer texts. Alas, the experiments also found that subjects could only sustain reading at high speeds with good comprehension for short bursts. With longer texts, the RSVP reading experience is monotonous and exhausting. It's another case of making reading harder, not easier, by removing elements that contribute to fluency. RSVP requires maintaining a constant focus on a single location, which is like having a staring contest with a book. If you're going to read this way, you'll have to factor in the time spent stopping to give your eyes and neck a rest. RSVP is also relentless: look away to scratch your nose, miss a bit of text, and the words are gone. Although the programs can be paused, rereading requires starting over or backing up, far more disruptive than the much maligned regressive eye movement. The irony of RSVP is that it turns text into a fast-fading signal like speech, creating a gratuitous modality mismatch. The net result is that people can only tolerate it for short stretches, which is not apparent if the product is only taken for a brief test run.

The free web-based RSVP programs are good for assessing the upper limits at which people can understand text. You can determine your personal redline, but if you try to read real texts at peak wpm, you'll blow that engine pretty quickly.

Like earlier speed-reading techniques, RSVP periodically gets repackaged for new generations of anxious readers. The producers of a 2014 RSVP app claimed to have achieved a breakthrough by incorporating "optimal recognition points." Each word is presented with a highlighted letter, said to be the location that allows the reader to recognize the word most quickly.

Although the product website has a section for the science behind the method, there aren't any studies showing that each word has an optimal recognition point, that it makes any difference which letter (or any letter at all) is highlighted, or that aligning words by "optimal recognition points" improves performance. Predictably, the website only allows a very limited, carefully controlled tryout of the procedure. The method retains the properties that make RSVP so unpleasant—fixation on a single location, relentless presentation, no regressive eye movements. The program repackages RSVP with a gimmick—the highlighted letter—of no known value.

At this time, I know of no method for changing text presentation that is an improvement on the standard ones. E-reading on a Kindle or tablet uses standard formatting with add-ons that can enhance the experience, but it's not a faster way to read. Everyone knows these devices make it all too easy to go off on interesting impulse expeditions, to the dictionary, to the author's website, to Wikipedia for background information. The new reading dilemma is whether to stay on track or go off road. The devices encourage Slow Reading, the analog of Slow Food, not speed reading, the junk-food approach.

Even if all speed-reading methods are ill motivated, to put it nonlitigiously, is it nonetheless possible that a person who diligently practiced them could improve their reading? Certainly. Some improvement could occur despite the methods, not because of them. The programs require people to allocate time for reading, focus attention on it, and set aside distractions—all good. There won't be much benefit if the texts are only skimmed, and no one will achieve super speeds. But improvements could occur simply because people are reading more and paying closer attention. Of course, the same could also be accomplished without buying a program or engaging in bogus exercises. You just read.

I'm no theologian, but neither was Evelyn Wood. If God values knowledge, and reading is an excellent way to acquire it, it follows that people who seek the favor of God should read a lot, not that Wood's methods will work. If God had wanted us to be speed readers, he might have given us a bigger retina with a lot more cones, and the ability to comprehend language a little faster too.

I've dwelled on the fallacies underlying these methods because the market for them will persist as long there are people who want to read faster, profiteers who prey on their aspirations and anxieties, and a lack of shared knowledge about why the techniques fail. Details about the perceptual span and RSVP aside, the deeper problem common to all such methods is that they have the relationship between eye movements and reading skill *backward*. They assume that reading skill results from efficient eye movements. Train your eyes to take in more information or read without having to move the eyes at all, and you will be a better, faster reader. However, efficient eye movements are the *result* of becoming a proficient reader, not the cause. People who acquire the knowledge and experience that underlie skilled reading do indeed move their eyes with greater efficiency, producing increases in reading speed. But we don't achieve this by practicing how to move our eyes. The same fallacy arises in some treatments for developmental dyslexics. Their eye movements are inefficient, to be sure. But training children to move their eyes like a good reader, as in some types of "vision therapy," is ineffective because erratic eye movements are a manifestation of an underlying reading problem that eye training doesn't address.

I began this chapter by showing how the visual system constrains how we read. That system isn't malleable in ways that are relevant to reading, and so reading-improvement schemes that focus on the eyes are misguided. Reading would be a different experience if our eyes were different: imagine a compound eye like an insect's, each unit with the acuity of a human eye (and imagine how text would have to be presented to take advantage of this remarkable organ!). Our eyes weren't designed for reading, however, and we have yet to figure out any method that markedly improves on what is in place.

If reading at megaspeeds is not feasible, does that mean reading can't be improved? Not at all.

The serious way to improve reading—how well we comprehend a text and, yes, speed and efficiency—is this (apologies, Michael Pollan):

Read. As much as possible. Mostly new stuff.

Which means:

Read. Reading skill depends on knowledge acquired from reading. Skilled readers know more about language, including many words and structures that occur in print but not in speech. They also have greater "background knowledge," familiarity with the structure and content of what is being read. We acquire this information in the act of reading itself—not by training our eyes to rotate in opposite directions, playing brain exercise games, or breathing diaphragmatically. Just reading.

As much as possible. Every time we read we update our knowledge of language. At a conscious level we read a text for its content: because it is a story or a textbook or a joke. At a subconscious level our brains automatically register information about the structure of language; the next chapter is all about this. Developing this elaborate linguistic network requires exposure to a large sample of texts. There are no special techniques to practice; you didn't miss the class when they were taught. You read.

Mostly new stuff. Knowledge of language expands through exposure to structures we do not already know. That may mean encountering unfamiliar words or familiar words used in novel ways. It may mean reading P. D. James, E. L. James, and Henry James because their use of language is so varied. The goal is not to make reading excessively difficult, merely different enough from what's come before to pick up something new without enormous effort. A large sample of texts in varied styles and genres will work, including some time spent just outside one's textual comfort zone.

Reading expands one's knowledge of language and the world in ways that increase reading skill, making it easier and more enjoyable to read. Increases in reading skill make it easier to consume the texts that feed this learning machinery. This feedback loop is the mechanism that leads to expertise.

It is not the eyes but what we know about language, print, and the world—knowledge that is easy to increase by reading—that determines reading skill. Where this expertise leads, the eyes will follow.

Time Your Reading

The text is from the first page of *Dr. Montessori's Own Handbook* by Maria Montessori, the educator who developed the "Montessori method" used in many preschools. Read at your normal rate for reading a text carefully without laboring over it. There is a comprehension question at the end. Use a watch or timing app.

Ready to start?

Start timer.

Recent years have seen a remarkable improvement in the conditions of child life. In all civilized countries, but especially in England, statistics show a decrease in infant mortality.

Related to this decrease in mortality a corresponding improvement is to be seen in the physical development of children; they are physically finer and more vigorous. It has been the diffusion, the popularization of science, which has brought about such notable advantages. Mothers have learned to welcome the dictates of modern hygiene and to put them into practice in bringing up their children. Many new social institutions have sprung up and have been perfected with the object of assisting children and protecting them during the period of physical growth.

In this way what is practically a new race is coming into being, a race more highly developed, finer and more robust; a race which will be capable of offering resistance to insidious disease.

What has science done to effect this? Science has suggested for us certain very simple rules by which the child has been restored as nearly as possible to conditions of a natural life, and an order and a guiding law have been given to the functions of the body. For example, it is science which suggested maternal feeding, the abolition of swaddling clothes, baths, life in the open air, exercise, simple short clothing, quiet and plenty of sleep. Rules were also laid down for the measurement of food adapting it rationally to the physiological needs of the child's life.

Yet with all this, science made no contribution that was entirely new. Mothers had always nursed their children, children had always been clothed, they had breathed and eaten before. The point is that the same physical acts which, performed blindly and without order, led to disease and death, when ordered *rationally* were the means of giving strength and life.

Stop timer.

Comprehension question (don't look back!):

What were three beneficial child-rearing practices that Montessori attributed to "science"?

To calculate reading speed:

1. Record the number of seconds it takes to read the text.
   309 words ÷ number of seconds = your words per second (wps).
2. Then, multiply wps by 60 to get words per minute (wpm).

Example: Say a person took 75 seconds to read the text.

309 ÷ 75 = 4.12 words per second

4.12 x 60 = 247 words per minute

CHAPTER 5

F u cn rd ths, u cn gt a gd jb n rdng rsch

LET'S DO SOME BIG DATA. We'll accumulate a massive hoard of small things, such as millions of tweets, thousands of personal profiles from an online dating site, or a respectable fraction of the 250 billion or so pictures posted on Facebook. Then we can run them through "deep learning" procedures that sift for patterns buried within. That will require a prodigious number of simple calculations on our big cluster of computers optimized for this task. The patterns will be too complex to accurately describe in words, but they will be represented as complex networks of statistical contingencies. We can then use this data structure in applications such as identifying faces or describing scenes. Your smartphone recognizes what you say to it using statistical regularities about spoken words derived using such procedures. I for one welcome our new data analytical overlords, who are already using these methods to perform amazing feats of cognitive complexity, with no apparent upper limits on what can be accomplished in sight.

Stylometry is dedicated to applying statistical analyses to questions about the content, style, and authorship of texts. It is surprising that quantitative methods, working only with occurrences of words and not what they mean, can reveal facts about writing that even the author of a text does not consciously know. Statistical profiles of texts have been used to out the authors of books published under pseudonyms, as when J. K. Rowling published an adult novel under the name Robert Galbraith. Similar methods showed that the articles containing falsified data published by a Dutch social psychologist in an odious case of scientific fraud exhibited distinctive statistical properties (the author overcompensated for his fakery by using more definitive and emotionally charged terms to characterize findings than in his nonfraudulent papers).

The most poignant application of these methods involved the British novelist and essayist Iris Murdoch. An Oxford litterateur who briefly studied philosophy with Ludwig Wittgenstein, she wrote a series of acclaimed novels about relationships among the intellectuals. The critical response to what would be her final novel, *Jackson's Dilemma* (1995), was unusually negative. A. S. Byatt, for example, compared the structure of the novel to "an Indian Rope Trick . . . in which all the people have no selves and therefore there is no story and no novel." Researchers created a statistical profile of the novel, again counting simple things such as the number of distinct words used, the frequencies of the words, and the number of words and clauses per sentence. The same analyses were also performed using two of Murdoch's highly regarded books. Whereas the two earlier novels were very similar in this statistical sense, they both differed from the final novel, which used a smaller vocabulary and simpler sentences.

The poignant part is that Murdoch was diagnosed with Alzheimer's disease in 1996 and died three years later (events dramatized in the 2001 movie *Iris*). The research showed that the final novel was very likely to have been written in the early stages of the disease, with its dreadful, inexorable decimation of language and other capacities.

These exercises may seem risible. Reducing *War and Peace* or J. K. Rowling books to a statistical profile is like characterizing a pointillist painting by counting the frequencies and locations of colored dots. Statistical patterns in the usage of words might reveal that one of Iris Murdoch's books was unlike the others but not much about its content—what the book is about, its themes, ideas, and events, its style and organization, whether an expression is literal or figurative, sincere or ironic, who the characters are and why they behave as they do. Truman Capote famously said of Jack Kerouac's *On the Road*, "That's not writing, it's typing." A literary critic might say that stylometry isn't literary analysis, it's counting.

I'm much more sympathetic, actually. The methods offer a novel way of looking at texts that complements traditional approaches and can uncover things that couldn't be known by other means. People have been tabulating the words in texts since the compilation of the first Bible concordance, a list of every word and where it appears in the book, in the thirteenth century. Stylometric methods are also the basis of automatic grading procedures, which assess undergraduate English essays about as reliably as instructors, and I'm guessing, because they're not telling, that the NSA uses them to find patterns of interest in the petabytes of data they collect from monitoring Internet and phone use. Whatever their intrinsic interest, utility, or limitations, these kinds of analyses are important because they show that, in addition to the other properties that give rise to how they are experienced, texts have an implicit statistical structure. The distributions of words and phrases—which words occur, how often, and in whose company—are echoes of the content. Stylometry, like other types of data mining, is the business of detecting and interpreting such echoes.

Becoming a skilled reader is Big Data for people. Humans are data hunters and gatherers. We respond to patterns in the environment. We register repetitions and novelties, similarities and differences, the way things vary and covary, and then how the things that covary covary. We experience the world as a three-dimensional space populated by objects and events because of its statistical regularities. Language, like the visual world, exhibits statistical regularities at many levels—phonology, morphology, words, word sequences, and relations between utterances and the contexts in which they occur, among others. Reading brings in regularities in how letters combine and how orthography relates to phonology and meaning. For people, the algorithm for analyzing these data is learning. Every time we use language, we also update our statistical representation of it. Our learning mechanisms are similar to some computational algorithms used in analyzing Big Data, which developed out of research on human learning.

The number of such learning events is uncountably large. They begin in utero: third-trimester fetuses are already learning about statistical properties of their mothers' speech that make it English rather than Russian. Language acquisition is driven by exposure to a massive amount of data, utterances that exhibit statistical regularities at many levels. Later, reading becomes an additional source of data about print and language.

This continual learning and updating occurs in the background as we pursue our main goals, producing and comprehending language for various purposes. Statistical learning takes place without conscious awareness or intention. It is a kind of implicit, subconscious learning, a complement to the explicit learning with conscious awareness that occurs with overt instruction or listening to a TED talk. The two types of learning are different but linked because both result in changes to the neural systems for long-term memory. Well-timed and targeted instruction is effective because it accelerates the acquisition of this enormous data structure. Explicit instruction and conscious effort are the visible tip of the iceberg; statistical learning is the mass below the surface.

In short, a reader has to be able to count, not in the sense of counting the number of pitches thrown by the sixth inning but rather in the sense of implicitly tracking statistical regularities in language and how they relate to the world as it is experienced. Learning to read is the process of acquiring the several types of statistical knowledge that support rapid and efficient comprehension, starting with phonological structure, orthographic structure, the mappings between orthography and phonology, vocabulary, and grammar. Deficits in any of these areas can seriously interfere with children's progress and adult proficiency. These types of knowledge differ—spelling is not the same as vocabulary or grammar—but they are all statistical, involving the frequencies and co-occurrences of the elements out of which they are composed.

People usually do not think of what they know as a multidimensional statistical matrix or of human learning as a lifelong Big Data project. A good way to get a sense of what's involved is by looking at the unique part of reading, the orthographic code.

Department of Redundancy Department

In the early 2000s, an odd-looking anonymous text began to ricochet around the Internet:

> Aoccdrnig to rscheearh at Cmabrigde Uinervtisy, it deosn't mttaer waht oredr the ltteers in a wrod are, the olny iprmoetnt tihng is taht the frist and lsat ltteer are at the rghit pclae. The rset can be a tatol mses and you can sitll raed it wouthit a porbelm. Tihs is bcuseae we do not raed ervey lteter by itslef but the wrod as a wlohe.

As the text circulated numerous variants appeared, including ones in other languages and writing systems. The text is perfect for social media. It fills a lot of blogspace. Some people think it reveals deep properties of the human mind relevant to creativity and corporate management strategies. It remains a popular object of curiosity, earning a page on Snopes.com, the fraud-detection website (which assesses its validity as undecided). The text went viral because it really can be read despite the many misspellings.

It's just that everything it says is false.

> There wasn't such a study at Cambridge University.
>
> It isn't true that only the first and last letters have to be in place.
>
> We don't read words "as a whole."
>
> And it's only because we *don't* that we can read TIHNG AS THING.

The Cmabrigde hoax caught on because it is interesting, not because it is true. It is a curious effect, being able to read the text without knowing how it is done. The description of the phenomenon was clever because it also allowed readers to experience it themselves. Give the author credit: it is hard to make a novel experience out of a quotidian activity like reading.

The actual explanation mainly involves knowledge of orthographic statistics. We learn about the patterns that occur (the words we know), ones that could occur (like MAVE or GLORP), and, equally important, ones that could not (TSIP, SITP, XPLK), except for special purposes (Mister Mxyzptlk, a foe of Superman; Randall Munroe's web comic xkcd). We learn how often patterns occur and in which combinations. This information is enormously helpful because the ones that occur are a small subset of the possible ones, which is the key to overcoming the distortions in the Cmabrigde text, not "reading words as wholes." And that is not only interesting, it is true.

Recognizing that a string of letters is a specific word requires ruling out that it is any other word. Alphabets are a potential problem because even a small number of letters can generate a very large number of spelling patterns. From twenty-six letters, 475,254 words can be formed that are one to four letters long. If five-letter words are allowed, the number jumps to over 12 million. Longer words take the numbers into light-year range. How can readers find the word NEEDLE in this enormous haystack of orthographic possibilities? We couldn't if all of these possibilities were allowed, but they aren't. The *Oxford English Dictionary* has entries for about 170,000 words in current use, and the vocabularies of people reading this book are on the order of 20,000 to 40,000 words (the estimate is imprecise because it depends on how "word" is defined). The fact that most letter combinations do not occur makes it easier to recognize the ones that do. The range of possibilities has already been severely restricted before a word is even read.

The spelling patterns that *are* used are a highly nonrandom sample of the millions of possibilities. Words are composed out of the smallish subset that can be pronounced in English (such as TR, EA, UN, LK, AVE, IST, OST, STR, and so on); many more letter combinations are ruled out because they can't be pronounced (e.g., LBATK, SKTP) or haven't been used (e.g., MAVE). Finally, an even smaller subset of the legal patterns are used very frequently. Although we know many thousands of words, surprisingly few of them account for most of what we read. The 150 most common words in the language account for about half of the words we read. The top 2,000 words account for about 90 percent. Word frequencies have a "long tail" distribution: a small number are used with very high frequency, with many others employed much less often.

Readers build neural structures that represent these statistical patterns and tune them every time a text is read. The mass of positive evidence about the limited set of patterns that do occur also carries information about patterns that do not. We do not have to be taught that the sequence TLKP is not a word because of its lack of similarity to patterns that do occur.

The nonrandom ways that letters combine mean that letters carry information about others that are likely to occur. This property is known as redundancy, which can be quantified using information theory, developed by mathematician Claude Shannon. It is a crucial concept.

Usually we think of redundancy in language as a bad thing. In writing we try to avoid redundant expressions such as EXTRADITE BACK and PIN NUMBER because the second word does not add information beyond what the first word conveys. Similarly, using accent marks to indicate syllabic stress in English would be redundant because the patterns are already highly predictable.

These excesses aside, redundancy is an intrinsic property of language, as orthography clearly illustrates. Redundancy decreases the uncertainty in how elements (letters, sounds, words, others) combine. That is how we avoid searching entire haystacks of possibilities. To illustrate, the letters WOR_ are redundant with respect to the final letter, which cannot be just any of the twenty-six letters; only K, M, E, D, N, and T form words. DOR_ only allows M, K, and Y; its redundancy regarding the final letter is greater because it places stronger constraints on what it could be.

The great benefit of a redundant code is that it is fault tolerant: word recognition doesn't collapse if a letter is obscured or difficult to recognize because it is scrawled. Redundancy allows us to cope with annoyances such as this:

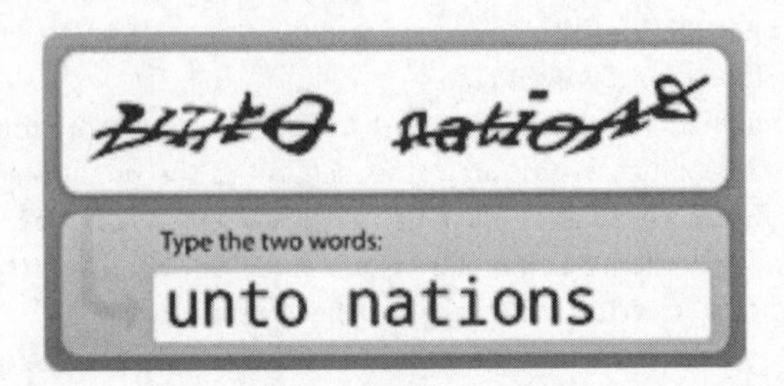

Captcha security systems rely on the fact that people are better at using knowledge of orthographic structure to overcome letter and number distortions than automated "bot" programs.

The degree of redundancy in written English is impressive. Consider just the vowels. A text can be readable even if most of the vowels are deleted, as in this line from an ad that was part of the New York City subway experience in the 1970s:

```
F u cn rd ths msg, u cn gt a gd jb
```

If all the vowels are deleted (as in Hebrew), many texts can still be read:

```
Th bsc dmnstrtn s tht txt s stll mr r lss lgbl whn th vwls
hv bn rmvd.*
```

There are limits, however:

```
n n ld rtcl tht s ndsrvdl bscr, rsrchr Mrln dms pntd t tht
rdblt drps ff rpdly f th vwllss txt nclds lss cmmn wrds sch
s tchbl, cntsn, cnfblt, nd bttrbr.**
```

Recognizing the less-familiar words requires more evidence about their spellings, and so vowel deletions are not tolerated well.

Orthographic redundancy allows us to recognize lightly camouflaged expletives like SH*T and F*CK, but the asterisk trick also fails with less common words such as BL*V**T* or CR*M*L*NT (BLOVIATE, CROMULENT). English is redundant but not to a degree that would allow vowels to be eliminated or marked with a placeholder, curse words and simple txt msgs aside.

Claude Shannon calculated that the alphabetic code for English is about 50 percent redundant. That does not mean that a text will be readable if 50 percent of the letters are deleted because the redundancy arises from dependencies among letters that are disturbed, in varying degrees, by deleting individual letters. Redundancy makes words and letters easier to identify. That's a good thing for readers, but also for archaeologists trying to decipher ancient texts and cryptographers trying to crack coded messages.

The statistical structure of written English is easy to illustrate, but these examples capture very little of what a reader knows. Peter Norvig, the head of research at Google, has posted some other simple orthographic statistics of interest, but even these only skim the surface. There is no way to describe

---

* The basic demonstration is that text is still more or less legible when the vowels have been removed.

** In an old article that is undeservedly obscure, researcher Marilyn Adams pointed out that readability drops off rapidly if the vowel-less text includes less common words such as TEACHABLE, CONTUSION, CONFABULATE, and BUTTERBEER.

orthographic structure in its entirety, but brains and computers capture it very well.

Varieties of Orthographic Expertise

Readers become orthographic experts by absorbing a lot of data, which is one reason why the sheer amount and variety of texts that children read is important. For a beginning reader, every word is a unique pattern. Major statistical patterns emerge as the child encounters a larger sample of words, and later, finer-grained dependencies such as the fact that syllables can both begin and end with ST, begin but not end with TR, end but not begin with BS, and SB can only occur across syllables (e.g., DISBAR). The path to orthographic expertise begins with practice practice practice but leads to more more more. Only a limited amount of spelling can be taught, and instruction typically ends by fourth grade. Orthographic expertise is not acquired through the years of deliberate practice required to become an expert at playing chess or the tuba. We don't study orthographic patterns in order to be able to read; we gain orthographic expertise by reading. In the course of gathering all that spelling data, a person can also enjoy some books.

Humans are statistical learners, but so are other species; we are just much better at it. Even pigeons, who are certifiably dull, can be taught to identify twenty-six letters presented in a single font, under favorable testing conditions. They require extensive training but will work for seed. That is pretty much it for pigeon reading.

Baboons, who can walk without also moving their heads, do somewhat better. In 2012 they achieved, though they did not know it, their fifteen minutes of fame when it was reported that scientists had a trained a few to read words. A study in *Science*, a glamour journal known for publishing novel but often sketchy findings, reported that baboons could learn to distinguish words (such as BRAKE) from pseudowords (could-have-been-words like BRONE). In human studies the task is called making a lexical decision (is it a word or a nonword?).

The baboons were taught to recognize a bunch of letter strings ("words"). In the test phase, they were presented with these words intermixed with novel letter strings ("nonwords"). The baboons' accuracy in telling them apart was well above chance. The authors thought that the study showed that baboons have the capacity to pick up on orthographic statistics involving combinations of letters, indicating that this part of reading draws on capacities that are not among our most advanced.

The question raised by such studies is whether the behavior resembles human performance in any significant way. After the "baboons can read!" stories broke, the Internet hive went to work and found that statistical properties of the particular words and nonwords in the experiment differed in a simpler way that the experimenters had not noticed. Some letters were much more likely to occur in the test words and others in the nonwords, which the baboons, not being stupid, figured out. The baboons were tracking frequencies of letters, not learning multiletter orthographic patterns, which was enough to allow them to perform at the reported levels with these stimuli. The behavior isn't trivial, but it isn't word recognition either (and someone should check with the pigeons to see how they would do). No animal species other than humans is known to have the capacity to learn statistical structures as complex as those in writing systems, and those baboons weren't getting even close. And learning that those orthographic patterns symbolize other things, like the sounds and meanings of words? That's another quantum leap.

At the other extreme are cases of extraordinary orthographic expertise. The masters of the orthographic universe are tournament Scrabble players. Scrabble is a lexical decision game. The player's task is to form high-scoring words that will fit into the current board. The player must know which letter combinations form words found in an official Scrabble dictionary. They also need to be able to determine which combinations are not words in order to spot when an opponent plays a phoney, the Scrabble term for a nonword that could pass for a word. To reach the highest skill levels, tournament players study word lists: two-letter words, three-letter words, words containing letters with the highest point values (J, Q, X, A), short Q words, longer Q words, words that contain many vowels (MIAOU, ZOEAE), and so on. Players memorize anagrams—all of the words that can be formed from specific groups of letters—and practice anagramming under time pressure.

A championship Scrabble player's orthographic expertise has a different character from a skilled reader's. Readers pick up orthographic statistics incidentally as a by-product of reading, whereas Scrabble expertise results from extended deliberate practice, like other specialized talents such as eating the most hot dogs in ten minutes. Readers need to know the meanings of words; Scrabble players do not. Knowing that ninety-five legal Scrabble words can be formed from the letters APTYKLI is helpful in playing the game; readers, in contrast, have to recognize patterns as particular words, not discover words in randomly ordered strings. Elite Scrabble players' expertise is a function of the amount of Scrabble-specific practice (studying word lists, analyzing previous games, time spent in competitive play) rather than general factors such as vocabulary size or verbal ability. Scrabble skills do not carry over to reading, but expert players are better at making lexical decisions for words presented vertically, as they can be in the game.

The championship Scrabble player's orthographic expertise is like a London taxi driver's exhaustive knowledge of the city's streets. A famous study of those taxi drivers showed that learning "The Knowledge" led to increases in the volume of gray matter (neurons, mainly) in a part of the hippocampus, a brain structure crucially involved in memory formation. I would expect to see similar anatomical effects in expert Scrabble players in brain areas that encode spellings and in areas that may be recruited to store all those facts.

The other orthographic experts are children who compete in spelling tournaments: "bees." They too engage in thousands of hours of deliberate practice memorizing the spellings of words. I have not found any studies of the impact of this knowledge on the children's reading or academic performance. The effects could be a little weird. Winning turns on knowledge of obscure words in the very long tail of the frequency distribution, such as STICHOMYTHIA and NUNATAK. Many of these are technical terms or borrowed from other languages, which certainly alters the statistics of the contestants' Big Orthographic Data. Unlike Scrabblists, spelling champions reportedly know the meanings of many of the words and have studied etymology and word formation (morphology). They read dictionaries, not lists of six-letter words containing J, K, or Q, suggesting that the knowledge they acquire has greater utility. The spellers may regret the recent redesign of the SAT, however, which no longer rewards knowledge of obscure words.

It should now be apparent that our knowledge of orthographic structure, though not in the Scrabble champion range, is what allows the Cmabrigde text to be read. Reordering the interior letters obviously does matter because the doctored text requires much more effort to read than a normal one. Because the code is redundant, however, words can be identified even when the spelling is distorted—just not *too* much. In addition to keeping the first and last letters in place, the interior letters cannot be arranged in just any random order, which the author must have realized because the person fudged, excluding patterns that would ruin the intended effect. They mainly used the simplest kind of scrambling, reversing the order of two letters, as in W*RO*D, OR*ED*R, and LET*ET*R (these are called bigram transpositions). More elaborate scrambling was used in only a few words, such as RSCHEEARH (as the word RESEARCH was misspelled in the original).

Bigram transpositions are particularly easy for people to override. FCUK is the logo of the clothing chain French Connection UK, which would not be memorable if the company name were French Connection US. In 2008, Jon Stewart had fun with the "Palin Talk Express," a bigram transposition pun on John McCain's "Plain Talk Express." Typing TEH for THE is such a common error that the misspelling has become a jargon word with its own meanings and grammar. Bigram transpositions are easy to spot because they create letter sequences that are illegal (like CMA and EDR at the end of a word) or very rare. They are easy to correct because the scrambled version (such as OREDR) is still closer to the actual word (ORDER) than to any other word. If the letters are randomly scrambled, as in LERETST, the string is no longer close enough to any word to get back to it easily.

A utility for generating Cmabrigde texts inadvertently proves that they are only readable if the alterations are limited to simple ones like bigram transpositions. Paste in a text, and it will return a scrambled version in which the first and last letters of words are retained, but the order of the other letters is randomly scrambled, in accord with the original claim. Unless a text is at the *Green Eggs and Ham* level, the scrambled versions are almost impenetrable. Here is how some Wikipedia turns out:

> Wrtinig is a mduiem of huamn couomcmtnaiin that rseteenprs lngaguae and eiomotn thgourh the iicsptrinon or rcioen-rdg of snigs and sbmlyos. In msot lneagaugs, wniirtg is a celnmmpoet to seecph or skoepn lgagnaue. Writnig is not a lganauge but a from of tlcoogenhy taht doeleevpd as tloos depeleovd with haumn seictoy.*

Our actual sensitivity to the order of letters is shaped by two competing demands. On one side, the exact order of letters matters; otherwise we would not be able to distinguish TAP, PAT, and APT, SALT and SLAT, and all the other cases in which multiple words are formed by reordering letters. On the other side, orthographic redundancy—the constraints on how letters combine and the fact that most letter sequences are not used—creates wiggle room, allowing minor distortions to be overridden. The only broad generalization is that the extent to which order information matters varies, and we adjust accordingly.

Orthographic redundancy is the main basis for reading the Cmabrigde text, but not the only one. The reader's task is not just to recognize misspelled words: it is to comprehend the meaningful, if factually mistaken, text in which the words occur. At this level the text is not a TATOL MSES. Part of the problem of comprehending it has been presolved by leaving almost half the words

* Writing is a medium of human communication that represents language and emotion through the inscription or recording of signs and symbols. In most languages, writing is a complement to speech or spoken language. Writing is not a language but a form of technology that developed as tools developed with human society.

intact (thirty-two out of sixty-six). People can also use background knowledge about such things as university names and research studies.

At last we can accurately state what the Cmabrigde paragraph shows: readers can recognize words when the first and last letters are in place, some interior letters are reordered, mainly by bigram transposition, and the words occur in a meaningful context in which many words are spelled correctly. Although the paragraph's specific claims were wrong, orthographic statistics are indeed important for skilled reading, not just spelling or Scrabble tournaments. Those Scrabble experts are probably great at reading scrambled texts, too.

Why Statistical Knowledge Is Powerful

Readers aren't just Big Data machines grinding away at texts. We think. We know things such as how species reproduce, the germ theory of disease, the functions of the departures from strict chronology in *Middlemarch*. Text statistics can serve valuable purposes, such as classifying texts for author recognition or comparison, essay grading, or predicting how well a book will sell. Some statistics, such as how often a president uses personal pronouns in speeches, are food for thought, and thoughtlessness. They provide evidence that can be used to answer interesting questions about language that can only be asked now that such methods exist. It still takes a human to formulate the questions, to determine which of the many possible statistical analyses would supply relevant information, and to interpret the data that come back. As previously undetected properties of texts are discovered, however, they will influence the kinds of questions that the human asks.

Text and spoken-language statistics aren't the answer to every question, but these data are enormously important nonetheless and loom especially large in learning to read and then becoming very skilled. We've looked at spelling, but readers gather data about the statistical properties of texts at multiple levels: letters, letter combinations, syllables, morphemes, words, sequences of words, phrases, and on to galaxies beyond. Each level consists of a finite set of elements that are combined to form a much larger set of elements at another level. The knowledge at each level is probabilistic: we know what is likely or unlikely to occur but only rarely exactly what has to occur. WOR_ can be followed by several letters, not just one; of the possible continuations, K, is more likely than E or M. TAKE YOUR can be followed by many words, but TIME is more likely than CHAI or PINEAPPLE. Reading THE KEYS TO THE CABINET, we know that a plural verb is likely to occur but not which one or when (e.g., ARE HERE, WERE HERE, IN THE KITCHEN ARE HERE); with a slight change, THE KEY TO THE CABINET, expectations shift to singular verbs.

The deep question is how all this probabilistic information can be useful. Comprehending a sentence requires resolving the ambiguities at each level to a high degree, though not always completely. We need to determine the actual letters and sequences of letters, not just the ones that are likely, and the same holds for words and sequences of words. If we are uncertain about what is happening at one level, how can it inform what is happening at another? Look again at that Captcha on page 90. If the first word is UNTO, we can figure out that the fuzzy second letter is an N. But the word can only be identified as UNTO if the letters have been recognized, including N. If there is uncertainty at both the word and letter levels, how can one inform the other? Which comes first, the letter or the word?

The brain's solution to this riddle can be illustrated by two familiar kinds of puzzles, Sudoku and crosswords. The mechanisms involved are easier to see in games, where they are closer to the conscious surface, than in reading, where they are executed without awareness and also far more complex.

Sudoku is an example of a constraint satisfaction problem, the formal term for what people usually mean by process of elimination. The rules of the game stipulate that the solution must satisfy several simultaneous conditions (concerning the assignments of numbers within subgrids, rows, and columns). These constraints are yoked: the solution to one affects the solutions to others. The puzzle cannot be completed by solving the subproblems separately, starting with one (perhaps the smaller subgrids) and then proceeding to others (the constraints on rows and columns), because the solution to one problem must be compatible with the others. To succeed the solver has to work back and forth between the subproblems; progress on one part affects progress on others. The puzzle is solved by converging on a solution that satisfies all of the constraints at once.

Rather than working stepwise from one level to the next higher one, the reader, like the puzzle solver, works interactively, moving back and forth between levels. As in the game, this works because the levels are not independent: one type (e.g., letters) is a component of another type (words), which is a component of yet another type (phrases), and so on. This nesting allows information from each level to reduce uncertainty at other levels. In the Captcha, figuring out something about the letters narrows the range of possibilities at the word level, which in turn constrains the possible component letters. Working back and forth between levels, one converges on the exact identities of the word and its letters.

Crossword puzzles (another constraint satisfaction problem but without the multiple nestings in Sudoku) illustrate a second property, the nonlinear way that probabilities are combined. Say that an across clue suggests several possible answers, but your confidence that any one of these answers is correct is fairly low. Say that the same holds for a down clue for an intersecting word: you come up with another set of candidate words, each with a low probability of being correct. Taken together, however, you observe that only two candidates, one across and one down, are consistent with each other (because they are the only ones in which the same letter occurs in the square where they intersect). Because the answers are not independent (they share the intersecting letter), the combination of two low-probability events greatly increases the probability that the two words are the correct answers.

Contrast this with the more familiar situation in which the probabilities of two independent events are combined. For example, the probability of being seven feet tall is very low, as is the probability of being born in Montclair, New Jersey. Since there is no connection between these events, combining them creates an even lower-probability one: a seven-foot-tall person from Montclair. This is not at all like the way we use statistics about language and print.

One last game to illustrate this mechanism: It is called "I'm Thinking of Something." The first clue:

I'm thinking of something yellow. What is it?

The clue is not very constraining because many things are yellow. The same would happen if the first clue had been

I'm thinking of a fruit.

or

I'm thinking of something round.

Many yellow things, many round things, many fruits each with a low probability of being the thing I'm thinking of.

Although each of these clues is not very constraining on its own, combining them makes it very likely that the answer is . . . lemon. (*Correct!*)

In this case, as in reading, the combination of low-probability clues yields a very high-probability answer because the properties are not independent. Fruits are objects that have color and shape. The answer could be wrong—I might have been thinking of a guava—but often one answer is far more likely than any other and thus usually correct. These are actual rather than metaphorical examples of nonlinear thinking. The answers—the identity of a letter or word, the puzzle solutions, the thing I'm thinking of, the interpretation of a sentence—result from the *nonlinear combination of probabilistic constraints.* The capacity to combine things we know in this way is one of the distinguishing features of human intelligence. In reading, the process occurs on a massively larger scale, without conscious awareness or effort.

These mechanisms are so powerful that a text can sometimes be read under conditions that are far more extreme than the Cmabrigde paragraph. For proof, avert your eyes from this book long enough to watch two minutes and fifty-five seconds of a television game show, *Wheel of Fortune*, in which contestants attempt to identify a phrase revealed one letter at a time (http://goo.gl/dKsDDe).

Back?

The winner guessed a familiar phrase ("I've got a good feeling about this") using several meager clues:

The puzzle category (phrase)

The number of words and their lengths

The apostrophe in the first word

The absence of an R (the previous contestant's incorrect guess)

The presence of one L (the winning contestant's guess) in a specific position

None of these clues is very informative on its own. But the combination of clues is compatible with one solution, which the contestant (who had practiced the game) could determine very quickly. That is not magic or luck; it is what happens when the properties of written language engage a powerful type of mental computation—and the contestant is a player.

CHAPTER 6

Becoming a Reader

TO THIS POINT I HAVE focused on properties of the written code that are the backbone of reading: characteristics of the symbols, what they represent, how they combine, and what we see when we look at them. Think of this as the phylogeny of reading: the defining characteristics of this species of language and their origins. Carrying the biological metaphor a step further, now it is time to look at ontogeny: how children acquire the skill.

How do children learn to read? It's a simple question, the most basic, interesting, and controversial question about reading, the focus of this chapter and the next, and, really, the subtext of this book. The question can be taken in two ways. The first is as a question about how children learn to read given what basic research tells us about the nature of reading (i.e., what is involved in comprehending written language) and about children—how the various types of knowledge that support skilled reading develop and come into alignment, like the characters finding their places in the tableau that will become the Seurat painting in Sondheim's *Sunday in the Park with George*. This answer is the same for all children. Cultural, economic, and educational circumstances obviously affect children's progress, but what they need to learn does not change.

The question can also refer to how children learn to read in practice, given cultural assumptions (for example, expectations about home involvement) and educational assumptions (about what needs to be taught and how). These issues are the subject of later chapters. Children's progress depends on experiences in the home, in preschool or child care, and in school. These experiences depend on what the adults in these settings believe about learning to read. As I'll show, in the United States and elsewhere, these beliefs are only loosely coupled to the conclusions about learning to read derived from basic research. The discrepancies between the two underlie the reading wars, the chronic state of disputation about how—or if—reading should be taught. These disagreements are exhausting, but they are not trivial. They are not easy to resolve because they arise from deep-seated differences between the cultures of science and education. Reading wars persist because, like other cross-cultural differences, they are deeply entrenched.

Skilled reading is a specialized type of expertise that only some people possess. So is plumbing. Professional plumbers demonstrate their expertise by passing licensing exams that assess things plumbers need to know (such as how to repair gas versus water pipes). All of the required skills are spelled out by the licensing agency. A would-be plumber acquires this expertise through a combination of instruction (classes) and practice (an extended apprenticeship). What if there were a licensing exam for skilled reading? If we can codify the skills required to become a plumber, can't the same be done for reading?

Proposed Requirements for Licensure as a Certified Skilled Reader

Prerequisite: All candidates must have achieved age-appropriate proficiency in a spoken language.

Reading-relevant expertise:

- Knowledge of orthographic structure and relations between written and spoken language, including how written words are pronounced and how spoken words are written
- Ability to recognize a large vocabulary of printed words quickly and accurately, including academic vocabulary mainly found in texts, such as ANALYSIS, ACQUISITION, and WHEREAS
- Extended knowledge of the meanings of vocabulary words, including multiple meanings and senses, and the ability to interpret them correctly in context
- Knowledge of the kinds of phrases and sentence structures in which words commonly occur, including literal and figurative expressions
- Ability to comprehend sentence structures of varying complexity as they are read, as well as the larger structures created from sequences of sentences (paragraphs, sections, articles, chapters, books), making necessary links within and across levels
- Ability to recognize lapses in comprehension and perform simple repairs

- Possession of background knowledge relevant to understanding a broad range of topics*
- Ability to vary depth of reading comprehension and textual engagement in accordance with goals

The candidate must also have completed an extended apprenticeship consisting of several thousand hours of eyes-on reading experience.

The licensing exam will consist of a practicum in which the candidate demonstrates these types of knowledge (such as the Organisation for Economic Co-operation and Development's Programme for International Student Assessment (PISA) reading test for English).

To qualify for master reader certification, a candidate must exhibit additional expertise, such as ability to infer the meanings of unfamiliar words using knowledge of morphological structure and distributional statistics; ability to analyze texts to achieve deeper understanding of content, style, and implications; familiarity with multiple types of texts (including narrative, expository, persuasive, and at least two others); and additional expertise, depending on the master reader's area of specialization.

Within each of the major areas of expertise, the potential licensee would need to demonstrate competence in a large number of subskills. The National Assessment of Educational Progress (NAEP) assessment for eighth graders incorporates thirty-two subskills, ranging from "use global understanding of the article to provide explanation" to "use understanding of character to interpret author's purpose." The Common Core standards for English language arts are, in essence, an attempt to codify the licensing requirements for reading at each grade level from K to 12. The requirements get very specific:

> [Individual demonstrates ability to] cite strong and thorough textual evidence to support analysis of what the text says explicitly as well as inferences drawn from the text, including determining where the text leaves matters uncertain. (Grade 11–12)

If I were asked to propose a standard for reading beyond the twelfth-grade level, it would be "Demonstrates ability to comprehend the Common Core standards for reading."

* Exact content may vary but should minimally include shared cultural knowledge concerning, for example, media, government, and social institutions; it may include deep expertise in specific areas such as music, guns, cars, sports, or Star Wars.

Having completed a rigorous course of training, engaged in extended apprentice reading, and passed the licensing exam, the certified skilled reader could be issued this document in formats suitable for framing and posting:

> A person in possession of this license shall be able to read and comprehend text presented by means of traditional paper formats, as well as screen-based media including smartphones, tablets, phablets, laptops, and Netflix subtitles. The reader may be called upon to read in school or employment settings and so must be capable of maintaining sufficient attention to comprehend long-form documents such as the *Telemarketing Employee's Handbook* and the thicker *Harry Potter* books. The licensee demonstrates proficiency at reading jacket blurbs and exhibits good judgment in deciding whether to wait until the book is available at the public library or to buy it in hardback, paperback, audio, or digital format, or perhaps secondhand if the shipping charge is not exorbitant. A licensed reader will have had several thousand hours of textual experience, excluding low-literacy activities such as browsing online dating profiles. A licensed reader may be called upon to read to children with the expectation that none of the important words will be skipped or changed. Finally, a licensed reader is certified as capable of reading on airplanes in coach even when it is hard to turn the page without elbowing the person in the next seat. Responsibilities of the licensed reader include updating personal language statistics without conscious awareness and maintaining a tidy orthographic network.

Although the license is a spoof, the list of skills is not. It accurately represents core components of proficient reading. (A few were omitted in the interest of not belaboring the joke.) The licensing requirements are outcomes, not developmental stages or modules in a reading curriculum to be mastered one by one. The licensing exercise shows that expert reading is not an inscrutable art: the major qualifications can be listed. Having specified what skilled readers know and do, we can ask how they get there.

Reading Develops

Reading skill follows a lifelong trajectory. The preparations begin in infancy, with learning about language and the world. Children typically learn to read during the first several years of formal schooling, but expertise continues to develop for an extended period lasting through late adolescence. Reading skills are further modified by how the activity figures in our adult lives. Learning to read is a complex problem because multiple overlapping subskills develop at the same time. Children vary in how rapidly they progress along the path to reading, but there is little skipping ahead because basic skills are prerequisites for more advanced ones.

Two important consequences follow from the fact that reading follows a developmental trajectory. First, the way to become a skilled reader is not by emulating one. Think of how a person gains athletic skill (substitute musical skill if you balk at sports analogies). It would be gravely hazardous for a coach to train novice divers to emulate what an Olympic champion does coming off the ten-meter platform. The athlete (musician) builds skills over time. Along the way the learner performs in ways that do not closely resemble the expert. Such transitional states are nonetheless necessary to move further along the trajectory toward expertise. The same is true of reading.

The second consequence is that what it means to be a skilled reader depends on where an individual is on this trajectory. A very good fourth-grade reader would make a poor eighth-grade reader. An eighth grader who scores at the proficient level on the NAEP would nonetheless be a weak reader by college-level standards. The behaviors that are characteristic of age-appropriate reading—and relevant to predicting and monitoring progress—also change over time. For a prereader, knowing the names of letters and being able to tell if words rhyme are important. By first grade, those skills no longer count for much because they have been learned. Now a good reader needs to master a different set of skills: being able to sound out words and read texts aloud with a modicum of fluency and sufficient linguistic and world knowledge to understand them. By fourth grade, reading aloud is over. Emphasis shifts to silent reading and comprehending texts in order to learn and experience new things. The demands on the reader continue to increase over the course of formal schooling, through high school and beyond and in many occupations. Thus, the goal posts do not stay in place for long.

I've said that reading and speech are like codependents in a long-term relationship. The relations between the parties change over time. Two esteemed reading researchers, Isabel Beck and Connie Juel, relate a story that illustrates this point. A teacher is reading the book *Make Way for Ducklings* to a class of first graders. The children's reactions show that they are following the story. They laugh in the right places. They show concern when Mrs. Mallard almost gets run over by a bicycle. At one point the narrative says, "There were sure to be foxes in the woods or turtles in the water, and she [Mrs. Mallard] was not going to raise a family where there might be foxes or turtles." When asked, the children can explain why Mrs. Mallard would not raise a family there even though the text does not explicitly say. Although the children are able to understand the text, few if any would be able to read it.

For most of childhood, speech is dominant. The origins of reading are in the acquisition of spoken language, well before a child has seen a single letter. Speech is the source of knowledge about language and a conduit for information about the world the child inhabits. Children's entry into reading and how fast they progress depend on their knowledge of speech. Because it is so crucial to becoming a reader, measures of prereaders' spoken language are strong predictors of later progress.

Reading lags behind from the start and continues to do so well after the onset of schooling. Becoming literate is the scenario by which reading catches up to speech and then surpasses it. With the acquisition of sufficient skill, reading takes over as the primary means by which linguistic knowledge continues to expand, because many words and sentence structures appear in print but rarely in speech, and as the primary means to acquire the specialized knowledge of a topic or type of text on which comprehension depends.

The earliest events on the path to reading are the acquisition of a lot of spoken language and a bit about print. The child learns a staggering amount of language during the first several years of life. This knowledge is deeply implicated in the two main challenges facing the beginning reader: figuring out how print represents speech and comprehending language represented in this new form.

First, phonology. A child who learns a spoken language can both produce and comprehend utterances. These two sides of the linguistic coin are somewhat different. Production involves using the vocal apparatus to generate complex sound waves. The act consists of a series of coordinated articulatory gestures involving the breathing cycle, jaw, lips, and tongue. The linguists Louis Goldstein and Cathy Browman depicted these movements as a gestural score, analogous to a musical score. In comprehension, those articulatory gestures are experienced as acoustic events analyzed using the auditory system. The components here are properties of sound waves, such as changes in frequency and intensity over time. The codes are different but linked: having been heard, a pseudoword such as "glorp" can be repeated, and having been uttered, it can be heard. The translation from one code to the other is mediated by phonology, an intermediate representation, neither pronunciation nor sound but rather a more abstract code shaped by both. Infants tune this system by babbling, producing sounds and hearing how they match the speech of others.

It takes infants less than a year to learn the inventory of sounds in the language to which they are exposed. It takes longer to learn the phonological forms of words. These representations change as the child's vocabulary grows and later when they learn about print. Infants' first words are acoustic blobs. The discovery of component parts, which is important for speech and later reading, begins as a by-product of vocabulary growth. If the only words a baby knew were "cat" and "mommy," they could be recognized as two distinct patterns without being analyzed into component parts, as with nonlinguistic sounds such as rushing water or a door slam. The child's first hundred or so words are treated as blobs because, like "cat" and "mommy," they are spaced pretty far apart, phonologically speaking (they include words such as bath, bowl, spoon, diaper, chair, duck, and milk). As vocabulary size increases, the overlap among words causes the parts to begin to emerge. Rhyming is fun but also increases the child's awareness of rimes, the part of the spoken syllable that overlaps (as in "hop-pop" and "said-bed"). The overlap across rhyming words also draws attention to the different initial sounds. The word "cat" gains neighbors such as "can," "car," and "could," helping the child to isolate the initial sound, the phoneme /k/. This process begins to take hold during the rapid growth in vocabulary that begins around eighteen months of age. A detailed theory of this differentiation process has been proposed by the developmental psycholinguist Amanda Walley, who calls it "lexical restructuring." It is a result of statistical learning, driven by the size of the child's vocabulary, how often words are heard, and patterns across words.

The smallest units of importance in reading an alphabet are phonemes, exemplified by the three sounds in "bat." Phonemes are abstractions because they are discrete, whereas the speech signal is continuous. From the statistics of spoken English, children develop rough approximations of some phonemes as clumps of sounds emerge from the contrasts across words. A child who learns that his name begins with "m" begins to develop a representation of that sound even though it is spoken as part of a single articulatory gesture, the syllable "mark." The name is no longer treated as an acoustic blob, but neither is it a string of phoneme beads. The invaluable illusion that speech consists of phonemes is only completed with further exposure to print, often starting with learning to spell and write one's name.

The development of phonological representations for words and their parts is a major step in learning a spoken language. The properties of these representations also play a critical role in reading, and impairments in phonological representation are usually observed in developmental reading and speech disorders. Phonological development is not the only factor involved in learning to read, but it is always an important part.

Why Do Letters Have Names?

Children gain some general awareness of the functions of print and its relation to speech from looking at books and being read to, but the first direct connections involve letters and their names. A child can learn letter names from alphabet books, talking toys, playing alphabet blocks with an adult, or watching *Sesame Street*, but everyone learns the alphabet song (to the tune of "Twinkle Twinkle Little Star" in English). Learning the song is a childhood rite of passage, right up there with jumping in puddles and finger painting. Parents are thrilled with a play activity that is literacy related and does not require cleanup. See YouTube for videos of toddlers offering cute renditions of the song in English as well as Russian, French, and other languages. The songs help toddlers learn the inventory of letters and their recitation order. Most of the world's alphabet songs teach children the names for letters: "ay" for A, "bee" for B, "em" for M, "double-u" for W in English; "ah," "beh," "veh" in Russian; and similarly in other languages. Some alphabet songs use sounds associated with the letters ("buh" instead of "bee," for example). Studies in English and other languages have consistently found that prereaders' knowledge of letter names is among the strongest predictors of their subsequent progress.

Alphabet songs may be common cross-culturally because they are a simple entrée to reading that is rewarding for both child and adult. But why letter names are so strongly connected to beginning reading achievement is a little puzzling. The names are not related to the sounds of the letters in a simple or consistent way: "bee" starts with the sound; "eff" does not. For vowels in English, one of several pronunciations is arbitrarily used as the name, which ensures that the letter (e.g., A) will only be pronounced like the name some of the time. If "em" were the sound of the letter M, the pronunciation of MARK would be "emmark." The names teach the child to associate a letter with a *syllable* (three for W), whereas their correspondence to *phonemes* is relevant in reading. Letter names also allow words to be spelled aloud, which is helpful and makes spelling bees possible, but children learn letter names well before they serve these functions. What else is special about letter names?

Among their many tasks, a beginning reader has to learn to recognize letters—not a trivial problem because of the variety of forms they can take (see Figure 6.1). The reader must be able to treat the visually similar patterns in the alphabet on the left as different letters but treat the visually dissimilar patterns on the right as the same. Letter recognition is a categorization problem: all of the variants must be treated as exemplars of the same letter. The problem is similar to deciding whether an object is a fruit. Similarity helps: many fruits look similar, tend to be similar in size and weight, and usually have a pleasant fruity smell or taste. (A lot of A's also look similar.) But then there are pineapples and durians, which do not look much like other fruits, and durians also smell stinky. (Some of those A's are pretty funky too.) Some fruits more closely resemble another type of object—bananas and boomerangs, for example. (Some A's look more like H's than other A's). Readers also use the contexts in

FIGURE 6.1. The letter recognition problem: treating the visually similar patterns in the alphabet on the top left as different letters, while also treating the visually dissimilar patterns on the top right as exemplars of the same letter. The contexts in which letters occur also provide cues (bottom). The alphabet is from Rumelhart and Siple (1974), Figure 2, p. 101. Copyright © American Psychological Association. Reprinted by permission.

which letters occur as cues: A's are graphical symbols that occur in words in positions where the letter A is required (Figure 6.1, bottom). Recognizing letters using these cues requires a lot of statistical learning on a lot of text, which isn't yet available to a young child but works for machine learning algorithms, which can now recognize even handwritten letters with high accuracy.

Children, however, can use another cue, the one thing that all of those A's definitely share: their name, "ay." Similarity is important in forming categories, but names provide the glue. The role of names in forming categories has been closely studied by researcher Gary Lupyan, who finds that having a category name makes it easier to form new categories and to assimilate new exemplars into existing categories, overcoming the limits of similarity. This theory extends easily to letters. It is not just that a rose by any other name would smell as sweet. It is that the name helps in determining whether a flower *is* a kind of rose rather than a peony or marigold.

In addition to its other functions, the alphabet song is important because it teaches the child the names of twenty-six categories of objects. That proves useful as they gather data about their visual properties, sounds, and combinatorial statistics, solving the letter-recognition problem.

Vocabulary: It's Not Just Number of Words

Children's production of their first words is a special period in language development, a delight to observe, the emergence of the ability to communicate verbally always a wondrous event but also a relief for parents who await the first words and word combinations as signs that language learning is on track. Although vocabulary acquisition has been much studied in English and other languages, only recently have researchers, educators, and policymakers recognized just how much young children's language varies. The sources of that variation, its impact on reading and school achievement, and how those effects can be modulated are now the focus of intense interest.

Why is vocabulary so important to becoming a reader? Children in the K–3 range encounter relatively few of the words they know in the books they read. A first grader can sit in the advanced group if she can read only a few hundred words, enough for *The Cat in the Hat*, which has 236. The child will need more words to read more advanced texts, but at the outset the factor that limits reading is knowledge of print, not vocabulary.

Vocabulary is important for other reasons that become apparent by considering what it means to know a word. Intuitively, knowledge of a vocabulary word is a learned association between form (a spoken or, later, written pattern) and meaning. This narrow sense of knowing a word underlies standardized tests of vocabulary development. The child hears a word such as "cup," for example, and has to point to the one of four pictured objects that corresponds to its meaning. Children who correctly point to the picture of a cup know the word but can nonetheless differ greatly in the information they associate with it, a dimension researcher Charles Perfetti termed "lexical quality." A word is a hub linking many types of information: its sound, pronunciation, and spelling; its multiple senses (e.g., drinking utensil, unit of measurement, trophy); the entities to which it refers (e.g., types of cups), as well as the sensory and perceptual properties of those entities, their functions, and how to use them; facts such as where they are made, bought, and kept and which is Mother's favorite; their grammatical functions (e.g., "cup" is both noun and verb) and how the word combines with others to form expressions such as "sippy cup" and "cuplike." The density of the information linked to a word reflects what the child knows about many things beyond a simple association between form and meaning.

Words are also tied to expectations about words that can occur with them. A cup is usually something that is acted upon rather than the initiator of an action (I HELD THE CUP, THE CUP DROPPED), whereas the opposite is true of a tornado. Having heard or read THE CUP DROPPED creates expectations about who or what caused the dropping and how or why it occurred. Although superficially similar, THE TORNADO STRUCK creates expectations about who the event affected, not who caused it. The noun and verb senses of words such as CUP have their own distributional properties. The verb occurs with words for things that can be cupped, such as hands, chin, or face; the noun with things that cups hold, such as coffee or flour. Much of our knowledge of the grammar of a language is tied to individual words in this way. These types of knowledge are probabilistic, of course. THE CUP HELD triggers expectations about what is likely to follow rather than the exact continuation. The cup could hold coffee or Brisk Lipton tea, or it could be held by a person or held in a museum because it was made of fur.

Our knowledge of a word is therefore not very much like a dictionary entry, unless your dictionary is endowed with the capacity to experience the world and track statistics about how often and in what linguistic and nonlinguistic contexts the word occurs. Vocabulary in this broader sense develops by adding hubs and building connections to these many types of knowledge, a process that continues through the lifespan, as when we add a word such as FRACKING or attach additional meanings to a word such as TWEET.

Vocabulary illustrates a general puzzle about language and reading: So Much to Learn, So Little Time. Children acquire a first language in a few years. Five-to six-year-old English learners have vocabularies in the range of 2,500 to 5,000 words, and they add about 3,000 words per year (about eight words a day) for the first several years of schooling. How is this possible? Whereas it is arguable whether grammatical knowledge is innate, vocabulary definitely has to be learned because languages form words differently and use words to pick out concepts in idiosyncratic ways. Parents and other caregivers do provide some explicit vocabulary instruction (such as the names of animals and letters), especially at the start, and this plays an important role, but they do not teach thousands of words. The puzzle is how children learn so many words (and so much about them) in so little time.

This question assumes great importance because of the substantial individual differences in vocabulary size and quality that are present at school entry. Elementary school language arts curricula routinely include vocabulary-building activities, and researchers have identified effective teaching procedures. The sticking point is the sheer number of words involved. There are not enough minutes in a school year to teach children thousands of words or eliminate vocabulary disparities by such methods alone.

Like learning orthographic structure and letter categories, learning vocabulary is a Big Data problem solved with a small amount of timely instruction and a lot of statistical learning. The beauty part is that statistical learning incorporates a mechanism for expanding vocabulary without explicit instruction or deliberate practice. The mechanism relies on the fact that words that are similar in meaning tend to occur in similar linguistic environments. Consider just the words that precede or follow a word such as LION. They are also likely to occur with TIGER. Two nouns referring to big cats that share many traits tend to be used with the same words because, for example, most of the adjectives that apply to one also apply to the other. The same happens for verbs with similar meanings, such as TOSS, THROW, and FLING. In just the first several years of life, children are exposed to millions of words in utterances used for communicative purposes in relevant real-world settings. Having accrued statistics about words such as LION and TIGER allows the listener (or, later, the reader) to infer much about the meaning of a new word such as LYNX when it occurs in those contexts—*without being told.*

The key idea underlying this process was described in 1957 by British linguist John Firth, who coined the aphorism "You shall know a word by the company it keeps." A person can infer much about the meaning of a new word because it occurs in the same contexts as familiar words. It then takes very little additional information to learn what is special about, say, a lynx. Whether this process works very well could not be determined for many years because it was not possible to analyze the statistical patterns in large samples of language. Moreover, Noam Chomsky had emphasized the generativity of language, the fact that we can produce and comprehend novel combinations of words, even ones that bear little resemblance to previous utterances. Language is so flexible and open-ended that combinatorial statistics might be of little use. Interest in the issue was revived by a landmark 1997 study by Thomas Landauer and Susan Dumais at the dawn of the Big Data era. By then it was becoming possible to conduct the relevant statistical analyses of text and speech. Such analyses show that much can be learned about words from relatively simple statistics such as trigrams (three-word sequences). A common word such as LION is both preceded and followed by huge numbers of different words. However, this is another long-tail situation. A relatively small proportion of the words that precede or follow LION do so repeatedly (for example, LION is frequently preceded by COWARDLY, and KING is a frequent follower). The high-frequency patterns allow inferences about the meanings of novel words. Landauer and Dumais established that because word combinations exhibit strong statistical regularities and children are exposed to sufficiently large samples of utterances, they are prepped for many words when they finally occur, as in the LION/TIGER/LYNX example.

This theory is a breakthrough because it explains how children can rapidly acquire large bodies of knowledge such as vocabulary with only intermittent instruction or correction. Children go from 0 to 1,000 words in about thirty-six months (though the rate varies, as we'll see). The learning curve is exponential, not linear: the first fifty or so words are learned slowly, but then the rate increases rapidly. According to the statistical learning theory, the increase occurs as the child accumulates sufficient linguistic and experiential data to bootstrap the meanings of more and more words once they occur. This account of word learning also explains a phenomenon called fast mapping. Young children can learn new words extremely rapidly, sometimes with a single exposure in a meaningful context. One explanation is that children are born knowing all of the concepts that words could potentially label. The statistical learning account is that children are continuously gathering data about the structure of language and the world in which it is used, which prepares the way for adding new words very rapidly.

Statistical learning provides a plausible answer to the So Much to Learn, So Little Time puzzle. It is not subject to the limitations of explicit instruction because it works automatically in the background as language is used. However, it also explains why readers benefit from vocabulary instruction, as many studies have demonstrated, even though only a small percentage of words can be taught. The impact of instruction extends beyond the word itself because it causes updates in the statistical network that are relevant to many other words, as in Landauer and Dumais's theory. Computational analyses of the performance of various learning procedures suggest that a combination of many implicit learning events and a smaller number of well-timed and targeted explicit learning trials is highly efficient.

Reading to Children: Necessary but Not Sufficient

Although screens are making inroads—a *Goodnight Moon* app!—the most significant reading-related activity for young children is still being read to, which is as American as apple pie, or as pie was before the revolt against the carbs. I am referring to the traditional activity in which an experienced reader and a child interact, not a video of someone reading a story aloud. Reading to children has obvious benefits, and being the rare issue that poses no risk of offending anyone, it's the ideal option for a celebrity public service announcement or presidential photo opportunity.

I am not Scrooge. Reading to children is important. It socializes them into the experience of reading and engages them in the activity before they can read. They learn about books and their parts, their variegated sizes, shapes, colors, and textures, and the arrangements of the letters, some of which they will exult in recognizing. They hear wonderful/scary/surprising/funny stories and see drawings of all sorts; they learn about other places, people, and things; they get the attention and usually the affection of another person. The experiences can help to kindle the child's interest in reading and boost their motivation to learn. All good. But let me explain what goes on. I will not ruin anyone's memories of reading books with Grandma or to their children. I am going to show that what children get from being read to is both more and less than usually assumed.

Like reading itself, reading to children encompasses several activities. Reading *to* children is a misnomer inasmuch as people usually read *with* them, talking, digressing, and asking and answering questions (that is why it is sometimes called "shared reading"). How adults read to children depends on when (a few minutes before bed, during a block of quality time), where (at home with a single child, in preschool with a group), why (for entertainment or instruction, to induce sleep), by whom (the reader's language background, literacy level, motivation; their cultural norms), and other circumstances (the first versus the nth reading of *Pat the Bunny*). All of these permutations have positive aspects, of course. However, our culture's emphasis on the importance of reading to children creates the impression that it plays the same role in learning to read as speaking to children plays in their learning to talk.

That's not correct. Whereas talking with children guarantees that they will learn to speak (in the absence of pathological interference), reading to children does not guarantee that they will read. Children learn a spoken language through exposure and use, but reading requires systematic guidance and feedback, more than occurs in casual reading to children. In short, reading to children is not the same as teaching children to read. I emphasize this point because the mantra about reading to children makes it seem that this is all that is required. A child who has difficulty learning to read therefore must not have been read to enough. Among the first questions that will be asked of the parents of a child who is struggling is whether they read to the child and if there are books in the home. Reading to children is important but not sufficient; children benefit from it, some quite a lot, but it neither obviates the role of instruction nor vaccinates against dyslexia. Children who are read to until the cow jumps over the moon can still have difficulty becoming readers.

The apparent exceptions to the rule? Children who do learn to read from being read to or who "teach themselves." This phenomenon is intriguing but hard to study. I regularly ask the students in my classes when and how they learned to read. It is not a scientific study; I am mainly interested in getting them to connect their experiences to the research we're examining. A few people will report learning to read before starting school, having been taught by a family member or "figured it out" on their own. The late Martin Gardner, the polymath who wrote a popular column in *Scientific American* for many years, was of the latter sort, according to his *New York Times* obituary: "Unbeknownst to his mother at the time, he learned to read by looking at the words on the page as she read him L. Frank Baum's Oz books."

When people describe learning to read on their own, my thoughts naturally turn to Kanzi, a bonobo (a primate closely related to the common chimpanzee) who was the star subject in an animal language-learning experiment in the 1980s. Several bonobos and common chimpanzees were taught to communicate using a display of lighted keys that corresponded to words such as STRAWBERRY and GO. The scientists who conducted these studies thought the apes' use of the keyboard was like the child's use of language, but others disagreed. For one thing, it took many hours of training for an ape to learn the association between a key and a referent such as strawberries, in contrast to the fast-mapping seen in children's word learning. Kanzi, however, was reported to learn these key signs without extensive training. It is true that Kanzi did not receive many hours of direct training, but he did not learn the system on his own. He learned it both literally and figuratively on his mother's back: she carried him while *she* was being trained. Kanzi's *Times* obituary might someday read, "Unbeknownst to Kanzi's mother, he learned to 'talk' by looking at the signs on the keyboard as she was taught to use them."

Reading with children mainly focuses on the content and the shared experience, with varying emphasis on properties of the written code. Research has clearly established that incorporating attention to print into reading to preschoolers has beneficial effects. Taken to an extreme, reading to children with a focus on print is tantamount to a large amount of one-on-one instruction. For an adult and a child of suitable temperaments, extended interactions of this type may indeed result in the child beginning to read. As in the classroom, how much guidance and feedback are needed depends on other characteristics of the child and the environment. The more advantages a child has, the less instruction is required. Martin Gardner, who grew up to be a puzzle aficionado and author of over seventy books, may well have required little more than the pairing of his mother's speech to the corresponding words on the page. We do not know how little Alan Turing, the genius mathematician of *The Imitation Game*, cracked the orthographic code, but it surely would have been fascinating to watch.

Such exceptions notwithstanding, reading to children rarely involves teaching them to read. The activity contributes to the child's progress but is not all that is required. Some children do learn to read with minimal instruction, but unlike learning to talk, it is a mistake to assume that they will.

Reading to children serves a lesser-known function that is at least as important as introducing them to print: expanding their knowledge of spoken language. Adult speech influences children's language acquisition more than any other experience. Reading to children is also a form of speaking to them that promotes linguistic development in special ways. When we read aloud to children, we are speaking in a manner unlike everyday speech. Even picture books for the youngest children contain vocabulary and grammatical structures that occur rarely, if at all, in other speech to children. Reading to children is a type of linguistic enrichment.

Remember the duckling sentence? "There were sure to be foxes in the woods or turtles in the water, and she [Mrs. Mallard] was not going to raise a family where there might be foxes or turtles." Many children find out about foxes and turtles through the books they are read rather than through everyday speech or direct experience (all the more so for Sneetches and Wild Things). The word MALLARD (the referent for SHE in the quotation) is unlikely to come up in speech to children outside a duck blind. At thirty words, the sentence is longer and more syntactically complex than children's own speech and the speech they hear from adults. As the sentence illustrates, children's books are a unique source of data for the language learner.

Eric Carle's *The Hungry Caterpillar* is among the most popular children's books ever. It's a board book for very young children, printed on heavy cardboard that is food and bodily-fluid resistant. Its subject matter includes an apple, pears, plums, strawberries, oranges, a piece of chocolate cake, an ice-cream cone, a pickle, a slice of Swiss cheese, a piece of salami, a lollipop, a piece of cherry pie, a sausage, a cupcake, a slice of watermelon, and a green leaf, only some of which a child may know about from other sources. The first sentence, "In the light of the moon a little egg lay on a leaf," begins with a complex preposed prepositional phrase, a syntactic structure with a literary rather than spoken character.

Although it is obvious that adults learn many kinds of specialized vocabulary through reading, the process begins much earlier, with stories read to prereaders. The linkage between vocabulary and reading works in both directions. Larger, richer vocabulary facilitates learning to read and skilled reading, but reading also promotes vocabulary development. This bidirectional relation holds for syntax as well. Children learn grammar from the utterances they hear, which include texts that are read to them, which in turn include syntactic structures that are rarely used in other types of speech. Familiarity with a broader range of syntactic constructions facilitates beginning reading and gaining skill. You talk better too.

Reading to children is an area in which differences in economic circumstances affect child development. A low-income family will be less able to buy books and more likely to live in a neighborhood with fewer public libraries, which serve larger populations and contain fewer books which are in worse physical condition than those in middle-class areas. The laudable Reach Out and Read program founded by a team of pediatricians and educators has distributed millions of books to millions of children. Providing reading materials is an important step, but the impact depends on how they are used. And that depends on other factors correlated with socioeconomic status, such as whether there is a parent or caregiver available to read to the child and that person's reading and language abilities.

One final caution about reading to children. The act takes different forms and serves several functions. It does not require sticking closely to the text, but the language-building benefits I've described do. The reader has the option to recast a text in a more casual, talky idiom that simplifies vocabulary and syntax. What parent doesn't attempt to freestyle some of the time? The impact on the child changes significantly, however, if a reader consistently recasts the text, as may occur because both parties enjoy it, or the adult's knowledge of English is limited, or they speak a dialect very different from that of the text. Telling a story in one's own argot may enhance the experience in some respects while undercutting other potential linguistic benefits.

Finishing the /h/-/a/-/t/

This extensive run-up brings children to the brink of reading. The transition into reading typically occurs after the onset of formal reading instruction, with the push to move from reading-related activities to being able to read. In the past, this transition typically occurred late in first grade. Kindergarten is sometimes said to be the new first grade in many schools, but it is not clear whether children are passing major milestones earlier. Within the pre-K to grade two range, the timing of this transition varies and is not strongly predictive of long-term reading attainment. The onset of reading can be delayed by environmental factors such as a lack of competent instruction or events that disrupt school attendance whose longer-term impact depends on whether and how quickly they are ameliorated. It can also be delayed by reading or language impairments and other developmental anomalies. Although some late starters become good readers, delays in attaining basic skills in favorable environments beyond late in grade two merit closer examination and active response. Delays in identifying struggling readers are hazardous because earlier interventions are more successful.

The exact age at which the child begins formal schooling does not appear to be critical in the United States. School systems use calendar cutoff dates to establish the minimum age at which a child can enroll. Children who barely make the cutoff will be almost a year younger than children who barely missed it the previous year. Fred Morrison, a reading researcher at the University of Michigan, conducted very clever studies that made use of this calendar quirk. He and his colleagues looked at three groups. Older kindergarteners and younger first graders were close in chronological age but fell on different sides of the calendar cutoff. (The older kindergarteners also included children whose parents had "redshirted" them on the view that an extra year would give them a developmental edge on younger children in the same grade.) Older first graders were kindergarteners from the previous year who had just missed the cut. Children's reading performance was assessed at the start and end of the school year. The main question was whether reading achievement was related to age or amount of schooling. Younger and older first graders made similar progress, with no advantage for later enrollment. Younger first graders made far more progress than older kindergarteners, even though the two groups were very close in age. Thus, time in school mattered more than age, evidence that argues against delayed enrollment.

If things are going well, children entering kindergarten have acquired extensive knowledge of spoken language, including vocabulary and grammar. They have begun to discover the internal structure of words, including some phonemes. They've learned their letters and letter names and understand that print is a way of representing words they know from speech. Prereaders can usually recognize and pronounce a few words accurately (e.g., their own names, brand names such as Coke or Disney written in their trademark fonts and colors), but the process isn't much different from recognizing a can of Coke or a Disney princess. Prereaders can use bits they have learned about letters to correctly guess some words but will also make some howlers. Asked to pick out the word "bat" from a few printed words, the child may know enough to go for one that starts with the /b/ but then pick BUTTERFLY.

Some years ago Philip Gough proposed "a simple view of reading," an account of the child's transition into the activity. Gough was being a little coy. The simple view reflects a deep insight about learning to read. He observed that early reading has two components: print knowledge and comprehension. Beginning readers can already comprehend spoken language. They will be able to read if they can just gain access to language from print. Their task is to build a new circuit linking the visual code to existing neural systems for language (Figure 6.2). Since print, sound, and meaning are all involved, I will use the broader term "basic skills," a narrow but essential set of procedures that allow written words to be recognized and understood efficiently. Basic skills and comprehension are not independent; both are tied to spoken language, and they shape each other over the course of learning to read. The beginning reader gets the comprehension part for free, but basic skills require instruction and practice.

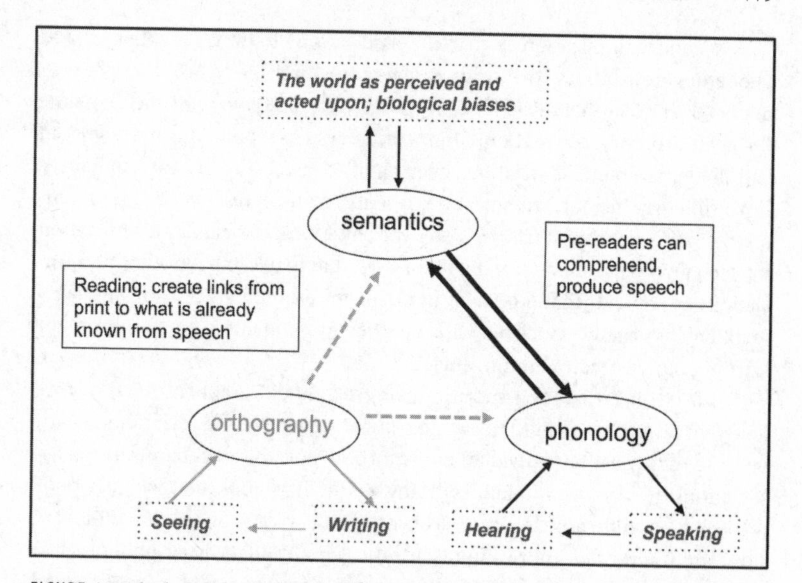

FIGURE 6.2. The logical problem of learning to read. The child has to link print to existing knowledge of spoken language. In principle, print could be linked to semantics (associating BOOK with its meaning) or phonology (associating it with "book"). Each of the codes in the ovals develops from perceiving and acting (e.g., recognizing letters but also writing them).

A child who acquires basic skills can recognize and pronounce words, the entry point for reading. The overt pronunciations (or, later, internal phonological codes) that the child generates can be comprehended by the spoken-language system, establishing the necessary bridge between print and speech. Much depends on this breakthrough, but it can be difficult to achieve. In the best case it requires effective instruction and guidance; in the worst case, children have extreme difficulty that seriously interferes with progress.

The initial hurdle is grasping the alphabetic principle, whereby units in the written code (graphemes) represent units in the spoken language (phonemes). Several potential obstacles arise. First, as I've already noted, phonemes are an abstraction that depends in part on exposure to print. The child's task isn't as simple as learning how graphemes relate to phonemes that are already in place; rather, they are solidifying phonemic representations by learning about print. The same happens going the other way: what's learned about phonemes shapes the learning of orthographic units. This interdependence means that a problem in learning one of the codes will also affect the other.

A second complication is that the associations between graphemes and phonemes are arbitrary. The forms of letters are unrelated to how they are pronounced. The fact that the letter B, for example, is the symbol for the phoneme /b/ is an arbitrary convention. Our writing system would still have worked had the letter-sound associations been shuffled, as occurs in other alphabets. In Cyrillic writing, for example, В represents /v/, Н represents /n/, and Р represents /r/. (You heard it from МАРК.) The child learns a default pronunciation for each grapheme. Of course it will turn out that they can have other pronunciations in context (that B is silent in COMB, for example), but knowing that a grapheme is usually associated with a particular phoneme is a prerequisite for learning more complex contingencies.

A third issue is that the associations are cross modal: graphemes (vision) are linked to phonemes (sound). Learning any set of arbitrary associations is hard because they must be individually memorized, but it is even harder if the associations involve two modalities. Many studies have looked at whether poor readers have difficulty learning arbitrary associations. Such difficulties are commonly reported; more important, the cross-modal visual-phonological associations are hardest and, in many cases, the only type that is impaired. As with other skills, impairments in learning arbitrary associations vary in severity; for many children, the impairment only affects learning when the associations are difficult, as when they cross modalities. The impact can be increased by other risk factors, such as an attentional impairment, or mitigated by protective factors, such as high phonological awareness and effective instruction.

It should now be clear why becoming alphabetic is a major hurdle that requires instruction, feedback, and practice. The child has to think phonemically, which involves both phonology and orthography, and learn arbitrary cross-modal associations between graphemes and phonemes. This step can be disrupted by conditions that affect any of the several moving parts involved. The amount of instruction required depends on the child and how they are taught.

Although it is called the alphabetic principle, the term is a rather high-level description of behavior that isn't much like learning a rule or other broad generalization. Children do not go from staring blankly at printed words to the realization that they consist of graphemes that correspond to phonemes, resulting in the wholesale reorganization of orthography and phonology. The developmental sequence instead involves accumulating knowledge of the statistical structures of orthography and phonology and the mappings between them.

The alphabetic principle is helpful as a label for a transition that becomes apparent as the child is acquiring this information: at some point they start treating graphemes as corresponding to phonemes. Computational models of statistical learning have shown how a gradual process can give rise to what looks like a sudden insight. At the neural level, many small changes can occur over time before there is an observable change in behavior, which can be perceived as an abrupt transition. This gradual learning process is accelerated by explicit instruction.

Children who can treat spoken words as consisting of phonemes are said to have acquired "phonemic awareness." This term is also a little infelicitous. Children aren't always "aware" that spoken words consist of phonemes in the sense that they are aware that each hand has five fingers. Children only have to be able to use this knowledge, in learning how the written and spoken codes are related. They demonstrate their knowledge by performing "phonemic awareness" tasks such as judging that two words begin (or end) with the same sound. That is different from being aware enough to explain what a phoneme is or what it means to "begin with the same sound."

It is also important to appreciate the extent to which "phonemic" awareness depends on spelling. Children can demonstrate that, too. One of the "phonemic awareness" tasks is called phoneme deletion: I ask you to say the word "bat" without the "b." You should say "at." Say "step" without the "p"; you say "ste," the three sounds that are left over. The task is more difficult than deciding if words begin or end with the same sound and usually can't be performed until the child has begun school. Landerl, Wimmer, and Frith did a very clever thing. They had children perform this task—which only involves listening to words and saying a response, no reading or writing—but they included some words that are spelled with silent letters. Say "lamb" without the "m." The answer? Based on the sound it should be "la," the leftover consonant and vowel. Eight-year-old children who were good readers often responded "lab." Say "sword" without the "s"? They often responded "word." In these cases the spelling of the word clearly influenced responses. "Phonemic" awareness is not just about sounds; it's also about spelling.

Learning how orthography and phonology are related is a hurdle in all alphabetic writing systems. What is different about English compared to languages such as Finnish and Italian is that having grasped that graphemes relate to phonemes, the child will soon make another discovery: the correspondences between them are inconsistent. If B-A-T is "bat," then C-A-T is "cat," and C-A-P is "cap," which is very cool and works for a lot of words! But then C-A-R is "car," which doesn't sound like "cat," and CINDERELLA starts with the same sound as SAT not CAT. The alphabetic principle? That and ten cents will get the child to everything else they need to learn.

CHAPTER 7

Reading: The Eternal Triangle

A CHILD WHO HAS GAINED a basic understanding of the relations between print and sound can get on with the task of learning to read words. I emphasize word recognition because it is the part that is unique to reading, not because it is sufficient. Comprehension involves other types of knowledge—spoken-language and topic knowledge are the most important—but all is contingent on recognizing and comprehending sequences of words accurately and efficiently.

Establishing the bridge between written and spoken language requires gaining sufficient knowledge of print-sound mappings to be able to generate correct pronunciations. This problem is harder in English than in many other writing systems, to be sure. Writing systems vary along a dimension called orthographic depth: how consistently and transparently units in orthography and phonology correspond. English is "deep" because of its inconsistencies; words that are spelled similarly often have different pronunciations. Finnish, in contrast, is one of many "shallow" alphabets with grapheme-phoneme correspondences that are completely consistent or nearly so. Finnish children typically learn these associations before the start of formal schooling, whereas for English speakers the task has hardly begun. Our poor children are subjected to a classic bait and switch: we emphasize the connections between letters and sounds and barrage them with alphabet books, apps, and videos; then, after they have finally gotten the alphabetic idea, the first words they read include HAVE, GIVE, SOME, ARE, WAS, SAID, WHO, WHAT, WHERE, DONE, LAUGH, and other "exceptions." These words cannot be avoided because they are among the most common. Dr. Seuss wrote *Green Eggs and Ham* using only fifty words, about 20 percent of them irregular. The myriad inconsistencies in the spelling and pronunciation of English are the stuff of doggerel and slam poetry and fodder for spelling-reform hobbyists.

I've said that it is necessary to learn the correspondences between print and sound in order to gain access to what is known from speech. But given the inconsistencies in the writing system, one might ask, Why bother? Perhaps a word could be recognized from its spelling, with help from the context in which it occurs. Children could learn to associate spelling patterns with meanings, cutting out the phonological middleman. Recognizing the word BOOK would be like recognizing the object. If words are treated only as visual patterns, the fact that DOSE, POSE, and LOSE inexplicably fail to rhyme is irrelevant. The insight that writing could represent speech was an epochal event in human history, but we aren't obligated to use that information when we read.

The options are illustrated in Figure 7.1, which shows a triangle with orthography, phonology, and semantics as vertices. The reader's most basic task is computing the meaning of a word from print. Every writing system allows this to be done in two ways. A word's written code (whether letters, characters, or indentations on clay tablets) can be directly associated with its meaning, often called the direct or visual route. The written code can also be associated with a pronunciation, which can then be used to access the word's meaning, as occurs in comprehending speech. (All but the youngest readers use the mental, phonological code rather than overt pronunciation for this purpose.) This is the phonologically mediated route. Words appear in contexts that can yield expectations about semantics (e.g., that the word will refer to an animal) or about likely words (e.g., that it may be CAT or BIRD but not TIGER or BEAR); context is also needed to disambiguate words with multiple meanings (e.g., TO TIRE versus THE TIRE), and words have to be integrated with prior context for sentences to be comprehended. Contextual information can be used with both pathways, doesn't favor one over the other, and doesn't greatly alter the procedures I'm about to describe; therefore, it's omitted from the figures.

How these pathways are used in reading underlies debates about reading instruction dating from the nineteenth century. The phonological pathway requires knowing how print relates to sound, the focus of "phonics" instruction. The visual pathway obviates the need for phonics. For reading scientists the evidence that the phonological pathway is used in reading and especially important in beginning reading is about as close to conclusive as research on complex human behavior can get. The opposing view, that using phonology is an inefficient strategy used by poor readers, is deeply embedded in educational theory and practice.

These are much-studied questions of intrinsic scientific interest. There are three angles to consider:

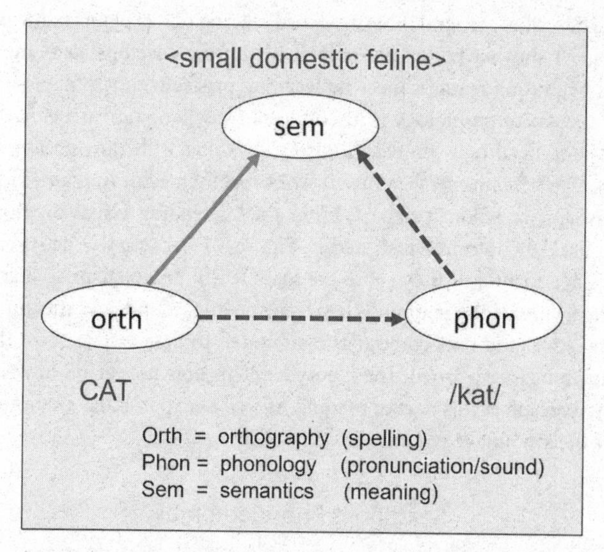

FIGURE 7.1. Two potential paths from print to meaning: orthography → semantics, direct associations between spellings of words and their meanings; orthography → phonology → semantics, computing the phonological code from print and using that to get to meaning.

1. Whether people use phonology in reading and how its use is affected by variables such as reading skill or type of writing system is a question of fact. I'll summarize some representative findings that reading scientists consider decisive. They show that readers use phonological information and that among younger readers, good readers are better able to use phonology than poor readers, allowing them to depend less on guessing words from context. These findings hold across writing systems, with some differences in detail.

2. These findings raise an important hypothetical question: Would English be easier to learn to read and would more people become skilled readers if the writing system were shallow like Finnish? Those Finns do score highly on the cross-national assessments. The Finns' educational system is impressive, but could it be that they merely got a better draw in the writing system lottery? How did English get the way it is, and couldn't it be fixed?

3. I'll describe a general theory of word reading that explains many findings. It shows why the phonological pathway develops more rapidly in beginning readers and why learning proceeds much more slowly if the use of phonology is discouraged (as when phonics instruction is superficial or withheld). It also shows that with the development of expertise, the two pathways work together, achieving an efficient division of labor. It departs from the "is reading visual or phonological" dichotomy illustrated in Figure 7.1 because the paths aren't independent and they share the load. It will be no surprise that the theory treats the mappings between spelling, sound, and meaning as statistical and uses computational models to demonstrate how these mappings are learned. The theory explains how properties of writing systems affect this part of reading, as well as experiential factors such as the amount of reading and type of instruction.

Reading Is Phondamental

Whether skilled reading is visual or phonological or some combination of the two has been a central question in the study of reading. In Chapter 2, I described several phenomena that seem to demonstrate the use of phonology in silent reading, but a skeptic could argue that how people understand jokes or movie dialogue is not relevant to normal skilled reading. Most educators have assumed that reading is purely visual and that phonology is something that poor readers fall back on.

Rhetoric and intuition carried the day for many years because decisive evidence was hard to obtain. For reading researchers, the issue was settled in the late twentieth century by several types of findings that converged on the same conclusion: phonological information is an essential element of skilled reading in every language and writing system; impairments in the use of this information are typical of poor readers and dyslexics. The claim is not that phonics is all there is to reading. Rather, it is that use of the phonological pathway is an essential component of skilled reading. For most children it requires instruction, hence phonics. Having assimilated knowledge of how print and sound are related within the networks that support fluent word reading enables the child to move ahead with other components of reading.

Evidence from Skilled Readers

The breakthrough experiments were conducted by the late Guy Van Orden, whose findings have been replicated many times in English and other

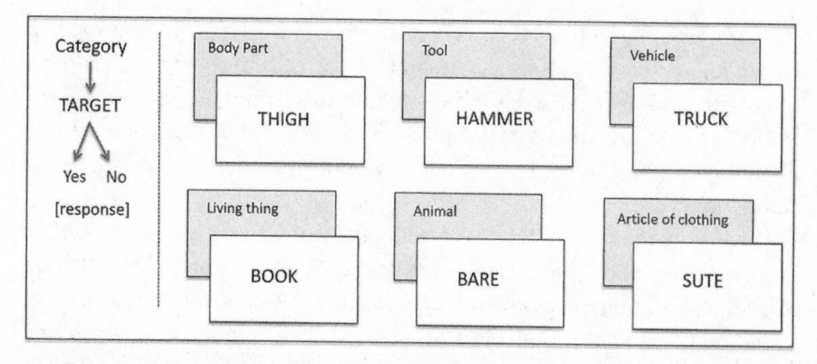

FIGURE 7.2. Main conditions in the Van Orden experiments. On each trial, subjects saw the name of a category (such as BODY PART) followed by a target word or nonword. Their task was to decide if the target was the word for a member of the category (such as THIGH). On some trials the target was a word or nonword that sounded like a category member (as in the BARE and SUTE examples).

languages. Van Orden's findings ran so counter to the received wisdom of the era (the late 1980s) that he initially had great difficulty publishing them. His first, landmark article eventually appeared in a minor psychology journal.

The subjects in these studies were college students who were good readers. The setup was simple (Figure 7.2). Each trial began with presentation of the name of a category. Then a letter string (the target) appeared in the middle of the screen. The task was to press a yes key if the target was a word for a member of the category (such as THIGH when the category is *body part*) or the no key if it was not (such as BOOK when the category is *living thing*). Such experiments include many trials of each type in random order. I've created a toy version to use with your friends and family. Note that the experiment does not involve reading aloud: subjects silently read the words on the screen and indicate their yes/no decisions by pressing response keys.

The crucial element of the design is the inclusion of trials such as

category: *animal*   target: BARE

BARE is a homophone for BEAR, an animal. What should happen? If college students read visually, it should be equally easy to respond no to BARE and to BOAT (the "control" stimulus: not a member of the category, not a homophone, similar spelling complexity). If subjects generate phonological codes in silent reading, then BARE should be more difficult to reject because it sounds like BEAR and will activate its semantics.

A yes response on trials such as *animal*: BARE is a "false positive" error. Van Orden found that subjects usually responded correctly, even on the BARE-type trials. However, they made far more errors in this condition than the others. Subjects would only make a false positive to BARE if they had computed its phonological code. Hence he concluded that skilled silent reading involves the computation of phonology.

What happened? The subject is reading the words in a normal manner. As on other trials, the subject reads BARE and computes its phonological code (orthography→phonology), which activates the meanings associated with it, including the animal and unadorned senses of "bair." On most trials subjects manage to avoid making the error by recognizing that the animal is not spelled BARE, but their responses are slowed. On other trials the fact that the phonology and semantics elicited by BARE are consistent with BEAR tips subjects over to an erroneous yes response.

People make these errors even when the target stimulus isn't even a word, as in *article of clothing*: SUTE. The nonword SUTE is a "pseudo" homophone of SUIT. If subjects use the visual pathway in reading, it should be easy to reject SUTE because it has no meaning and doesn't look like a familiar spelling. Errors would only occur if subjects activated its phonological code, which matches SUIT, activating SUIT semantics. They do indeed reliably make more erroneous "yes" responses on SUTE-type trials compared to nonwords like SOTE. Correct "no" responses for pseudohomophones are also slower than in other conditions.

The effects are compelling because subjects use phonology even though it results in *poorer* performance in this kind of experiment. Continuing to do something one would be better off avoiding strongly suggests the behavior is not under volitional control. That is why the inducement to "suppress subvocalization" in order to speed read is a sham.

For every novel finding, especially ones that challenge conventional wisdom, it is always necessary to assess alternative interpretations. Do the errors occur just because the spellings of wrong answers such as BARE and SUTE overlap with the correct answers BEAR and SUIT? No: the overlap was the same across conditions, including the controls (e.g., SOTE). Is it that the subjects catch on to the point of the experiment? No: the effect occurs even if there is a single pseudohomophone embedded in a long list of other items. Do subjects use phonology only on some types of trials, perhaps the ones with homophones or pseudohomophones? No: the trials are presented in random order. Subjects cannot adjust their strategy because they do not know what kind of target will appear next.

It also has to be shown that new findings can be reliably reproduced in other labs with other materials and subjects. In some areas of psychology and other sciences, there is concern about the number of published findings that are themselves unreplicable "false positives." The Van Orden effect is eminently replicable. A person can even reproduce it at home.

Finally, it is also important to examine the issue using other methods because the results could be due to idiosyncrasies such as the use of homophones and pseudohomophones. Additional studies can also address a concern often raised by educators: that the results were obtained under artificial "laboratory conditions" rather than when people were reading naturally. Both concerns were addressed in what is now an extensive array of studies using different materials, tasks, subjects, writing systems, and languages, and when people read single words or longer texts. Some examples: Phonological effects are obtained in eye-tracking experiments in which people just read texts at their own pace. They occur in silent reading of limericks and tongue twisters, such as THE TAXIS DELIVERED THE TOURISTS DIRECTLY TO THE TAVERN, compared to non–tongue twisters, such as THE CABS HAULED THE VISITORS STRAIGHT TO THE RESTAURANT, findings that will not surprise anyone who reads poetry. Complementary effects occur in spelling: phonologically based errors such as homophone substitutions (e.g., typing THE BARE CUB) occur more often than would be expected if errors were just based on spelling. Proofreaders are less likely to catch a homophone of an intended word because it sounds like the correct word. And so on.

Evidence from Children

What about younger readers? The Van Orden effect is also seen with children; tellingly, the effect is larger with more advanced readers because they have better knowledge of spelling-sound correspondences, a good thing that only gets them into trouble when the animal is BARE. A pair of studies from Charles Perfetti's lab illustrate the kinds of research that yielded a clearer picture of phonology, context, and early reading. In the first, Perfetti studied third-and fifth-grade children who had been identified as good or poor readers using a standardized test of reading comprehension. The task was to read words and nonwords (such as NUST) aloud, one at a time, with the speed and accuracy of responses recorded. In both grades, common (high-frequency) words were easier to read aloud than less common ones, with nonwords being the hardest. More important, reading aloud was related to children's comprehension abilities. The children read aloud the most common words equally well, but the poor comprehenders did much worse on the less familiar ones (low-frequency words and nonwords). Thus poor comprehenders read some words aloud as well as good comprehenders, but not enough. Later studies showed that these effects also occurred when the words appeared in stories.

Is skill in reading aloud causally related to comprehending texts, or is this an example of correlated but unconnected events, such as the .9586 correlation between per capita consumption of mozzarella cheese and number of civil engineering doctorates from 2000 to 2009? Unlike mozzarella and engineering, reading aloud and comprehension are closely connected in English. Children who struggle when reading texts aloud do not become good readers if left to read silently; their dysfluency merely becomes inaudible. Reading aloud and silent comprehension are causally connected because they both make use of the phonology→semantics pathway in the triangle in Figure 7.1. In reading aloud, the computed phonological code drives the production of overt speech. In silent reading it is used in computing semantics, as in the Van Orden experiments.

The second Perfetti experiment examined how more and less skilled readers read words that appeared in sentences. Many studies focus on reading individual words, but what if everything changes when words are read in their normal setting, a meaningful context? Fourth graders read two-sentence contexts that ended in a target word that they read aloud. The experimenters created contexts in which the target word was either highly predictable, not predictable but semantically plausible, or surprising (highly unexpected but still meaningful).

The main results were like the ones from the first experiment. Good readers were faster and more accurate overall, as expected. Poor readers depended much more on context than good readers. The two groups performed about equally well when words were highly predictable; however, poor readers performed much worse when words were not predictable. Perfetti also showed that when asked to do so, good readers were better at predicting the words in these stories than poor readers.

So this is the picture:

Poor readers have more difficulty decoding—using phonological codes to recognize words—especially ones that are used less often. Because their decoding skills are poor, they have to rely more on guessing words from the context. This is an ineffective strategy because they also have more difficulty reading the context words and are poor guessers.

Good readers, in contrast, are better at decoding words and therefore less dependent on context. They do not have to rely on the inefficient strategy of predicting words from context, even though they are better at it than poor readers. Instead of guessing which word will fit, a good reader rapidly identifies each word and integrates it with what has come before.

These experiments are over thirty years old, and the findings have been confirmed many times. Yet they are still at odds with how children are taught to read, a reflection of the disconnection between reading science and educational practice that is the topic of Chapter 11.

Blame English?

If learning how orthography represents phonology is so important, written English seems like a disaster. Readers of English might be entitled to resent the fact that shallow alphabetic writing systems have few if any contrasting words such as MINT/PINT, DIES/DIET, and DONE/BONE/GONE; silent letters; letters such as C, G, and Y that represent two phonemes; homographs like WIND, DOVE, and PERMIT; or one-off words like YACHT and COLONEL. Even THE, the most common word in the language, has two pronunciations, as heard in this Katy Perry lyric: "I've got the eye of the tiger."

These are only a few of the $10^{\text{umpteenth}}$ orthographic oddities, of course. Millions of people manage to learn this system, but perhaps the task would be easier and more individuals would attain higher skills if the peculiarities could be ironed out. It has happened before: many shallow alphabets resulted from royal or governmental ukase, as in the adoption of the Korean alphabet during the reign of the fifteenth-century monarch Sejong, an event honored with a national holiday.

Two questions have to be asked. First, why is written English the way it is and what would work better? Second, does it matter? Does the depth of the writing system have an impact on literacy? Is it harder to learn to read English than Finnish or Korean?

The sources of the irregularities in written English are buried in the history of the language and writing system. Since using words does not require knowing their histories, they seem arbitrary, especially to a beginning reader. In fact, the system adheres to several design principles. The problem is that they often conflict.

Like other alphabetic writing systems, written English incorporates the principle that units in the written code correspond to units in the spoken one. Fidelity to the principle is compromised, however, by changes in the pronunciations of words over time. Consider BEEN, which is homophonic with BEAN in most British dialects but with BIN in most American dialects. The duration of the vowel has been shortened, as in many irregularly pronounced words (e.g., HAVE, GIVE, SAID), probably from pressure to produce speech quickly with fluency. Pronunciations change more rapidly than spellings, causing words that once had predictable spelling-sound correspondences to move to the dark side. In principle, these irregularities could be corrected by changing the spellings, for example, BEEN→ BIN and SAID→SED, the kind of fix that self-appointed spelling reformers advocate. The revised spellings would eliminate the irregularities, but at what cost?

Here a second design principle comes into play, the idea of maintaining continuity in how words are written despite differences in pronunciation, which would be sacrificed if spellings were "simplified." Brits and Americans would spell even more words differently than they already do (e.g., COLOR/COLOUR, ORGANIZE/ORGANISE), and Canadians would have to take sides. And what about the regional dialects in each of these countries? Speakers in the American South would be entitled to drop the final G's and apostrophes from the spellings of words like HAVIN and GOIN, and GETTING would be spelled GITTN. Texas would demand its own spelling system. Learning to spell is hard enough; learning alternative spellings of the same word would take the complexity of the task over the top. The idea that spellings could be revised to keep them in alignment with pronunciation is as realistic as instituting pronunciation reform to ensure that everyone says words the same way.

BEEN and SAID also illustrate a third principle that creates seeming inconsistencies. The spelling of BEEN reflects the fact that it is a form of the verb BE, a connection that would be lost if it were spelled BIN. Similarly, the fact that SAID is the past tense of SAY would no longer be visible if it were spelled SED. In these cases, the writing system preserves information about *morphological* relations between words at the price of maintaining the alphabetic principle. You remember morphemes: the minimal units of meaning analogous to phonemes and graphemes.

Morphemes such as CAT or -LY make similar phonological and semantic contributions to the words in which they occur, but strange things happen over the many-year history of a language. Consider the words SIGN, SIGNAL, and SIGNATURE. All of them contain the same root morpheme ("lemma"). SIGN, however, is one of the words that gives English its shady reputation. The silent G seems utterly gratuitous because the word could have been spelled SINE (like WINE and FINE), or perhaps SYNE or SCIEN. Seems like something even Bob the Builder could fix.

Again, however, the best that can be achieved is a compromise among competing demands. SIGN most likely came into English from the French SIGNE, in which the G is also silent but not irregular: it's the rule in that language (the G is also silent in the French SIGNATURE and many other words). In English, SIGN can't be pronounced with the G articulated because spoken syllables can't end with the phonemes /g/-/n/. That wasn't a problem for later words such as SIGNAL and SIGNATURE, because the /g/ and /n/ occur in different syllables. How then to spell the word "sign"? Spelling it SINE would remove the connection to its morphological relatives while creating a false connection to unrelated meanings of SINE. Spelling it SIGN retains the clear connection to its morphological relatives, but at the cost of including the silent G. What's a community of language users to do?

According to the *Oxford English Dictionary*, they experimented with many spellings, among them SINE, SEGN, and SYNE, for a couple of hundred years, gradually settling on SIGN. Preserving information about morphological structure won out over spelling-sound consistency. In linguistics, group decision making that takes place over a long period, as occurred in determining the spelling of SIGN, is called diachronic change. When the process takes place in a much shorter time frame, it is called Wikipedia. Seriously, diachronic change is like the hive mind working on a problem but without the Internet.

SIGN is an oddity, but it is not arbitrary. It could even be argued, as the linguists Noam Chomsky and Morris Halle famously did, that using the spelling SIGN for the pronunciation "sine" was the optimal solution. The spelling preserves valuable morphological information, and the reader can easily determine that the G must be silent by other means: it can't be /i/g/n/ (three phonemes) because that sequence does not occur in English phonology. Moreover, since "sine" corresponds to a word the child knows from speech but /s/i/g/n/ (the four-phoneme pronunciation) does not, the G must be silent, as also occurs in REIGN, PHLEGM, MALIGN, and other words.

That is their story, and they've stuck to it. It may work better for linguists who study the structure and history of languages and for advanced readers than for beginners. The spellings of many words are a compromise between the conflicting demands of the alphabetic and morphological principles. Small comfort for a child who is learning to read them.

Skilled readers are aware of the morphological relations among words and use this knowledge in reading and writing, as when we try to deduce the meaning of an unfamiliar word such as SUPERNUMERARY or CONNECTIONIST. The only catch is that morphology too is only quasiregular. Morphemes often make similar contributions to the meanings of words, but not always. For example, a REDHEAD is a person with red hair, but a person with very dark hair is a BRUNETTE not a BLACKHEAD, which is a kind of facial blemish, and a DEADHEAD is something completely different.

Many spelling errors result from morphological misanalyses. Question: Which is the proper spelling of the word meaning a tiny amount: MINISCULE or MINUSCULE? The presence of the morpheme MINI-(meaning small) favors the wrong answer: the correct spelling is MINUSCULE, derived from the Latin MINUS (meaning less). SUPERCEDE seems to pattern closely with PRECEDE, RECEDE, and CONCEDE, but the word originated from another source, and the correct spelling is SUPERSEDE.

Using a single spelling for homophones such as SINE and SIGN would result in a loss of information about the *similarities* between morphologically related words. However, it would also result in a loss of information about the *differences* between unrelated words by increasing the number of homonyms—words such as WATCH and STALK that have a single spelling with a single pronunciation associated with two or more unrelated meanings (e.g., to watch, the watch). English has settled on a comfortable number of such words, which are rarely misinterpreted except when intended (as in wordplay). Using a single spelling simplifies one part of the system (the spelling-sound mappings) only by increasing complexity elsewhere (loss of morphological clues, more homonyms to disambiguate).

Borrowing from other languages is the other major source of spelling-sound inconsistencies. Words whose pronunciations were transparent in the source language often create anomalies when embedded in English. Some examples:

Apostrophe, chlorophyll, aesthete (Greek)
Algae, equus, ficus (Latin)
Mesa, amigo, chipotle (Spanish)
Bruschetta, cello, cupola (Italian)
Chagrin, croissant, amateur (French)
Bagel, schnoz, nudnik (Yiddish)

These words are excellent and English would be poorer without them, but they contribute to the chaos (from the Greek KHAOS) in English spelling and pronunciation. English is highly receptive to words from other languages; no one is guarding the linguistic borders, unlike countries with protectionist linguistic policies enforced by government agencies (e.g., in France, L'Académie Française; in Quebec, L'Office Québécois de la Langue Française). *Chacun à son goût.*

Put these factors together, and a quasiregular system comes out: most of the spelling-sound correspondences are consistent, but the orthography admits many deviations arising from multiple sources. Leaving aside the utter impracticality of implementing any major spelling reform—the US is a country that could not agree to adopt the metric system—representing English with a shallow orthography would create a modern mismatch between language and writing system.

Is It a Virtue to Be Shallow?

The endless kvetching about English spelling assumes the system's properties make it harder to learn and use, but is that true? Several English-speaking countries (Singapore, New Zealand, Australia, and Canada, Quebec exception noted) consistently score well above the US on cross-national assessments, so other factors are involved. Still, perhaps performance would be even better in English-speaking countries if the language were not written as it is.

Direct evidence comes from studies of learning to read in English and in shallow alphabetic writing systems. Some of the most interesting work took advantage of the linguistic environment in Wales. At the time, children could enroll in schools in which the language of instruction was either Welsh or English. The two languages are structurally and historically unrelated. Whereas English is deep, Welsh is shallow. The children in the studies were similar in ethnicity, culture, and socioeconomic status and attended schools that were otherwise alike.

The results clearly showed that by age seven, Welsh readers read words and pseudowords aloud about twice as well as English readers. This advantage for shallow alphabets has also been documented in Italian, Spanish, German, Finnish, Serbo-Croatian, Turkish, and other languages. As the title of a paper comparing English, Welsh, and Albanian put it, "Learning to Read in Albanian: A Skill Easily Acquired." Researchers have concluded that all of these languages are easier to learn to read than English.

Not so fast. These studies went off the rails because they equated "reading aloud" with "reading." Most reading is silent, and the goal is comprehension. In the Welsh-English studies, a funny thing happened: the English readers were poorer at reading aloud but scored *higher* on comprehension tests. As the researchers dryly noted, "This result suggests that a transparent [shallow] orthography does not confer any advantages as far as reading comprehension is concerned. As comprehension is clearly the goal of reading, this finding is potentially reassuring for teachers of English."

Reading aloud is closely related to comprehension in English but not in the shallow alphabetic writing systems. In English, correctly pronouncing exceptions such as HAVE, GIVE, and SOME indicates that the child knows the words and has learned how the spellings map to phonology, which is used in silent reading. Children who are progressing more slowly make more errors generating phonological codes for such words, which interferes with comprehension.

In shallow orthographies, children can correctly read aloud words that are not in their vocabularies—a good trick, but not correlated with comprehension. A Finnish child can pronounce OMISTUSHALUINEN and AVIOLIITTONEUVOLA without understanding what they mean (roughly, "acquisitive" and "marriage guidance center"). In fact, people can accurately read entire texts aloud in shallow orthographies without knowing the language. I know this because I proved it at my bar mitzvah. With the vowels included, Hebrew is shallow, and I learned to read it aloud with great accuracy and enthusiasm, though without understanding much of what I was saying. Luckily for children like me, Hebrew is a good bar mitzvah language, as are Finnish, Albanian, Welsh, Italian, and the other shallow alphabetic orthographies. Avoid Danish, however, which is deep, like English.

Just as written English would be harder to read if it were shallow, languages such as Finnish and Albanian would be harder to read if they were written like English. The languages with shallow orthographies have complex inflectional systems in which morphemes indicate properties such as tense, number, gender, and case. English, in contrast, has one of the simplest inflectional systems. Spelling-sound inconsistencies like the ones in English would be intolerable in those languages.

In Finnish, for example, nouns are inflected for number (as in English) but also for case, which reflects the noun's function in a sentence: whether the noun is the agent that causes an action, the object of an action, a location, and so on. Finnish has fifteen noun cases. Children do not just learn the singular HOUSE and plural HOUSES; they learn the set of inflected forms that express INTO/IN/TO/AT/FROM/WITH A HOUSE, among others. And that is only one slice of Finnish inflectional morphology. The language is also agglutinative, the property that yields words that require multiword phrases in English, such as ISTAHTAISINKOHAN (I wonder if I should sit down for a while).

The exact form of an inflection often depends on the preceding phoneme. These already complex contingencies would be horrendously difficult to learn and use if the writing system included consonants such as C and G that each represent two phonemes or vowels with multiple pronunciations. In short, it's a good thing those Welsh and Finnish children can learn the pronunciations of letters very quickly, because they'll be spending extra time studying inflectional morphology. Grammatical instruction, which continues for years, relies on reading texts.

Because the inflectional system is simple, English also has many more monosyllabic words than these languages. The exception words that are the focus of so much instructional Sturm und Drang are mostly short, high-frequency, monosyllabic words such as HAVE, GIVE, SAID, BEEN, and the others. Thus the irregulars are tucked among words that are easy to learn and remember.

In short, there is no free orthographic lunch. Simplicity in one area trades off with complexity in another. English learners just have to suck it up. That's what they do in Finland.

Reading the Triangle Way

Written English, as we know and love it: How did we ever learn to read it?

The beginning reader's fundamental challenge is learning how to recognize and comprehend written words. Most accounts of how this is accomplished start with the observation that several procedures are required because of the writing system's idiosyncrasies. The phonological pathway will work for words with simple, rule-governed spelling-sound correspondences, but telling apart homophones such as BARE and BEAR will also require input from the visual side or context. A rule that correctly specifies the pronunciations of GAVE, SAVE, and PAVE will generate a mispronunciation of HAVE; thus words with irregular pronunciations must be read "by sight." But then sometimes, it's thought, words only have to be read closely enough to determine if they match a strong prediction based on the prior context. Beginning readers are encouraged to develop several strategies for reading words because they can't know in advance which one will work for any given word.

Sounds good until one tries to specify how these procedures work. Although phonology is clearly used in skilled reading, it is disconcerting that no one knows what the pronunciation rules are or how they can be learned. That there are such rules seems obvious from simple patterns such as BAT-MAT-CAT/BAT-BAG-BAD/BAT-BIT-BUG and from the fact that we can pronounce novel strings like YAT on the first try. But these are simple consonant-vowel-consonant syllables. With other kinds of syllables, it is harder to say what the rules are. Words in the COOK/LOOK/BOOK neighborhood seem to obey a rule governing the pronunciation of-OOK, which makes SPOOK an exception. But SPOOK is fine if grouped with SPOON and SPOOL. BOOK is rule governed if grouped with LOOK and COOK but irregular compared to BOON, BOOM, and BOOT. Which -OWN words are rule governed and which are "exceptions": CLOWN, TOWN, and FROWN or OWN, FLOWN, and BLOWN? Worse, however they are stated, the number of rules is so large that only a small subset can be taught, surely the bane of every teacher charged with providing phonics instruction. How the child catches on to the rest is a mystery.

Reading words "by sight" is an equally unworkable proposition. Until the child has developed a large reading vocabulary, the associations between spelling and meaning are arbitrary and thus must be memorized, which requires extended practice. That is an accurate description of the traditional method for learning to read Chinese, which was to memorize characters through many hours of laboriously writing them, but not of learning to read words in English. And guessing words from context? Works occasionally but hopeless as a general strategy. These common assumptions about how children read make it hard to see how they manage to succeed.

An alternative approach emerged when researchers began to develop theories of learning based on how it occurs in the brain. Although we are far from understanding how the brain works, many of its fundamental properties are known. Behavior results from the activity of massive numbers of neurons, interconnected to form huge networks in which each neuron is affected by the thousands of neurons with which it links. Things that we know and do correspond to patterns of activity in such networks; learning is the process of developing and tuning them.

These concepts have been explored using computational models that perform simplified versions of tasks such as learning to recognize and pronounce words. The goal is not to build a robot that reads but to arrive at a theory of a complex human behavior such as reading that brings out its essential characteristics, using the models as tools. The models are a synthesis of research in several fields. Neuroscience provides evidence about properties of the brain that underlie behavior; computer science—especially the subfield called machine learning—is a source of ideas about learning and knowledge representation; cognitive science provides evidence about complex behavior and its components. The computational models address the causes of behavior at a mechanistic level that transcends the limits of intuition, like theories in physics that transcend our experience of the physical world.

My colleagues and I have used such models to study how children learn to read, the bases of differences in reading skill, the causes of dyslexia, the effectiveness of different types of interventions, the effects of brain injury on reading, the way reading occurs in different types of writing systems, the relations between word, face, and object recognition, and other topics. I want to explain their operation in enough detail to convey the basic concepts and how they changed our thinking about phenomena such as learning to read, without serving up too much to digest. What I present here is necessarily simplified, but it is not hand-waving: these are working computer programs that can be understood by looking under the hood. Many such models have been studied, and the details about how they work are available all over the Internet. The deeper principles that govern the design of the models are spelled out in our theoretical papers, as are their limitations. I had no role in inventing this computational framework; I am a member of the user group that recognized that the concepts were relevant to fundamental questions about reading and language. The inventors were people such as Geoff Hinton, one of the true heirs to Alan Turing, who is also responsible for more recent breakthroughs in the training of the larger, more complex networks used in deep learning. The models I'll describe are far simpler but more closely related to human behavior.

The general framework is shown in Figure 7.3. Words are represented by orthographic, phonological, and semantic codes. Each code is itself learned from perception and action, constrained by human biology. Orthographic knowledge, for example, results from perceiving letters and writing them. The phonological code develops from listening and speaking. Semantics develops from perceiving and acting upon the world. Properties of the neural systems that underlie these codes and the nature of the mappings between them also shape what is learned. Reading involves using orthography to compute the semantic and phonological codes for words, which can then be used in performing various tasks: comprehending words, pronouncing them aloud, and deciding if a letter string is a legal Scrabble word, among others. The same network is used in performing other tasks such as computing a word's meaning or spelling after hearing it or generating a word's pronunciation or spelling from its meaning.

We began by looking at how people learn the mappings from orthography to phonology, an interesting problem in English. We then moved to models of how the meanings of words are computed from print, focusing on the roles of the two pathways in Figure 7.1 How the models perform these tasks could then be related to the corresponding human behavior and its brain basis.

Basic properties of these neural networks that are relevant to reading include the following.

Representations

A network consists of groups (or "layers") of neuron-like units representing orthography, phonology, and semantics. For example, the spellings of words are represented by patterns of activation over the units in the orthographic layer (e.g., which are on or off). The units correspond to letters or features of letters. Each word's spelling is represented by a unique pattern of on and off units. Think of the card section at a college football game, where letters are formed by changing which side of the card is shown by each of the many cardholders.

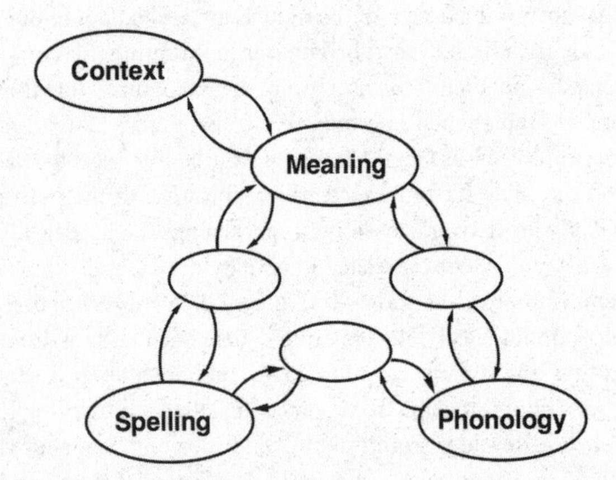

FIGURE 7.3. General "triangle" framework for models of reading and other uses of words. The implemented reading models differed in details and focus. Though greatly simplified, they incorporate basic principles thought to govern many kinds of behavior and their brain bases. Adapted from Seidenberg and McClelland (1989), Figure 1, p. 526. Copyright © American Psychological Association. Reprinted by permission.

The same approach is used to represent the phonological and semantic codes for words. Units in the phonological layer can correspond to phonemes or their component features. The meanings of words can be represented, in much simplified form, as units representing features such as small, yellow, animate, and winged. In real life we think these representations are learned from experience; they were built into these models in order to focus on how the system learns the mappings. The model is given the capacity to represent strings of letters, for example, but has to learn the combinations of letters that form words and how they relate to phonology and semantics.

These multiunit "distributed" representations have important properties. First, a small number of units is used to represent a much larger set of patterns, analogous to the way an alphabet can represent every word in a language. Second, the representation encodes pattern similarity. The orthographic representations of similarly spelled words will overlap more than those for words with dissimilar spellings, and the same goes for phonology and semantics. This is important because learning about one pattern (say, the pronunciation of MUST) will also benefit the learning of overlapping patterns (e.g., DUST, RUST).

Connectivity

A network is formed by connecting units in one layer to units in others. The pathway from orthography to phonology, or phonology to semantics, is actually a large number of connections between the units in these layers. The network architecture typically includes layers of intermediate "hidden" units that allow the models to learn more complex mappings. Patterns of activation over a hidden layer eventually come to represent the deeper structure underlying the relations between codes. Our models employed one of these layers on each pathway; a powerful deep learning network might use a dozen or more.

Activation

Units can be "activated," by analogy to neurons "firing," the sequence of events that includes passing electrical and chemical signals to other neurons, causing action potentials that entrain many other such events. The process of reading the word GAVE begins with the activation of units in the orthographic layer that represent this spelling pattern. Activation passes to the hidden layers and to the phonological and semantic units. The amount of activation that is forwarded depends on the strengths of the connections between units. Think of each connection as having a number attached—a "weight"—that modulates the flow of activation. The analogy here is to neurons affecting each other via synapses with variable strengths. The activation of each unit in the phonological layer, for example, is determined by the amount of activation coming in along all the connections leading into it.

Learning

The task of reading a word can now be described as

> Given a spelling pattern as "input," activate the pattern that represents its semantics as "output."

The learning problem can be described as

> Find a set of connection weights that allows the network to perform this task correctly for many words.

The weights are initially set to random values. A spelling pattern such as GAVE will therefore initially produce a random phonological pattern. The weights are changed on the basis of a prespecified learning algorithm. Many such procedures have been explored, varying in how closely they correspond to learning as it occurs in the brain. The general idea is that the weights are adjusted based on experience (when the learner is a computational model, the term is "training") and feedback about performance. Gradually the model converges on weights that allow the network to perform a task accurately for a large number of patterns. The learning algorithm is an efficient way to determine how much to adjust each weight to improve performance.

In the most commonly used procedure, called "backpropagation," the weights are adjusted based on the discrepancy ("error") between the phonological pattern the model produced for GAVE and the correct pattern. The powerful feature of this procedure is that weights that are more responsible for discrepancies get changed more than weights that are less responsible. That is, the size of the adjustment to a weight is proportional to the amount it contributes to the total error. On a GAVE trial, the weights are adjusted in a way that will reduce the error on this word.

The tricky part is that the model is trained on a large set of words—say, the ones a first grader is learning to read—not just GAVE. The model has to find *one* set of weights that will produce correct output for *all* words. Changing the weights in a way that is helpful for producing "gave" benefits overlapping words such as GATE and SAVE. However, that also moves the weights away from values that are optimal for pronouncing HAVE or GIVE. Conversely, training on HAVE moves the weights closer to what is needed for "have" but away from what is optimal for GAVE and its friends.

After the model is trained on many words, the values of the weights represent a compromise, having been pushed and pulled by the statistical properties of written English. The model is not taught rules for pronouncing words in English or told which words are exceptions. Rather, the subregularities that are hidden among the many mappings between spelling and phonology are discovered as a by-product of reducing the error on each learning trial.

We have explored many such models, configured to examine many parts of the reading system and how they develop. The models behave in amazing ways—like children who are learning to read. Here are a few of them.

The Same Procedure for "Rule-Governed" Words and "Exceptions"

We started with a model that included only the orthography→phonology pathway. It learned to produce phonological codes for about 3,000 common monosyllabic words with 97 percent accuracy. (The words that it missed were low-frequency oddballs like GUILE and DEBT. Those words require help from the orthography→semantics→phonology pathway, which we had not included.) The words included both "rule-governed" ones and "exceptions." Given the spelling pattern GAVE, it produced "gave"; given HAVE, it produced "have." It learned SAID and PAID, PINT and MINT, BLOWN and FLOWN, CLOWN and TOWN, DOSE, POSE, and LOSE—a good portion of the monosyllabic words in the language. They were all pronounced by the same procedure, using the same set of weighted connections between units, which encode the jumble of mappings from orthography to phonology. This was a surprising result, contradicting the strong intuition that words such as MINT are pronounced by rule and ones like PINT must be memorized.

How is this possible? The architecture of the model—the layers of units, distributed representations, learning procedure, and so on—allows it to pick up on statistical patterns that hold across words, many of which are not visible to the naked eye. Moreover, because the same weights are used for all words, what is learned about one word affects performance on all the others. That creates savings: having learned the -AVE → "ave" pattern from words in the SAVE, GAVE, PAVE neighborhood makes it easier to learn words that share that structure, such as BRAVE and SATE.

No Sight Words

Words with atypical pronunciations such as HAVE and GIVE are usually treated as sight words that must be memorized. However, HAVE's pronunciation is not arbitrary; it overlaps with HAD, HAS, HAVING, HAVEN'T, HIVE, and other words. Thus, what the child learns about words such as HAVE carries over to other words and vice versa. Moreover, what is learned about HAVE is relevant to many other words in which the vowel has an atypical pronunciation (such as GIVE and COME). The model can represent these partial regularities because the learning algorithm ferrets them out. According to the model, the child can learn a large number of "exceptions" because they are *not* arbitrary patterns that must be individually memorized.

Although the model doesn't learn "sight words" by memorizing them, once it has learned a word very well, it behaves *as though* it is read "by sight." This property is also relevant to children's learning but hard to intuit. Early in training, the model's performance on a word such as HAVE is affected by its overlap with other words. It is not memorized "by sight." However, like most "sight words," HAVE is a very high-frequency word. The model learns it relatively quickly because common words are trained more often than less common ones. The word is eventually learned so well that other words cease to exert any detectable influence on it. That is the model's version of reading "by sight." The unintuitive part is that the model still represents the components of the word, which continue to influence the processing of other, less common words and performance on novel words like MAVE. So, is HAVE read "as a whole," or do the parts matter? This is like asking if light is a wave or a particle. The answer is, it isn't one or the other; it's both.

Degrees of Consistency

A character in the 1980s movie *Stand by Me* says to his friends, "Alright, alright, Mickey's a mouse, Donald's a duck, Pluto's a dog. What's Goofy?"

When it comes to reading, the conversation goes like this:

> MUST is rule governed (because all of the -UST words rhyme).
>
> HAVE is an exception (because it conflicts with the other -AVE words).
>
> What's GAVE?

On traditional accounts, GAVE is rule governed. It rhymes with the other -AVE words, it follows the rule of silent -E, and its pronunciation is highly predictable. If other factors such as length are held equal, words such as GAVE should be as easy to read as other rule-governed words, such as MUST, and both will be easier than exceptions like HAVE. In the reading network, however, the same weights are used for all words. Performance on a word such as GAVE takes a slight hit because the weights also have to accommodate HAVE. MUST, in contrast, is in the clear because it does not have any close irregular neighbors. The prediction, then, is that it should be harder to read GAVE aloud than MUST, even though both are "rule governed."

The experiments were done, and the answers are in. Rule-governed but inconsistent words such as GAVE are harder to read than "pure" rule-governed words such as MUST. For skilled readers, "harder" means that the word is pronounced correctly but a little more slowly. The impact is larger for lower-frequency words such as PINT and AISLE. For beginning readers (and adults who are poor readers), the effects extend to higher-frequency words as well. The general phenomenon is that pronunciation is affected by the degree of consistency in the mapping between spelling and sound across words. "Rule-governed" words and "exceptions" occupy points on a statistical continuum that includes many degrees of spelling-sound consistency. This clarifies why it is so difficult to list the rules for pronouncing written English: *the system isn't rule governed; it's statistical.*

A New Account of Generalization

Language is said to be rule governed because we can generalize, applying what we know to new cases. Knowing the rule for generating the past tense in

English, we can be confident that the past tense of GLORP is GLORPED. If the models lack rules, how can they pronounce nonwords such as MAVE, NUST, and BRONE? The answer: the same way they pronounce words. The model doesn't know whether an input string of letters is a word or nonword; it just takes spelling patterns in and pumps phonological patterns out. NUST can be pronounced correctly because the model was exposed to overlapping items such as MUST, JUST, and NUT. In effect, the model "knows" the pronunciations of novel words before they have been encountered—the same principle as in learning vocabulary. The idea that generalization could arise from the behavior of a neural network rather than from rules was both a major conceptual breakthrough achieved by the researchers who developed this framework in the 1980s and a challenge to linguistic orthodoxy.

These two phenomena—consistency effects and generalization—show that these models do not simply memorize the words that are trained. The models develop representations of the parts of words. Consistency effects arise from patterns that hold across words containing these parts. Because the models develop representations of the parts of words, they can pronounce new words that combine them in novel ways.

The models illustrate the perils of relying on intuitions about how reading works. Before they were developed it would have been absurd to claim that words such as HAVE and GAVE are learned, read, and pronounced by the same mechanism, which is also used in sounding out novel words like MAVE. The claim was counterintuitive, and no known procedure worked this way. (The claim is still nonsensical if one isn't familiar with this type of system.) Yet that is what the implemented models do. Having accounted for existing data, the models made numerous novel predictions about typical and atypical development, impact of brain injury, and many other issues that have guided subsequent research.

Division of Labor

Later models focused on computing the meanings of words from print using an architecture in which meaning was determined by activation coming in from both visual (orthography → semantics) and phonological (orthography → phonology → semantics) pathways in the triangle. This was the big departure from the "visual versus phonological" dichotomy that Figure 7.1 illustrates. The network again had to find a set of weights that would allow accurate performance on many words. The question was how it would make use of the two pathways as it learned.

Because children can already comprehend many spoken words before they begin to read, we first trained the phonology → semantics pathway for a large subset of words. Once it had learned many of these mappings, we brought in the other parts of the triangle, orthography → phonology and orthography → semantics, and reading "instruction" commenced. On reading trials, the model had to compute the meaning of a word from its spelling. It also continued to have some "listening" trials, computing the meanings of words from phonology. The model was not told which pathways to use for the reading task; rather, it converged on a solution based on feedback about how well it was doing on each word, adjusting the weights in all parts of the network to improve performance.

The division of labor between the two pathways changed over time. The model initially relied mainly on the orthography → phonology → semantics part of the network. The phonology → semantics mapping was known for many words, and the model could learn the orthography → phonology mappings relatively quickly. The orthography → semantics part took longer to develop because the mappings were arbitrary associations. With additional training, the orthography → semantics part began to contribute more, such as activation that disambiguated homophones like BEAR/BARE. This development progression matches up well with neuroimaging studies showing that the orthography → phonology pathway normally develops more rapidly in children, with the orthography → semantics system coming online with additional experience.

After the model had learned almost all of the several thousand words in the training set, both pathways were contributing to the semantic patterns for most words. However, the division of labor—how much each part contributed—depended on properties of the words. The division was more complicated than simply using the orthography → semantics pathway for "exceptions" and the orthography → phonology → semantics pathway for "rule-governed" words. For example, YACHT-type words that have unusual spellings and pronunciations typically require greater input from the orthography → semantics side; however, if YACHT is one of your frequently used words, the orthography → phonology → semantics pathway carries more of the load. We used the trained model to simulate a variety of behavioral phenomena, including the Van Orden effect.

We ran the model several times, changing small variables to see if it would converge on different solutions for the reading task. It has long been thought that there are two types of readers, those who read visually and others who read phonologically, but it has been difficult to find clear evidence from behavioral studies. Division of labor between the two pathways changes with the development of reading skill. Skill is determined in part by amount of reading experience, which varies widely. The effect of less reading experience—in people and the model—is to produce behavior that would normally be seen in a younger person or an adult who reads at a younger level. At a given level of experience, however, the division of labor did not differ greatly.

These typical outcomes can be skewed by changing how the model is trained. Our models could use both visual and phonological pathways. However, some types of reading instruction assume that using the phonological part is a cause of poor reading. We therefore ran versions in which only the orthography → semantics pathway was trained. These models did not receive any "phonics," that is, training in the mappings between orthography and phonology. The orthography → semantics models did learn the set of mappings eventually; however, it took a massive number of training trials. The results suggested that "sight" reading is feasible in theory but takes far longer to achieve than reading with both pathways. Since children are expected to master this part of reading by the start of grade four, taking the extra time to learn using an inefficient method is not a good idea. The practical implication is that teachers who withhold phonics instruction to avoid inculcating poor reading practices are having the opposite effect, making it harder for children to acquire basic word-recognition skills.

Our research suggests that people do not all read in exactly the same manner, but nor do they devise radically different strategies. The basic architecture of the reading system is the same with some variation in the division of labor between the components due to experiential factors, such as the amount of reading experience and type of instruction, and possibly variation in the neural architecture. The impact of writing systems on reading is similar: the architecture is the same, but the division of labor varies. The most interesting evidence comes from studies using the same type of model to learn to read Chinese, which required greater involvement of the orthography → semantics side of the triangle, a finding consistent with behavioral and neuroimaging evidence. The difference is only a matter of degree, however: as with English words, both pathways contribute for most characters. Given the radical differences between the writing systems, the extent to which the solutions for Chinese and English overlap is perhaps the more surprising result.

Of course there is more to reading than computing the meanings and pronunciations of isolated words. We have to imagine the kind of model I've described embedded in a procedure that processes sequences of words, tracking the sentence-and text-level statistics discussed earlier. Such sequential models exist and have been useful in understanding how children learn vocabulary and grammar and how sentences are comprehended and produced. Readers also make use of content knowledge they have accumulated and stored in long-term memory. It has been easier to study isolated words, and more complex models can be difficult to keep under control. But basic word-reading skills are important, and for a while it wasn't clear how a child could actually acquire them. Our models have many limitations, but they have achieved at least one important advance: they've shown how learning to read words is possible!

And the implications for reading instruction? Are children little statistical learning machines who will learn to read if exposed to a sufficient amount of text? Have we proved that children really do learn to read "on their own"?

Not at all. We've shown why a moderate amount of instruction can have much broader impact, as in learning vocabulary. The models embody the statistical learning idea that what is learned about one word affects many others. In the future it should be possible to determine which specific instructional events will have the biggest overall impact given the current state of a child's knowledge. Explicit instruction in phonics and vocabulary could then be more efficient. It will take additional research to determine how these modeling results can inform educational practices. At this point we only know that what a teacher thinks children are learning from a lesson about sight words, phonics, or vocabulary may differ from how their brains respond to it. Bringing the two into closer alignment could have a powerful impact on learning, but we still need to find out.

CHAPTER 8

Dyslexia and Its Discontents

FOR MANY PEOPLE, LEARNING TO read is an uncomplicated affair. Reading happens because children and the family members and caregivers who raise them engage in activities that are part of the cultural fabric: talking and reading together, singing the alphabet song, playing word and language games, attending a preschool or day care program, going to school, reading. Children whose participation in these activities is limited or of poor quality—because of a lack of resources in the home, school, or community or other unfavorable circumstances—are at risk for reading failure. When a child does not become a reader because of such circumstances, it is appalling but not inexplicable.

For some people, learning to read is surprisingly difficult. It is surprising when a healthy child who engages in these activities nonetheless struggles with reading. It is surprising when the child's behavior is otherwise age appropriate, or even advanced. It is surprising when a child's struggle to read has no discernible cause.

These surprising reading difficulties invite the inference that they are caused by a condition—a developmental disorder—that affects learning to read. The condition is often called "dyslexia," although other terms such as "reading disability" are also used. Dyslexia has been studied since the early twentieth century and achieved After School Special status in 1972, but confusion about its characteristics and causes persists. Behavioral, neurobiological, and genetic evidence now unquestionably shows that some children develop in ways that greatly interfere with learning to read even when other circumstances are favorable. However, the road to that awareness has been long and twisted, marked by numerous controversies, some still in progress. Shifts in the understanding of the disorder and its causes have created confusion about how the condition should be defined, identified, and treated, what it should be called, and even whether it exists.

My goal is to describe current thinking about dyslexia and related impairments and to explain why the topic remains controversial to some. The basic characteristics and causes of dyslexia are understood well enough to inform identification and treatment; many questions remain, but they are mainly known unknowns, in the immortal words of Donald Rumsfeld, that require additional research along paths that are in place, in my view.

Although the basic facts and future directions are clear, dyslexia also illustrates inherent properties of developmental disorders that make them difficult to reduce to simple definitions. Moreover, the gaps in knowledge create a real-world dilemma: how to reliably identify children with reading impairments so that they can obtain timely and appropriate help even though our understanding is incomplete. Finally, dyslexia occupies an odd niche in the cultural ecosystem because it is both fascinating and complicated. Dyslexia is a disorder of everyday life with no obvious cause that can afflict anyone, and often does. Most of us know someone who is dyslexic; many people identify as mild or high-functioning dyslexics. On the research side, our understanding of reading impairments has increased in depth but also complexity. Research led not to the discovery of the underlying cause of reading impairments but rather to recognition that there is no single cause. The combination of a high-interest topic and a complex etiology creates an opening for self-appointed experts who have popularized simplistic theories of dyslexia whose appeal does not turn on being true. You will find a lot of that in the self-help section of the bookstore; we will be over in the science area.

The Rough Guide to Dyslexia

My description of dyslexia takes the form of a guide to the major issues, findings, and open questions. It is patterned after a concise travel guide that covers the major can't-miss sites without going far off the beaten path. I'll also provide some tips for avoiding the topical equivalents of tourist traps and purse snatchings. Here I focus on behavior; I take up the neural and genetic causes in the next chapter.

To get oriented to our destination, dyslexia refers to impaired reading. The components of this complex behavior can be examined to determine which are affected. For example, one child might be poor at reading words quickly and accurately or sounding out unfamiliar words. Another might read words accurately but nonetheless comprehend texts poorly. The causes of such deficits can be addressed at multiple levels, ranging from immediate, or "proximal," causes that are closest to the impaired behavior to underlying, or "distal," causes, such as genetic anomalies that affect brain development and later reading.

The proximal causes are impairments in cognitive, perceptual, or motor functions that affect components of reading. Dysfluencies in reading words, for example, can arise from several causes: slow recognition of letters or letter combinations, impaired ability to link spelling and phonology, lack of expertise due to insufficient practice, among others.

Of course, we can dig deeper. Identifying a deficit in recognizing letter patterns or linking letters and sounds is useful for a teacher or reading specialist trying to help a struggling reader, but these behaviors have causes as well. Does the impairment derive from an auditory or visual perceptual problem? A learning or language problem? An attentional deficit? Several minor deficits that are important only when they occur together? Identifying these underlying causes might help that teacher or reading specialist to be more effective with that reader.

Such accounts still explain one kind of behavior (reading impairment) in terms of other kinds of behavior (e.g., perception, learning, attention). The character of the explanation changes once we turn to the brain bases of reading impairments: anomalies in the structure and functioning of the neural circuits that underlie the observed behavioral deficits. Finally, properties of these circuits are themselves determined to a large degree by genetics in the broad sense: human genetic endowment, how it varies across individuals, and the manner in which it guides development and behavior under the modulating influence of the environment throughout the lifespan. Genetics can contribute to understanding reading disorders by specifying mechanisms that influence brain development in ways that underlie the proximal causes. For example, impaired phonological processing, which is strongly implicated in developmental dyslexia, appears to arise from the action of genes that influence brain development in regions important for spoken language.

We know most about the behavioral characteristics of dyslexia and their proximal causes. Table 8.1 lists the behavioral deficits commonly seen in dyslexic children. This information provides a solid basis for identifying children who are dyslexic or at risk for dyslexia and responding appropriately. The greatest immediate gains in ameliorating reading impairments could be achieved by making better use of this information. Dyslexia is also associated with several types of anomalies in the structure and function of the neural systems for reading. That information is itself useful—it demonstrates that dyslexics are not struggling with reading just because they aren't trying hard enough—but we cannot as yet reliably determine from brain imaging and related tools whether a person is dyslexic or a superreader. That said, some of

TABLE 8.1. Characteristics of Dyslexic Readers

These behaviors are typical of children who are struggling with reading. The exact profile varies by child and age, and depends on the types and severities of underlying deficits. The tested behaviors involve basic reading skills and other types of knowledge that affect comprehension. As the child gets older, the effects of an underlying deficit may be altered by instructional practices and remediation, and by protective factors including IQ, higher SES environment, and personality traits such as tenacity, resilience, and intrinsic motivation.

Phonology
Impaired performance on phonemic tasks:
- deletion (say "split" without the "p")
- matching (do "bat" and "let" end with the same sound?)
- blending (what is the word when you combine the sounds "b", "a", and "t"?)

May exhibit residual deficits on other phonological measures (e.g., letter names, sounds)

Reading aloud
Performance is slow, dysfluent, error-prone
- particular difficulty with irregularly pronounced words (SAID, ONCE, GIVE)
- in shallower orthographies, pronunciations are accurate but slow
- errors often based on incorrectly guessing word from initial letter and context

Poor at sounding out pseudowords (NUST, MAVE, TRASK)

Processing speed
Slow at naming familiar digits, colors, objects (Rapid Automatized Naming task)
Difficulty reading words fluently, not just accurately

Orthography
Limited knowledge of orthographic structure;
- difficulty distinguishing legal from illegal letter strings (DORN VS DOLR)

Weak knowledge of word spellings
- misspellings
- misidentifications (RAIN OR RANE?)
- dysfluency in generating spellings

Working memory
Deficits on "verbal working memory" tasks
- nonword repetition ("fragistat," "contoban")
- "memory span" tasks

Language
Limits on vocabulary size and lexical quality
Familiarity with a narrower range of sentence structures and expressions
Difficulty reading texts aloud with appropriate stress, intonation

the brain evidence speaks to specific issues in ways that go beyond what we can deduce from behavior. For example, the debates about whether reading is visual or phonological might have been avoided had we known that orthography and phonology become deeply intertwined at the neural level.

What Is Dyslexia?

Dyslexia is a condition that gives rise to poor reading, but some poor readers are not "dyslexic." The term refers to reading that is impaired due to developmental neural and genetic anomalies that affect this skill, a focus that excludes many other factors that can also lead to poor reading: personal characteristics, such as motivation and perseverance, and personal circumstances, such as the home, school, and community environments. Children may opt out of reading because it seems unimportant, because they lack parental encouragement and support, or because they are facing social pressures. They may not learn to read because they attend a poor school or a series of such schools because they are homeless. Every person's reading skills reflect the influence of countless such factors. Dyslexia focuses on reading impairments that are neurobiological and genetic in origin rather than wholly environmental and on capacities that are close to reading (e.g., language, vision, learning, memory) rather than general (e.g., personality traits that have broad effects on behavior). Although the distinction is clear, in reality these various factors coexist, influencing one another during development.

I have to emphasize that children who are struggling to read are entitled to help regardless of the cause; it is not that only some types of poor reading qualify for serious attention. All relevant factors have to be considered in evaluating the basis for a reading problem and determining how to address it. Achieving a deeper understanding of dyslexia, however, also requires distinguishing between causes that have different origins and effects.

How is dyslexia defined? Getting an answer requires deciding whom to ask. There is no keeper of the royal definition. Definitions vary because they are proposed by individuals and groups whose orientations, expertise, and goals differ enormously and because concepts have changed as more has been learned. Medical and educational organizations and government agencies periodically offer definitions to meet their needs. The American Academy of Pediatrics provides guidelines for its members (excellent but technical). The US Department of Education has issued definitions of reading and learning disabilities as demanded by legislation. For example, the Individuals with Disabilities Education Act (IDEA) ensures access to free, appropriate public education. That requires defining learning, reading, language, and other disabilities. The definitions used in such legislation are written in legalese, with its distinctive mix of specificity and opacity.

The *Diagnostic and Statistical Manual of Mental Disorders* (DSM) published by the American Psychiatric Association is the standard source for the definition and classification of behavioral disorders. The characterization of developmental learning disabilities, including reading, in the current version (DSM-5) is idiosyncratic and does not align well with the research literature. It defines a general category, "Specific Learning Disabilities," with manifestations in reading, writing, or math. The manual serves important clinical and administrative functions, affecting diagnosis, treatment recommendations, access to services, and how the costs are borne, but it is not the best source for information about the nature and causes of these conditions. Researchers have their own accounts, which are more nuanced and based on a better understanding of reading and development, but they are less rigidly codified than the DSM-5 and therefore not good enough for government work.

Both researchers and clinicians are moving toward distinguishing two general types of developmental reading impairments, which parallel the two components in Philip Gough's "simple view." The term "dyslexia" is being reserved for children whose difficulties center on acquiring basic skills—the print part that is specific to reading. These difficulties interfere with comprehending texts of course, but the child's spoken-language comprehension is within the age-appropriate range. Other children's reading difficulties appear to originate in the language component. Some of these children attain adequate basic skills but still comprehend texts poorly. These children have spoken-language impairments that carry over into reading.

The fact that definitions of dyslexia differ is more than just a nuisance; it has called into question whether the condition exists.

Neurologists working at the turn of the twentieth century introduced the concept of dyslexia as a neurodevelopmental disorder that interferes with reading. It was construed as a selective impairment affecting reading (and sometimes writing) but not other capacities.

Later research focused on identifying the defining characteristics of the impairment and developing explicit definitions and diagnostic criteria. The initial definition of dyslexia was, in essence, unexplained reading difficulty, meaning poor reading that is not due to factors such as lack of opportunity, inadequate instruction, or "frank" neuropathology. In 1968 the World Federation of Neurology defined dyslexia, based on such exclusionary criteria, as "a disorder manifested by difficulty learning to read, despite conventional instruction, adequate intelligence and sociocultural opportunity. It is dependent upon fundamental cognitive disabilities which are frequently of constitutional origin." This definition helped to validate dyslexia as a developmental disorder and incorporated the idea that the category should exclude children whose reading deficit arises from factors such as intellectual disability. However, it did not speak to the causes of the impairment or how to diagnose and treat it.

The "adequate intelligence" clause masked several complications that subsequently came to light. Distinguishing reading difficulties that result from a general intellectual impairment from reading-specific ones seemed important. A general impairment might well affect reading but also many other kinds of behavior. Dyslexia, in contrast, was viewed as a narrower impairment with reading-specific causes and effects.

Although the intention was clear, the concept of "adequate intelligence" for reading was not. The US Department of Education definitions of the 1960s and 1970s excluded children with "mental retardation" (their term). Later definitions emphasized discrepancies between IQ and reading performance. A higher-IQ child whose reading was poor would be categorized as dyslexic, but a lower-IQ child who read at the same level would not. Reading ability and intelligence fall along continua, but the discrepancy criterion required setting boundaries: below a certain level on reading, above a certain level on IQ, or a discrepancy between reading and IQ performance that exceeded a certain size.

Any such procedure will be imperfect because the cutoffs are based on utility rather than true discontinuity. Given the imprecision in how intelligence and reading are assessed, a difference of a few points can be meaningless yet influence how children are categorized and thus their educational placement and access to resources. Children whose scores fall close to the cutoff but on opposite sides will be categorized differently; behaviorally, however, they will be more similar to each other than to most of the others in their respective categories. And where should the cutoffs be placed? The inherent arbitrariness of the boundary between dyslexic and nondyslexic is responsible for the lack of consensus about the incidence of the condition, which is said to affect anywhere from a small proportion of the population up to 15 or 20 percent.

The cutoffs were initially set so as to identify children for whom the discrepancy was relatively large. This strategy was not unreasonable at first. Researchers were trying to understand the nature and causes of reading impairments; thus they began with seemingly clear cases: the high-IQ, high-functioning child with an inexplicable void where the reading should have been. The understanding gained from looking at such cases could then be used in tackling more complex ones.

Later behavior genetics studies supported this logic. These are the famous twin studies that estimate the heritability of characteristics such as height or religiosity. The heritability of a trait is not a fixed quantity because the genetic influence is modulated by environmental factors. Estimates of the genetic impact on variability in reading skill illustrate this point. The genetic component is greater for higher-IQ children than for lower-IQ children. If a child with a higher IQ nonetheless reads poorly, the deficit is likely to reflect genetic variation relevant to reading, not IQ. If the goal is to identify the neurodevelopmental causes of reading impairments, the higher-IQ children are a good place to start. Treatment specialists also focused on such children because they seemed most likely to benefit from intervention. The high-IQ child with a highly specific reading impairment eventually became the dyslexic stereotype.

The use of discrepancies between IQ and reading to identify dyslexics is now recognized as a flawed procedure. IQ (that is, whatever IQ tests measure) is related to early reading achievement: correlations between nonverbal IQ (measured using tests that do not depend heavily on language) and children's reading fall in the 0.3 to 0.6 range, accounting for around 10 to 30 percent of the variability in scores. These correlations are modest but not negligible; they mean that fewer children in the higher-IQ range will be poor readers.

Higher IQ acts as a protective factor against falling into the dyslexic range, but it is less relevant to getting out. For children who are poor readers, IQ is not a strong predictor of intervention responses or longer-term outcomes. Moreover, the behavioral characteristics of poor readers are very similar across a wide IQ range. Patterns of brain activity for higher-and lower-IQ poor readers are nearly indistinguishable, and both differ from those seen in good readers. Within this broad range of IQs, poor readers struggle in the same ways, need help in the same areas, and respond similarly to interventions. In short, the skills that pose difficulties for children are not closely related to the skills that IQ tests measure. The primary question is about children's reading—whether it is below age-expected levels—not their intelligence.

Does Dyslexia Exist?

If dyslexia cannot be defined by a discrepancy between reading and IQ, who is dyslexic? Dyslexics are children (and, later, adults) whose reading is at the low end of a normal distribution. Reading skill results from a combination of dimensional factors (that is, ones that vary in degree), yielding a bell-shaped curve. The reading difficulties of the children in the lower tail are severe and require special attention. "Dyslexia" refers to these children. Viewed this way, dyslexia is on a continuum with "normal" reading. All children face the same challenges in learning to read, but dyslexics have more difficulty with the essential components. The accumulation of deficits in these areas and their multiplicative effects on each other affect performance to the point where the child is categorized as "dyslexic" under a prevailing standard.

For researchers, this approach is liberating because we can look at the entire range of performance without having to presort individuals into categories. We can investigate how and why behaviors vary along this continuum, excluding only the individuals whose reading problems are clearly secondary to other conditions, such as a hearing impairment or very low IQ. (These individuals can be studied separately.) Researchers still sometimes focus on children with substantial discrepancies between IQ and reading skill as a strategy for addressing some questions, but it is not assumed that impairments are limited to these individuals.

If dyslexia only refers to readers at the low end of a continuum, is it a valid diagnostic or descriptive category? This is a serious question, but it can be easily misinterpreted with potentially detrimental effects. If a skill is unevenly distributed across a population, there necessarily must be individuals who perform at the lower end. I am at the extreme lower end of the distribution for singing on key. Children who have this condition are asked to mouth the words, not sent for treatment. If reading skill also falls on a continuum, what is the justification for labeling the poorest performers as having a developmental disorder?

Such observations have led some researchers and educators to question whether the concept of dyslexia is valid. Dyslexia can be seen as an example of the creeping medicalization and pathologizing of behavior that falls within the normal range but happens to occupy the lower end: restless leg syndrome, caffeine dependence, Internet gaming disorder. Is there a meaningful boundary between normal and disordered behavior in such cases? The behaviors may be extreme, but are they pathological?

Each case has to be judged on its own merits. In the case of reading, the dyslexia category picks out the individuals whose performance requires special attention. Children learn to read at different rates. Many beginning readers who have fallen behind others in their age group are not dyslexic and will catch up with ordinary support and sufficient effort. Dyslexics do not "grow out" of the condition, though some eventually become good readers. How far an individual progresses depends on the severity of the impairment and circumstances such as the availability and timing of quality interventions, protective factors such as intelligence and persistence, and a favorable environment. Children who are dyslexic require extended attention focused on ameliorating problems before they multiply. Reading impairments are far easier to address in children than adults, making it critical to recognize them early. In short, it is important to identify such children because their needs are indeed special. Dyslexia is thus the *purposeful* medicalization of normal behavior: both part of a normal distribution *and* a treatable condition. There is a label for failing readers but not for atonal singers because one of these conditions is more important to well-being than the other.

If this view of dyslexia still seems disconcerting, consider how it compares to other diagnostic categories whose validity is unquestioned. The mumps is a disease with a specific cause, infection by the mumps virus, that produces specific symptoms. The diagnosis entails little ambiguity, and the boundary between afflicted and unafflicted individuals is clear. Dyslexia is not like the mumps. Obesity and hypertension are also medically recognized conditions; they are not like the mumps either. They *are* like dyslexia, however. Weight varies on a continuum. Some people are unhealthily heavy for their size or age. Interested parties are free to offer definitions and guidelines, but the boundary between heavy and obese is fuzzy rather than categorical and subject to cultural norms to boot. Similarly, the primary symptom of hypertension is high blood pressure, also measured on a continuum. The boundary between having or not having the condition is neither fixed nor categorical but rather determined by conventions that periodically change.

Diagnostic categories such as obesity and hypertension exist because they pick out conditions that are unhealthful and require explicit, targeted treatment. The extreme manifestations of these conditions may also have somewhat different etiologies than the rest of the distribution, which can inform how they are treated. Obesity and hypertension cannot be diagnosed as definitively as the mumps, but that does not invalidate the categories, which serve important functions. The same is true of dyslexia.

This view does not sit well with many reading educators. The concern runs as follows: dyslexia is a label for poor reading; the reading abilities of the children in K–4 classrooms usually span a wide range; teachers are sensitive to the varied needs of their students and routinely address them—it's their job. Labeling some poor readers as dyslexic isn't of any practical use. Moreover, if the identification of a child as dyslexic triggers access to additional resources (e.g., one-on-one instruction), it creates incentives to apply the label in a broad manner that undermines whatever validity it might have had. (This has been a major concern in Great Britain.) Better to leave it to teachers to pursue the traditional goal of helping every child become a reader.

Skepticism about dyslexia has become the focus of debate in the United States and Great Britain. Here is Richard Allington, an American reading-education expert and former president of the International Reading (now Literacy) Association:

> "There is no such thing as a learning disability or dyslexia," Allington told [educators in Baltimore County, Maryland], citing research and his own 45-year experience of never finding anyone he couldn't teach to read.

And here's Allington again, speaking of "excuses" that get in the way of reading:

> Some of the biggest excuses are learning and attention disorders such as dyslexia or Attention Deficit Hyperactivity Disorder [ADHD], Allington said.
>
> "[They] provide excuses for not bothering to provide high-quality reading instruction to students," he said. But "they don't exist. Teachers and schools create those [disorders]. . . . Doctors and parents think they exist because they've been convinced by people at schools that they do exist. It's a psychological tool that schools use to protect themselves from looking like it's their fault."

Whereas a 2009 cover article in the journal *Science* announced, "Dyslexia: A New Synergy Between Education and Cognitive Neuroscience," a 2008 article in a prominent educational theory journal asked, "Does dyslexia exist?" Keith Stanovich, a researcher who has long attempted to link research and educational practice, had asked and answered the same question years earlier, in 1994.

Extreme skepticism about dyslexia is misguided and contrary to the best interests of the child, in my view.

1. It encourages treating symptoms without regard to underlying causes. The skeptics' approach is narrowly pragmatic. Low reading achievement is the problem, they say; delving into how the child arrived at this point is a waste of time, akin to second-guessing history. In contrast, the modern concept of dyslexia and approaches to identification and remediation based on it emphasize the need to assess the causes of impaired performance. Poor reading fluency is a symptom much like a severe headache: it can arise from a multitude of internal and external factors. The effectiveness of the treatment depends on whether it addresses the relevant causes. In both cases merely treating the symptom is a risky approach and potentially disastrous.

2. It places unrealistic demands on teachers. The needs of these children extend beyond what a classroom teacher can provide. Standard classroom practices have not been successful for these children. Providing

more of what the teacher is already doing is not an adequate response. Too often the recommendation for dyslexic children (and their parents) is to try harder and to do more of what works for other children, even though they have fallen behind precisely because what works for other children has not worked for them. Moreover, it is simply untrue that teachers have the skills to deal with deep-seated reading difficulties. Few will have had relevant training; they are more likely to have been taught that the condition does not exist or that the child will grow out of it. Nor is it realistic to think that teachers can provide adequate intervention in addition to performing their other classroom duties. As with obesity and hypertension, people at the extreme low end of this distribution require more extensive, specialized intervention.

3. Extreme skepticism creates additional barriers to intervention. Dyslexia is an invisible condition: a child reads poorly for reasons that are buried in the nervous system. The impaired behavior does not command an urgent response because it does not differ from that of many other children who eventually succeed at reading. Dyslexia contrasts with conditions such as autism in which the symptoms are overt and command attention. We do not expect teachers to be autism experts or to fully address these children's needs in addition to managing a classroom. The same should hold true with regard to children with acute reading difficulties. Identifying children who are dyslexic or at risk for the condition reduces the likelihood that the problem will be overlooked until they are older, when it is harder to ameliorate and other negative consequences are apt to have kicked in.

Overdiagnosis and overtreatment are systemic problems, in parts of the US health-care system. For dyslexia, however, the problems are underdiagnosis and undertreatment. Parents routinely struggle with educational and medical authorities to gain recognition of their child's reading problem and access to effective help. Medical practitioners often view poor reading as an educational issue and therefore the purview of educational experts. Educators who accept that dyslexia exists may view it as a medical condition that they do not have the training, responsibility, or authority to treat. The dyslexia label helps prevent the child from falling between these cracks. Defining dyslexia is less important than identifying children with reading difficulties so that they can be addressed. Extreme skepticism increases the already high burden of proof that affected individuals and their parents face.

And the assertions that dyslexia and ADHD were invented by schools to excuse their failures or that any child with sufficient motivation and adequate instruction can learn to read? They are shameless fictions that are hurtful to parents and potentially harmful to their children.

How Can Dyslexics Be Identified?

Let's go with the genetic, neurobiological, and behavioral evidence that dyslexia exists and turn to a serious issue: how to identify affected individuals. Like other developmental disorders, dyslexia has several characteristics that present challenges for parents who want to know why their child is struggling, for reading experts who want to help them, and for researchers trying to figure out what's going on.

Dyslexia is an emergent condition: it develops. The best "treatment" is prevention, taking action to keep the child off that path. Doing so requires identifying children who are at risk when they are young, which is hard. The behaviors of four-to-six-year-olds who are at risk for dyslexia are not very different from those of children who are not. Typically developing children learn to read at different rates for constitutional and environmental reasons. Some hares get off to a fast start; some tortoises start off more slowly but soon catch up. During a window of uncertainty, the tortoises closely resemble children who are at risk for serious reading problems. The fact that many of these similarly behaving children do become good readers is why parents are often mistakenly assured that their (dyslexic) child will catch up without extra help.

Given this behavioral overlap, the absence of other definitive markers for the condition is unfortunate. Unlike the mumps example, there is no dyslexia virus. Unlike cystic fibrosis or Huntington's disease, which are single-gene disorders, dyslexia is polygenic (influenced by more than one gene). Simple behavioral criteria such as discrepancies between IQ and reading fail to sort children appropriately. Some children fall behind only because they have not had adequate instruction or opportunities or encouragement to read. Definitive identification of children as dyslexic may not be possible until they are ten or eleven, which is far too late for an intervention to have the best chance of succeeding. Moreover, many would not have fallen so far behind had help been available earlier. Technologies such as neuroimaging or behavioral tests based on brain evidence may eventually fill this diagnostic gap, but the science is not there yet. The two crucial issues then are identifying young children for whom early intervention is essential despite these ambiguities and intervening effectively.

Reading researchers devised an elegant solution to these problems, known as Response to Intervention (RTI). It is widely utilized in the US and other countries and was incorporated in the 2004 reauthorization of the IDEA. Instead of trying to diagnose dyslexia by reading the behavioral tea leaves of small children, RTI uses a multistep procedure. First, all K–1 children are screened to determine who is at risk for reading difficulties. Such screenings, as adopted in many states, only require gathering data on a few short, simple tests of basic prereading skills; information about any other behavioral problems is also obtained. A family history of reading or language difficulties is a strong risk indicator. The screening procedure will pull in children at risk for dyslexia as well as some tortoises. The children's progress in their regular classrooms is then assessed frequently using short tests of key skills. Classroom teachers who are supposed to have received additional training in effective instructional practices and strategies for targeting areas that need the most work provide this "Tier 1" intervention.

For some children—those who just needed more time on task with helpful feedback, for example—this will be enough to bring them up to speed. Children who do not make adequate progress will move on to a "Tier 2" intervention, consisting of additional instruction conducted one-on-one or in small groups. The child's response to the intervention again determines what happens next. Children who still have not made adequate progress move to "Tier 3," which can take several forms, depending on local resources and policies. Often it entails placement in a special education classroom.

RTI, then, uses the child's own behavior to trigger interventions, which can nip an incipient reading difficulty in the bud. It also incorporates a procedure for identifying children with a deep-seated reading impairment: failure to benefit sufficiently from Tier 1 and 2 interventions. Tier 3 is reserved for those children who demonstrably do not thrive in the general education setting. Under this approach, the inclusionary criterion for reading disability is poor response to high-quality instruction and intervention.

RTI is a thoughtful, logical, well-designed program. It has only one flaw: it has to be implemented in real-world environments that are often inhospitable. How well the RTI program works depends on how well it is implemented, which is left to the school district or system to decide. Implementation depends on funding, mainly for personnel, which varies enormously. Tier 1 may be left to the classroom teacher, or there may be funding for supplemental staff. Tier 2 may entail group tutoring by a lightly trained paraprofessional or an expensive one-on-one program with teachers who have received specialized training. If Tier 3 is special education, the classroom will be populated by children with issues that span a broad range, and the teacher is almost certain to have received little advanced training related to reading. Once placed in special education, the reading-impaired child is unlikely to return to a general classroom.

The deeper problem is that the effectiveness of RTI is undercut by the disagreements about how reading works, how it should be taught, and how reading problems should be addressed that pervade reading education. RTI is predicated on the idea that reading difficulties can be identified by observing who fails to benefit from quality instruction. The approach does not work, however, if the child's instruction has been inadequate. The extent to which American public schools provide quality instruction is in question. Many teachers will have been inculcated with beliefs about reading, learning, and development, shared by their principals and school system authorities, that promote ineffective practices, as I document in later chapters.

RTI is a good approach, and it succeeds in schools that can mount a program faithful to its tenets. The logic of identifying children at risk for reading failure, closely monitoring their progress, and tying interventions to their strengths and weaknesses is solid and much preferable to gambling that a child will "catch up." No harm is done if a tortoise who would have caught up anyway receives additional instruction. Taking this preventive approach, rather than trying to help an older child or adult who reads poorly, makes sense for the same reasons as employing preventive medicine to reduce the incidence of diseases that are difficult and expensive to treat. However, RTI is not effective as implemented in many schools, leading to disenchantment with the approach. The consequent need for parents to seek quality intervention outside the school contributes to disparities in educational opportunity, favoring those who can afford it.

The Drama of the Dyslexic Child

Defining dyslexia is like trying to define language. It can be done, and some definitions are better than others, but all of them telescope a complex concept into a few words. The best "definitions" of these concepts are fact-based theories that answer important questions. For language, they are questions such as, What do people know when they know a language, and how is this knowledge acquired and used? For dyslexia they are questions such as, What are the characteristics and causes of reading impairments, and how can they be prevented or ameliorated?

For an eight-or nine-year-old who is reading well below grade level, the problem areas are not hard to spot. Start by looking at children whose reading is on course. The better readers are well on their way to recognizing a large number of words quickly and accurately without much thought or effort—almost reflexively, which is the goal. They can figure out unfamiliar words by sounding them out or by recognizing that they have parts that occur in other known words or in the same contexts as known words, all of which can be facilitated by knowledge of the text's topic. They read sentences aloud fluently with appropriate intonation. They have begun to read texts independently, honing their skills and expanding their knowledge of language and content in the process. Better readers do not progress equally rapidly in every area, but they are moving ahead on all fronts. A nine-year-old who is moving into fourth grade is expected to be doing so.

Dyslexics are still struggling to recognize and pronounce words efficiently, impeding progress in other areas. Words have to be read with sufficient fluency to achieve the forward momentum needed to comprehend phrases, sentences, and larger blocks of text. The child's problems quickly multiply because they cannot obtain the amount and variety of experience on which skilled reading depends. Reading difficulties undermine motivation and encourage opting out, creating a nasty feedback loop.

Identifying these children's reading difficulties is *easy*. Determining their causes, which is crucial to preventing or ameliorating them, is much harder.

99 Problems but Reading Isn't the Only One

Most children are not identified as having reading impairments until they have failed enough, to the point where even the most optimistic teacher can see that they will not simply "catch up." Nine or ten years is already a long time for a child to have been living with a brain that is developing in atypical ways. Usually reading isn't all that is tangled up by this point.

Other behaviors can be affected because reading impairments result from anomalies in capacities that are not just used for reading. Moreover, every act of reading is also an act of visual perception, learning, memory, language, and the rest. The goal may be comprehending a text, but the bonus is the tuning and refining of these shared systems, which then tend to correlate. Reading also has broad effects on behavior because of its role in learning. Even reading and math, which may seem like nonoverlapping magisteria (Stephen J. Gould's phrase) are deeply intertwined, from the verbal description of symbols, concepts, and procedures, to word problems, to curricula that emphasize connecting mathematical operations to real-world situations. The connections between reading and other tasks and types of expertise make teasing apart cause and effect a daunting challenge. Repeatedly it has been found that atypical behaviors that were initially seen as causes of dyslexia are actually consequences of it.

Reading-Specific Impairment Without a Reading-Specific Cause

The concept of a "reading-specific" impairment is a non sequitur. Reading is a task; the impairment is not in the task itself but in one or more of the capacities involved. It's like riding a bike. There can't be a bicycle-riding-specific impairment, only impairments in the enabling capacities (or inadequate opportunity or instruction). Blowing out a knee will definitely interfere with bike riding but also with many other knee-dependent activities. Similarly, a low-vision problem that affects letter recognition will also affect recognition of objects and faces.

An underlying deficit can nonetheless have a much bigger impact on reading than on other behavior. Although created out of existing parts, reading is a unique system with an elaborate infrastructure. Think of just the properties of letters. The associations between letters and sounds are both arbitrary (P could have represented /r/) and systematic (across words, P usually represents /p/). These properties exploit capacities to learn both arbitrary associations and statistical regularities that, while not specific to reading, are combined in a unique, interlocking manner, which is overlaid on a complex letter-recognition process. An underlying deficit might interfere with this arrangement with little effect on anything else, creating the appearance of a reading-specific impairment. A phonological impairment that affects reading, for example, may have only subtle effects on producing or comprehending speech.

Same Problem, Different Causes

Poor readers tend to have similar problem areas but not necessarily for the same reasons. For example, dyslexics are almost always weak at sounding out new words, which cuts off a primary learning mechanism. The process of piecing together the pronunciation of an unfamiliar spelling pattern can be impeded by a variety of disturbances that are not easily discerned from the behavior itself. This is a good example of an area in which neuroimaging and other brain data are beginning to permit etiologies to be identified with greater precision, allowing interventions to be focused more effectively.

Behaviors Change

Just as the characteristics of skilled reading change with development, so do the characteristics of poor reading. For example, the number of letters children can name at age four is a good indicator of how they are progressing. By age eight, this is not informative because even poor readers eventually learn their letters. The underlying deficit remains, but its impact shifts from letter naming to more complex tasks, such as pronouncing groups of letters. These shifting behavioral profiles can be frustrating for people who seek a simple diagnostic checklist, but it is something to get over: dyslexia is a moving target and has to be approached as such. Theories of dyslexia address how behavior develops and changes; assessments of reading performance are scored using norms for what is age appropriate. Think of it as a checklist with a time dimension and a probability and degree of severity associated with each item.

Comorbid Conditions

Dyslexia frequently occurs with other developmental disorders. The conditions are "comorbid," a medical term that also brings to mind unrelated, unpleasant meanings of "morbid," so I will just refer to co-occurring conditions. The most common co-occurring conditions are speech and language disorders, ADHD, and math impairments. The conditions mainly co-occur because language, reading, and math make use of many of the same cognitive capacities. Deficits in these shared capacities can create comorbidity. In this case, the impairments do not just co-occur; they overlap.

Co-occurring conditions can also result from pleiotropy, in which a single genetic disorder has multiple independent effects. Fragile X syndrome, for example, results in cognitive and learning impairments, as well as distinctive physical features such as large ears, a long face, flat feet, and hyperflexible joints. Having large ears does not cause the cognitive impairments, even though they originate from the same genetic anomaly. Some conditions that co-occur with dyslexia may be of this type, but the genetics is more complicated than in fragile X and not yet clear. For example, dyslexia co-occurs with clumsiness more often than would be expected if the etiologies were independent. Clumsiness may be a pleiotropic characteristic unrelated to the reading impairment, or the two may be more closely linked. An impairment in learning and performing integrated motor sequences might affect actions such as hopping and leaping but also speech production, affecting phonological development and therefore reading.

Comorbidity makes developmental disorders harder to identify. The fact that a child exhibits behaviors such as inattention, distractibility, or defiance of authority can easily mask an underlying reading problem. But then a conspicuous reading problem can also mask a co-occurring condition such as ADHD. A comprehensive clinical workup by a skilled clinician (a clinical neuropsychologist or speech-language pathologist specializing in developmental disorders including dyslexia), taken with information gathered from parents and teachers, can help to sort through these possibilities, but close attention must be paid.

Factors Interact

The factors that affect reading influence each other, making it hard to track their impact. To take a simple example, motivation and persistence matter, but protracted, visible reading failure due to a developmental disability will undermine them—that is, unless some other, protective factor comes into play, such as a favorable temperament or environment. A more complex but equally realistic example: Children's neurocognitive capacities develop over time; we are not born with adult brains but progress toward them. This extended developmental process is affected by experience. However, what a child can learn from experience depends on their current developmental state. Factors that continuously influence each other are a basic condition of development.

I've mentioned the challenges in studying and identifying dyslexia not to be gloomy but to inoculate against simplistic theories that reduce the condition to a single cause. In fact, researchers have powerful quantitative methods for analyzing complex phenomena in which multiple dimensional factors (that is, ones that vary in degree) interact over time, and these have been applied with some success to reading and its brain bases. Computational models can also be used for this purpose. It is a tangled web to unweave, but more has been learned about the condition in the last twenty years than in the previous hundred. Going forward we can assume the following:

- Dyslexia does not have a single cause. Numerous underlying anomalies—at genetic, neurodevelopmental, and cognitive levels—can interfere with complex skills such as reading.
- Underlying deficits vary in severity, in the behaviors they affect and to what degree, and in their malleability or persistence.
- Impaired reading can result from the co-occurrence of several relatively mild deficits that would not be debilitating in isolation. The effects of such deficits can be modulated by strengths in other areas (protective factors).
- The manifestations of these deficits change over time as children develop.

The condition may be complex, but the by-product of this deeper understanding is the realization that there are both more opportunities to prevent reading impairments and more ways to address them should they develop than earlier theories allowed.

The Natural History of Dyslexia

Dyslexia is a language-based disorder. It origins are anomalies in the development of speech (perception, production) and language (vocabulary, grammar) from birth. The decisive evidence comes from studies that tracked children's behavior from a young age, even the first day of life, long before their exposure to print. The logic of these prospective studies is simple: Follow the development of a group of infants or toddlers for several years until they begin reading, by which time some will exhibit difficulties, and others will not. Go back to the data collected when they were younger to determine whether any characteristics apparent then predict later reading—ones that have clear, causal links to reading development, not eye color or astrological sign.

Following the same children for several years takes many potential confounding factors out of the equation because the children are being compared to themselves at different times. It also removes the possibility that an impairment is a result of poor reading rather than a cause. Studies in several languages conclusively show that precursors of dyslexia can be detected in infants and toddlers. The known precursors involve speech and language.

The pioneering experiment was conducted by Hollis Scarborough in the late 1980s. Scarborough's innovation was to conduct a family risk study. A prospective longitudinal study will only succeed if a sufficient number of participants develop reading impairments. The probability that a child will develop dyslexia goes up markedly if there is a history of reading impairment in one or more first-degree relatives (parents or siblings). In Scarborough's study, high-risk children recruited from such families were compared to low-risk children without this family history. She began studying the children when they were thirty months old and followed them until they were five. The children performed simple tasks that assessed several aspects of language, along with IQ and a few other measures. Recordings of the children's spontaneous speech were used to track spoken-language development closely. At sixty months basic reading-readiness skills (e.g., knowledge of letter names and letter-sound associations) were also assessed.

The key finding was that children who developed reading difficulties by age five had exhibited spoken-language deficits at thirty months. The impact of the language deficit changed as the child grew older:

- At 2.5 years, the to-be dyslexic children produced sentences with simpler syntax and pronounced words less accurately than nondyslexics.
- By three to four years, deficits in vocabulary and phonology (rhyming, deciding if two words begin or end with the same sound) had emerged.
- At five years, just prior to school entry, the children were impaired on prereading skills (knowledge of letters, letter sounds, phonemic awareness) and vocabulary.

Here then is a glimpse of the natural history of dyslexia as it emerges from its origins in spoken language.

These findings were extended by other researchers and in other languages. The studies conducted by Maggie Snowling and Charles Hulme, two of the preeminent reading scientists of the modern era, are especially compelling. One study followed at-risk children from age three to thirteen. At three, children who would later develop dyslexia exhibited poor spoken-language vocabulary, expressive language, and grammar. The additional finding was that the weaknesses in spoken-language skills persisted. Even at thirteen, performance on phonological tasks was still impaired; vocabulary and syntax fell further behind age-expected levels, in part because these children read less. These results suggest that the language impairments do not fully resolve, although, as usual, individual outcomes vary. Maggie Bruck's landmark studies of adults who had been identified as dyslexic in childhood had earlier yielded the same conclusion.

About half of the at-risk children in the Snowling-Hulme study did not become dyslexic, meaning that their reading was above the "dyslexic" cutoff. However, these children's performance fell between the other groups on most measures: better than the dyslexics, worse than the nondyslexics from nonrisk families. The results are consistent with the view that dyslexics and nondyslexics fall on a behavioral continuum. The impact of the family risk is modulated by those interacting genetic and environmental risk and protective factors. A later study from this group found that variation in spoken-language skills detectable at three to four years of age was predictive of reading comprehension at age eight.

Precursors of dyslexia can be detected even in infants by measuring minute changes in electrical potentials ("brain waves") detected at the scalp. Picture babies wearing cute bathing caps with many wires attached, like cyber hair extensions. The electrical signals are faint echoes of brain activity. Researchers measure changes in these signals in response to pictures, sounds, faces, and other stimuli of interest to babies (these changes are called "evoked potentials"). Speech sounds elicit atypical evoked potentials in infants at familial risk for dyslexia. Infants who are not at familial risk also produce these anomalous responses if they are exposed to maternal smoking, a sobering reminder that adverse environmental conditions modify brain development.

These are impressive findings, but it wasn't known which infants went on to be dyslexic. That angle was addressed in a heroic longitudinal experiment conducted by Dennis Molfese and colleagues. Remarkably, anomalous evoked potentials in responses to speech stimuli collected within thirty-six hours of birth were strongly related to the same children's spoken-language skills at ages three and five and to reading impairments at age *eight*.

Are these effects correlational? Yes. The researchers did not directly measure the development of the reading pathways. However, the probability of a causal relation is very high. First, the data are from the same children at different times. Second, later language and reading levels were predicted by the *degree* to which the early evoked potentials differed from normal, tightening the connection. Third, theories of reading development based on other evidence explain the developmental path from early anomalies in responses to speech to the development of reading. The researchers did not have to concoct a theory to account for their data after they had collected it.

These studies highlight why it is so important to view reading developmentally. The behaviors that are associated with reading attainment change over time. Performance on simple phonemic awareness tasks, for example, is highly relevant for prereaders. These same measures are not strongly related to reading comprehension in eight-year-olds. Does that mean that phonemic skills are unimportant because they only have transient effects that are unrelated to what reading is really about, comprehension? No. They are crucially important at a point on a developmental trajectory, allowing the child to gain more advanced skills.

Learning to Read with a Phonological Impairment

The developmental anomalies that underlie dyslexia are initially manifested in children's spoken language and later come to bear on reading. Although the children's speech is not markedly different from nondyslexics', they have subtle phonological impairments that affect learning about print, sound, and the mappings between them. Our dyslexic model illustrates how this can occur.

The model (Figure 8.1) was similar to others I've described. An input layer of units represents orthography; an output layer represents phonology. There is also an intermediate layer of one hundred hidden units between orthography and phonology. The "cleanup" units are like a layer of hidden units

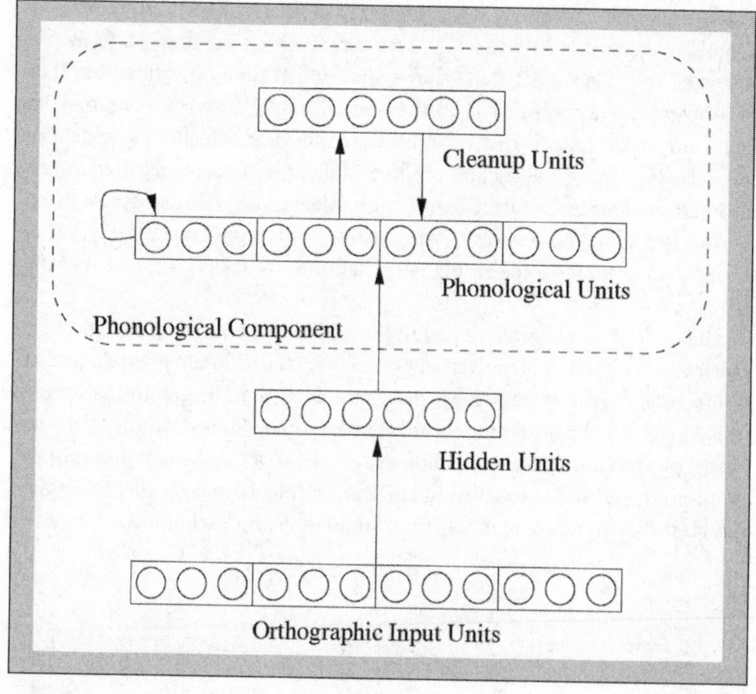

FIGURE 8.1. Schematic of the Harm and Seidenberg (1999) model of word reading and dyslexia. See text for descriptions of the parts. Adapted from Harm and Seidenberg (1999), Figure 9, p. 13. Copyright © American Psychological Association. Reprinted by permission.

connecting phonology to itself, creating a feedback loop. The model takes a spelling pattern as input and passes activation along the weighted connections to the hidden units and from there to units in the phonological layer. In simpler networks, processing stops there. In this network, activation is passed to the cleanup layer and then back to phonology. This process continues until the units in the phonological layer settle into a steady state (stop changing very much). How long the model takes to settle into the pattern is roughly related to how long it takes people to do the same thing.

The model first learned the phonological codes for several thousand words, to approximate the fact that prereaders know many words from speech. Then it was trained on the reading task: given a spelling pattern, produce the correct phonological code, for several thousand words. After this model learned the mappings, we looked at the patterns of activation over the hidden units (between orthography and phonology) for groups of words and nonwords. Consider the words EAT, MEAT, and TREAT, in which the digraph EA is pronounced "ee." Figure 8.2 (left) shows the hidden unit activations for these words and the nonword GEAT in the normal model after it was trained. For each unit, activations varied between –0.5 and +0.5; activation level is indicated by the size of the square; positive values are in black, negative in gray. The patterns are very similar because the model picked up the structure shared by the EAT-MEAT-TREAT words, the spelling -EAT pronounced "eat." That allowed it to pronounce the similarly spelled nonword GEAT without its being taught.

The dyslexic model was trained the same way but with a phonological impairment: the passing of activation between units within the phonological attractor was noisier—less precise, more variable—than in the normal model. This model learned most of the words but took many more training trials. The model produced the correct phonology for EAT, MEAT, and TREAT but mispronounced GEAT. Looking at the hidden unit representations (Figure 8.2, right), it is clear they were less precisely tuned than in the normal model. More units

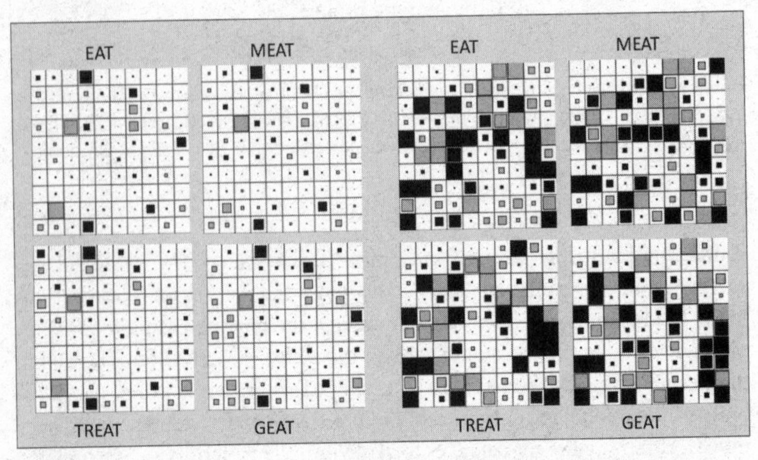

FIGURE 8.2. Snapshots of the activation levels of the hidden units for EAT, MEAT, and TREAT, and GEAT. *Left:* the unimpaired model; *right:* the "dyslexic" model. Both models learned to pronounce the words correctly, but the dyslexic model required many more training trials. The unimpaired model correctly generalized to GEAT, but the dyslexic model did not. The boxes show the relative activation levels of the one hundred hidden units in the orthography → phonology pathway for each item. Adapted from Harm & Seidenberg (1999), Figure 15, p. 25. Copyright © American Psychological Association. Reprinted by permission.

were activated and more of them were at extreme values. The patterns also overlap less than in the nondyslexic model.

The dyslexic model was behaving like a child who learns words individually with little carryover from similar words. That takes more learning trials. Having failed to pick up on the -EAT pattern, the model (and, we think, the child) cannot "sound out" a novel word like GEAT.

Let's be clear: This model is highly simplified. It does not capture details of the child's experience. We did not teach the alphabet song and then gradually introduce words. The model doesn't even know that words have meanings. And yet it suggests the answer to an essential question: Why should a minor phonological aberration have a heavy impact on learning to read? Because the impairment interferes with discovering the phonological components of words, causing words that overlap to be treated as less alike. The model failed to acquire the alphabetic principle—its knowledge that graphemes correspond to phonemes was partial at best—but more important, the impairment affected the discovery of statistical regularities involving other units, such as the rime (e.g., -EAT) and many others, that also determine pronunciations in English. The model's behavior is consistent with two basic characteristics of dyslexic readers: they learn words more slowly, and they are poor at sounding out unfamiliar words and nonwords. The results suggest that the dyslexic brain develops in ways that interfere with discovering commonalities among words involving orthography and phonology. The model also demonstrated that a phonological deficit could interfere with reading while having little impact on speech perception. Reading is the "unnatural" task and more vulnerable to phonological impairment.

When Michael Harm and I did this work in the 1990s, it was an advance to show how several behavioral characteristics of dyslexia could arise from a phonological deficit. However, the connection to children's behavior was circumstantial. Relevant brain evidence didn't yet exist. The model's behavior corresponded to dyslexics' in some important ways, but that did not ensure that the underlying mechanisms were the same. Although the model appeared to provide a strong test of causality—we trained a phonologically impaired model, it interfered with word learning in recognizable ways—a computational model is not a child. Every model is a simplification and therefore wrong at some level of detail, a fact of modeling life that requires proceeding cautiously. Our findings awaited stronger evidence.

Now that there is much more information about brain structure, function, and development, the noise hypothesis is looking good—better, in fact, than we had reason to anticipate in 1999. Behavioral evidence that dyslexics treat patterns that share structure as "too different" has been accumulating. Consider this behavioral study, conducted by Tyler Perrachione for his 2012 PhD thesis at MIT: The participants were young adults who were either dyslexic or typical readers. While positioned in a magnetic resonance imaging (MRI) scanner, they listened to recordings of a series of words and indicated when a word matched a picture that was displayed. That task was just to keep them listening attentively. The data of interest were the levels of brain activation elicited by the spoken words, as indexed by the BOLD (blood-oxygen-level-dependent) response that is the primary measure in functional MRI (fMRI) studies. In some blocks of trials, all words were spoken by a single person; in other blocks they were spoken by four different people, whose voices alternated randomly from word to word.

For all participants, listening to words elicited activation in areas involved in speech comprehension. Nondyslexic readers showed an *adaptation effect*: activation in these areas was reduced when they listened to words spoken by the same person compared to several speakers. Activation is reduced because the listener picks up on characteristics of an individual's speech. Adaptation does not occur when the speaking voice keeps changing. The crucial finding was that dyslexics showed much smaller adaptation effects. Activation levels in the single- and multiple-speaker conditions were very similar. Thus, dyslexics treated words that were similar to each other because they were spoken by a single speaker as more distinct—less similar—than did the nondyslexics. As in our model, dyslexics did not pick up on the overlap between words (here, carried by qualities of the speaker's voice). This is particularly interesting because dyslexics were showing *greater* sensitivity to voice qualities—how even a single speaker's speech varies—than the nondyslexics, which is contrary to the usual pattern of dyslexics performing more poorly than nondyslexics.

Perrachione then replicated this finding using other kinds of stimuli. Dyslexics showed smaller adaptation effects compared to nondyslexics with spoken words (listening to repetitions of the same word versus a series of different words spoken by a single speaker) and for printed words (repetition of the same word versus a series of different words). Although the dyslexics consistently showed smaller adaptation effects in these experiments, the degree of adaptation varied somewhat. Perrachione also reported analyses showing that the amount of adaptation was positively correlated with performance on a common standardized test of sounding out novel words (called "word attack").

Similar effects have been observed in studies conducted in other labs, in other languages, and with other stimulus materials. It appears that the effect we observed in the dyslexic model is probably more general than we knew, involving both linguistic and nonlinguistic stimuli and both visual and auditory modalities in at least some individuals. Dyslexics treat words such as EAT, MEAT, and TREAT as "too different," but the same happens with repeated examples of a single spoken or written word. The effects we studied arose from noisy processing in the phonological system, but it is easy to imagine similar effects in brain areas more closely tuned to orthography or other information.

In the real world of a child learning to read and write, such effects would make the tasks much more difficult. In reading, as in spoken language, a person has to be able to generalize, applying what has been learned to new instances. The alternative, treating every instance as different, is untenable. Consider, for example, having to learn the plural of every noun separately: Book-books, beer-beers, and so on, ad (almost) infinitum. Generalizing requires picking up on the salient characteristics of a pattern and ignoring what is irrelevant, such as font variation. Whether the initial letter is in upper or lower case also isn't relevant for the plural, but whether the initial letter is B is (because the plural of DEER is DEER). The behavioral and neuroimaging studies suggest that dyslexics are, to some unknown extent, failing to generalize, treating instances as "too different." That may be due to "noise," as in our model, or other mechanisms such as failures to inhibit irrelevant information.

The main impact of this type of condition would be having to learn important things multiple times. Our dyslexic model learned the EAT → "eat" mapping for each word in that neighborhood without much help from the others, which was not efficient and caused it to mispronounce GEAT. Or go back to the letter categorization problem. Imagine that the most efficient, most easily learned solution is to use a single attractor to recognize all of the K's, because they don't vary a whole lot. Then imagine that under some conditions (such as "noisy processing"), the brain solves the problem by partitioning the Ks along some secondary dimension, such as serif versus sans-serif font (K versus K). Now there are two K-processors. Words containing K's like Kite and Kite would also be handled differently. Worse, the pronunciations of letters and letter patterns would have to be learned twice. Of course, if the same thing were happening on the phonology side, there would also be multiple copies (e.g., attractors for high- versus low-pitched pronunciations of "kite"). That might also complicate saying the word, which requires the speaker to select one phonological representation to articulate. Since the system is interactive, copies on the phonological side would feed back on orthography. Failing to represent the shared structure and ignore what's irrelevant would make this part of reading very difficult.

The example is a cartoonish simplification suitable for a barroom napkin. No one knows how the brain recognizes the K's, but it is likely to involve millions of neurons with a very elaborate, changing connectivity structure. The solution wouldn't then be much like the attractor dynamics in an artificial neural network with a few hundred units and simple connectivity. But the cartoon version illustrates something seen in the behavioral and neuroimaging data: failures to benefit from shared structure including exact repetitions and, what may be the same thing, oversensitivity to irrelevant information.

The noise hypothesis is exciting because it is being studied at the multiple levels that the science now allows: behavior; computational mechanisms; brain structure, function, and chemistry; genetic influences on development. It must be emphasized that these findings are suggestive but far from established, and that "noise" does not preclude additional causes. We know nothing about how many dyslexics or nondyslexics exhibit such behaviors, how the severity of the putative impairment varies, whether different brain areas can be affected, whether children who fail to generalize about, say, speaker voices exhibit any of the other atypical behaviors that occur, and whether such impairments are direct causes of reading problems. We examined noise in the phonological component of our model, but other evidence suggests that it may occur in the mappings from orthography to phonology or in early visual processing. In the model noise was a parameter that could be set at a value that produced the dyslexic effects, but at much higher levels the model could hardly learn at all. Noise is interesting because it could affect many kinds of behavior and thus explain several of the proximal deficits thought to underlie dyslexia; however, it is a very general concept that needs to be more closely tied to specific neurodevelopmental mechanisms. Some recent evidence on that score is discussed in the next chapter.

The Phonological Umbrella

Phonological deficits are prominent in dyslexia, and their impact on reading has been extensively documented in multiple languages and writing systems. However, dyslexics also exhibit other, apparently unrelated deficits. Current thinking holds that the etiology of dyslexia involves multiple risk and protective factors that vary in degree and that no single factor is either necessary or sufficient to cause the disorder. Whether a child falls into the dyslexic spectrum depends on their configuration of strengths and weaknesses. There's little doubt that outcomes are determined by interactions among multiple factors that vary in degree. A major question remains, however: How many deficits are involved?

The behaviors in Table 8.1 certainly look as though they arise from several impairments. There's phonology, but also a speed deficit, a working memory deficit, an orthographic deficit, language deficits, and so on, as determined by performance on specific behavioral tasks. Whether these are distinct deficits is an open question. One of the keys is to look at how the diagnostic tasks are performed, not what they are said to measure. The tasks all involve phonology in some way. Hearing a nonword and then saying it ("working memory deficit"). Seeing pictured objects and naming them ("speed" deficit, associative learning deficit). Hearing a word and then saying it with a phoneme deleted ("phonemic awareness" deficit). Vocabulary—word learning—also depends on phonology, as does comprehending sentences. Are there multiple kinds of deficits, or are these multiple manifestations of one or more phonological impairments? The answer isn't known, but here is a suggestive lead.

Some years ago a team of European researchers conducted an early neuroimaging study of dyslexics who were native speakers of Italian, French, or English. Italian is very shallow, English is deep, and French is closer to English. In all three languages/writing systems, the dyslexics showed atypical activation in a brain region involved in phonological processing. The behavioral manifestations of the impairment differed, however. The English dyslexics' reading aloud was slower and less accurate than nondyslexics'; words and nonwords were equally affected. The French results were much the same. The Italian results were different. First, because the writing system is shallow, words and nonwords were read with high accuracy; the impairment only affected response speed. Second, whereas words were read only a bit more slowly than normal, nonword reading was very slow and laborious. The underlying deficit was the same, but it produced a much larger effect on response speed in Italian than in the other languages. Later studies confirmed that in shallow orthographies, dyslexia is manifested in slower but more accurate responses than in English.

Thus, the same underlying phonological deficit can affect either speed or accuracy of reading aloud. In the Italian-French-English experiment, the manifestation depended on the writing system. In English, it depends on the task: the deficit affects speed on timed tasks, such as the rapid automatized naming (RAN) task discussed below, but accuracy on untimed tasks such as phoneme deletion or nonword repetition. There isn't a separate "speed" impairment; the impact of the phonological deficit depends on what the reader is asked to do and what is measured.

Do the other behavioral impairments listed in Table 8.1 derive from phonological anomalies? We have not modeled all of them, but it is not hard to connect the dots (or units). The common thread is that performance is impaired because the phonological deficit causes the reader to treat stimuli (such as instances of letters, phonemes, or words) as different rather than picking up on their commonalities. This results in less efficient learning, comprehension, production, and generalization.

Many of the common deficits specifically involve formulating spoken responses. For example, dyslexics perform more poorly on the RAN task, which requires naming pictured objects, such as chair, book, doll, and shoe, or naming the colors of a series of dots (red, blue, green, black). The child can easily recognize the objects and colors and knows the corresponding names, yet is slower or less accurate in naming them. This might occur if the child had developed multiple representations of the words' phonological codes. Producing a spoken response requires choosing among them, which is slower and more prone to error. A child who has difficulty repeating back a nonword such as "glomper" may have a related problem with producing speech, not memory.

There may be several types of phonological deficits, and dyslexia may have other, nonphonological causes as well. But "phonology" is a big umbrella covering all of the ways in which knowledge derived from pronunciation and sound is used in reading, speaking, and other tasks. Impaired phonology jeopardizes performance on all of them.

Is Dyslexia a "Desirable Difficulty"?

A condition such as dyslexia, caused by multiple interacting factors that vary in severity and malleability, gives rise to a broad range of behaviors. We should call it the dyslexia spectrum, which, like the autism spectrum, includes individuals whose patterns of strengths and deficits differ markedly, as do relevant conditions in the home, school, and community. The corollary is that individual outcomes also vary a great deal.

Much of the discussion around dyslexia focuses on negatives: underlying impairments or deficits that make reading difficult, inadequacies in institutional responses to nascent reading difficulties, how debilitating the condition is for many people. This "negative" emphasis has a positive motivation: better understanding of the condition and its etiology will lead to more effective interventions that yield better outcomes for a greater percentage of people.

It is also important to recognize that the range of outcomes includes highly positive ones. For many children, timely, well-designed interventions that focus on relevant problem areas are effective. (The major challenges are making such programs widely available and determining what is different about "nonresponders" who continue to struggle.) Many dyslexics discover effective coping strategies or, better, are guided to them by capable specialists. Sometimes the strategies are as simple as obtaining additional time to complete assignments or exams. Supplemental resources, such as audio and video materials or shared class notes, can make a big difference. Other cognitive or personality attributes can ameliorate the impact of a reading problem. We know that such positive outcomes occur, though firm data about the distribution of outcomes and about the factors that make the most difference is scarce.

Given that the successes don't receive as much attention, it may be helpful for dyslexics, their families, and their teachers to hear about highly successful people who identify as dyslexic—celebrities, mega-rich Wall Street investors, venture capitalists. Such individuals convey the message that dyslexia does not preclude success more effectively than any research study could. We know little about the life histories of these individuals, and personal narratives aren't always reliable, especially when a person's position requires maintaining close control over personal information and public image. The message that dyslexics can be highly successful is valid in any case.

The apparent fact that some extraordinarily high-achieving individuals are dyslexic raises an interesting question: Could the conditions that cause a reading impairment have beneficial side effects, such as special talents? I can tell you that there isn't much hard evidence on the question because it is so difficult to study. No one has undertaken a large-scale prospective longitudinal study of individuals at risk for high achievement. Like reading, a person's life is a trajectory determined by interactions among a multiplicity of factors. The number of factors is even larger than in reading, however, and many are unrecognized, unpredictable, or misunderstood until, alas, too late. There are a lot of accidents of fate, too.

Such unknowns aside, could conditions that interfere with reading enhance other kinds of performance? Current evidence suggests the question is well worth pursuing. It's likely, for example, that at least some dyslexics show greater sensitivity to certain kinds of stimuli—speaking voices, faces, objects, and so on. Greater sensitivity to differences between exemplars of letters and sounds is bad for reading, but it's easy to imagine contexts in which it might be advantageous. Mel Blanc (Bugs Bunny) and June Foray (Rocky the Flying Squirrel) provided the voices for dozens of beloved animated characters. They were great at their jobs because they were able to create and maintain an amazing array of unusual, appealing, distinctive speaking voices; they also had to be able to hear those differences themselves. Would they have shown greater sensitivity to variation in how the same word was spoken by several speakers or to different words spoken by a single speaker? Would their vocal skills have had any impact on their reading or language comprehension? Positive or negative?

A font designer's job is to attend to variation in the "incidental" properties of letters that readers normally abstract away from. Would a font designer's expertise make it easier or harder to recognize letters or text that *is* **wr**ITTEN like TH*is*? Or is their expertise unrelated to properties of a dyslexic brain? Are such people any more or less prone to reading difficulties than others? We do not know, but the questions are not frivolous.

A small number of neuroimaging studies have found that dyslexics performed more poorly than nondyslexics on tasks involving print but were superior on tasks involving visuo-spatial processing of nonprint stimuli. In one study, for example, the task was to decide if a line drawing represented a possible 3-D object or an impossible figure, one that did not have a coherent 3-D interpretation. The subjects also performed a lexical decision task, deciding if strings of letters represented words or nonwords. Dyslexics were faster than nondyslexics on the impossible figures task (without making more errors) but worse at making lexical decisions. These behavioral differences were associated with interpretable differences in brain activity, with the dyslexics showing lower activation in left hemisphere circuits for reading on the lexical decision task and greater right hemisphere involvement on the visuo-spatial task. Thus, the dyslexics' brains were organized in a manner that was favorable for one task but disfavorable for the other. A very interesting result, but I wouldn't run with it quite yet: the number of subjects was too small to determine how often the pattern occurs in dyslexics or nondyslexics; the dyslexic advantage was small; and whether facility in identifying impossible figures carries over to tasks of greater consequence remains to be determined.

These are empirical questions that can be answered with additional research. In the meantime it would be a serious mistake to assume that because a dyslexic brain is organized differently, it necessarily confers other benefits. That doesn't follow. The brain isn't a zero-sum game in which a deficit in one area entails an advantage in another. At this point we know that dyslexia frequently co-occurs with ADHD, math impairments, and a couple of kinds of developmental language impairments. In fact, it is hard to find areas where dyslexics perform better than nondyslexics. That is why the findings of greater sensitivity to stimulus qualities are unusual. The conditions that cause dyslexia in some individuals might promote the development of other skills, even to an extraordinary level, and that is worth finding out. It is not a prominent characteristic of this population, however.

Knowing that a celebrity is dyslexic may be motivating for a child who is struggling; no harm in that. But matters take a darker turn when it is asked if dyslexia could be a "desirable difficulty," as Malcolm Gladwell did in his 2013 book *David and Goliath*. Dyslexia is a serious condition that is challenging at best and often debilitating. Gladwell acknowledges this and then asks, "You wouldn't wish dyslexia on your child. Or would you?"

My view is that his observations about dyslexia are so shallow they shouldn't merit serious attention. But this is an author whose books reach an enormous audience. This book is so well known it seems likely to create additional obstacles for dyslexics and their families. The usual responses to dyslexia are "He'll catch up" and "Try harder." Now there is "Not to worry: it could be an advantage." It therefore demands a corrective response.

Gladwell's Chapter 4 begins with the standard description of dyslexia as a serious liability, which is true but not interesting because it's widely known. The Gladwell touch is to ask whether the opposite might be true in some spectacular cases: people whose dyslexia launched them to stratospheric success. He acknowledges this doesn't occur for most dyslexics, but he builds the case that it probably does for some, and the payoffs are such that you just might even wish it upon your child.

Questions such as *Could dyslexia be a desirable difficulty?* and *Would you wish it on your child?* can't be empirically proved or disproved. They are value judgments that depend on how one weighs the evidence. It's important, then, to weigh the quality of the evidence that Gladwell provided.

It doesn't weigh much.

1. Gladwell relates dyslexia to the concept of "desirable difficulty," introduced by Robert Bjork, a UCLA psychologist. In his work the "difficulties" involve ways of structuring a student's experience (e.g., time spent on one type of problem before switching to another type) that may produce a longer learning curve but better mastery. Gladwell's idea is that dyslexia might be similar: a condition that makes reading more difficult but yields benefits. He describes a study of college and high school students that examined the impact of making the print harder to read by using smaller letters in the Comic Sans font at 60 percent grayscale. Assessments showed that students learned more from the degraded texts. The print manipulation created a desirable difficulty, apparently forcing students to expend more effort, which yielded better performance.

What is the connection to dyslexia? Dyslexics expend greater effort reading texts because they are poor readers. Perhaps they too attain desirable benefits.

These situations are unrelated. In one case, highly skilled readers (the college students were Princeton undergraduates) read laboriously because the texts had been degraded by the experimenters. They were able to use their excellent reading skills to accommodate the manipulation for the duration of the experiment. Seems like a fun but inconsequential challenge.

In the other case, dyslexics read normal texts laboriously because they are poor readers, the result of having learned to read with a brain that processes print *as though* it were degraded. That experiment has been going on their entire lives, and they don't have the option to discontinue.

Further, Bjork's idea is that "desirable difficulty" leads to better learning in the long run. Applied to dyslexia, that would predict that initial difficulties in learning to read would eventually result in more highly skilled reading. But that's obviously wrong; dyslexia makes it harder to become a skilled reader. Gladwell's idea is different from Bjork's: that the difficulty in one area, reading, would confer benefits elsewhere. But that too is contradicted by evidence that reading difficulties interfere with other kinds of learning.

2. In support of the idea that dyslexia might be a "desirable difficulty," Gladwell notes that "an extraordinarily high percentage of entrepreneurs are dyslexic." For evidence, he relies on a 2009 study from England. Let's dispense with this quickly. The study used an informal survey to identify people with dyslexic tendencies. The researcher noted that dyslexia cannot be diagnosed this way, unlike, say, a personality test that distinguishes introverts from extroverts. The checklist had twenty questions related to reading, spelling, memory, organizational skills, doing arithmetic in your head, and other tasks. For unexplained reasons, the researcher reported the data from a subset of the questions (difficulty with spelling, taking down and passing messages, and a couple of others). Whether this choice was made before or after she had examined the results is unclear.

The questionnaire was sent to 2,000 potential participants, identified as "entrepreneurs" and corporate managers. The low 7 percent return rate would normally invalidate the results because the few people who were willing to participate were unlike the 93 percent who declined. The returns yielded surveys from 102 entrepreneurs and 37 managers. The main data concern the "incidence of dyslexia," even though the author had noted that the checklist can't be used this way. Whereas only 3 of 37 corporate managers exhibited four or more dyslexic "traits," 36 of the 102 entrepreneurs did so. There was no way to assess whether people's responses accurately reflected their abilities. There were some other methodological problems and minor findings, which I'll skip.

The study is flawed at every step. Gladwell takes the results at face value because they were published, the results seem to support his story, and they would be interesting if they were true, showing no interest in the quality of the evidence. He then takes the results further, concluding that "an extraordinarily high percentage of entrepreneurs are dyslexic."

3. The remaining evidence comes from the personal histories of two individuals whose dyslexia enabled them to achieve extraordinary success. David Boies is a famous lawyer, one of the most successful, accomplished people at the top of a very high legal pyramid, who worked on a slew of historic cases, including the IBM and Microsoft antitrust cases, *Gore v. Bush*, and the overturning of California's Prop 8 ban on same-sex marriage. Gary Cohn is president of Goldman Sachs, the multinational investment firm.

I cannot address whether these individuals are or were dyslexic. I couldn't possibly know. I can address the characterizations of these individuals in the Gladwell book, however. The question is what can be concluded from the information that is provided there. To be clear, I'll refer to DB and GC, the characters in the book, not the actual individuals.

Case DB: The important feature of this narrative is that the behaviors that are described are neither unusual nor strong indicators of a reading impairment. DB reports not liking to read, which is true of so many children there is a term for them: "reluctant readers." This disinterest in reading may be related to his not having learned to read until third grade, which is within normal limits. Being a late starter because of disinterest in reading is not the same as being unable to learn.

DB's reading avoidance continued in law school. He read brief summaries of case studies rather than the entire casebook entries. He listened closely in class rather than taking notes, which he thinks helped him develop exceptional listening abilities and gave him more time to think about what was being said. These behaviors are again unremarkable. Dyslexics aren't the only people who take shortcuts like reading the Cliffs Notes/Sparknotes version instead of the book. College professors implore students to listen and respond to what they are saying and resist the urge to focus on note taking. DB sounds like an ideal student in that regard. Like many people who are excellent readers, DB feels that he reads more slowly and with greater effort than others. But people's beliefs about reading proficiency are skewed; what's perceived as slow and effortful can be ordinary careful reading.

These examples illustrate a common tendency to mistake normal behaviors for reading deficiency. Perhaps the plainest example is that DB also reports that his spelling was so poor that spellcheckers often could not come up with the correction. That is a characteristic of spelling software, not spelling impairment. It happened to me countless times in writing this book, and I am not dyslexic! Spelling is a recall task; reading is a recognition task. Recall is intrinsically harder than recognition. People can easily read words they routinely misspell.

The suggestion that DB succeeded by capitalizing on his listening and memory skills suggests an alternative narrative. DB presents as a person of multiple talents. People who have particularly good memory and listening skills are less dependent on acquiring information via text, which results in reading less often. Reading less works against gaining reading skill, which requires extended practice. Such persons can be weak readers, though not because of a constitutional deficit. Their compensatory abilities also buy them more time to bring their reading up to speed, which can happen when their jobs finally demand it.

Case GC: This individual is also dyslexic by self-report. Like DB, he describes a history of poor school performance, dislike of reading, and a period of indirection. Other details about his reading are sketchy, but two points stand out. First, he describes himself as an extremely poor reader who takes six hours to read a twenty-two-page document and notes that he won't be reading Gladwell's book because of the time and effort that it would require. Second, he relates the story of how he got a job on Wall Street. It involved bluffing an executive into giving him a job by falsely claiming that he knew a lot about options trading. During the week between the job offer and the first day of work, he reports picking up a copy of the Bible of options trading, *McMillan's Options as a Strategic Investment*, which he proceeded to read, describing in detail the great effort it required. The man was so successful at the job that he eventually rose to chief executive.

This narrative is contradictory. If GC reads twenty-two pages in six hours (3.67 pages per hour), there would not be enough hours in a week to read the McMillan book, which is over 1,000 pages long. Reading the entire book at that rate would have required reading continuously for about thirty-two hours a day.

Although it can't be determined from what's in the Gladwell book, these conflicting data could be reconciled in several ways. Maybe GC didn't need to read the whole book. Maybe he read a shorter version. Maybe he skipped around. Maybe he skimmed. Maybe he's a genius. Maybe he's a really good reader when he wants to be. Maybe he misremembered completing the book in nine days. It can't be determined. But the gaps and inconsistencies in what is reported undermine its evidential value.

I can summarize my concerns very simply. The information about these individuals is equally consistent with another interpretation: people can fall behind in reading because it requires more effort than other things that come much more easily. If so motivated, they can catch up because the poor reading is not due to a learning disability. The case studies illustrate the importance of reading skill, though not reading speed, in achieving extraordinary success in fields such as law and banking.

My guess—and it is only that—is that these individuals are "stealth dyslexics." This trendy term has no clinical definition, but it is being embraced by many superachievers who view themselves as having attained their success despite a well-hidden reading impairment. I would further conjecture that these are individuals whose reading is merely normal, in contrast to other skills that may be exceptional. They exhibit the discrepancy between high IQ and ordinary reading described by Bruce Pennington (page 323). They are not dyslexic, but they identify as such. They fell behind in reading for a time because they were far more interested in and rewarded for activities that came far more easily. They don't really feel badly about their reading; having overcome dyslexia is another of their impressive achievements.

Is dyslexia a desirable difficulty? Before deciding, it's necessary to look at the question in another way. Gladwell is interested in whether dyslexics are overrepresented among high achievers, such as tech industry entrepreneurs or lawyers who win Supreme Court cases, because that would indicate that something about being dyslexic confers an advantage. His question is, Given that someone is a superachiever, what is the probability that they are dyslexic? Although there are very high-achieving dyslexics, the evidence that the dyslexia occurs more often among overachievers than in the ordinary achiever population is poor.

Whatever the answer to that question turns out to be, it is essential to ask a second one. Given that one is dyslexic, whatever that may entail, what is the probability of becoming a superachiever? The answer to that is known because it is also true of nondyslexics: very, very small. If dyslexics were represented five times as often among superachievers, that would be very interesting, but it would have a negligible effect on the probability of becoming a member of that crowd. It would be analogous to holding several tickets in the Powerball lottery instead of one.

Would you wish dyslexia on your child? Not your child, no.

CHAPTER 9

Brain Bases of Reading

I used to think the human brain was the most fascinating part of the body. Then I realized, well, look what's telling me that.

—EMO PHILLIPS

BEHIND ITS MUNDANE FAÇADE, READING is an extraordinarily complex act. I've mainly described reading as a behavior that is the product of our capacities to see, hear, write, speak, learn, remember, and think. As if that were not enough, like every other behavior, reading has a neural basis, which adds another dimension to understanding it. Reading is accomplished by a brain mounted in the body of an organism that interacts with the world. We therefore cannot fully understand why reading has particular characteristics without knowing about the brain mechanisms that give rise to them.

Or perhaps not.

I opened the chapter with the Emo Phillips joke because, aside from being the best brain joke, it expresses an ambivalence that many of us feel about the brain. Our left, analytical brains say, "The brain is the most complex object on earth, the seat of intelligence, emotion, consciousness, action, and social interaction. 'How does the brain work' is one of the most difficult, important questions our brains have been able to formulate." The deep-seated sense that the brain provides the most basic explanation for behavior is similar to the intuition that particle physics provides the most basic explanation for physical reality.

Our right, holistic brains say, "I'm not very technically minded. I do care about basic characteristics of reading and what is special about it, but what really matters is the experience of reading itself. Particle physics doesn't tell us much about the experience of the physical world either. I'm glad that Lefty, my conjoined twin, is interested in the machinery, but that is a very specialized kind of knowledge only a left brain could like. For a right brain like me, what difference could it possibly make what gets done where? My feeling is—and I say this as the feeling hemisphere—if it can't be explained using metaphors and simple syntax, it's not of any use to me. Terms like 'superior temporal sulcus' and 'perisylvian areas' send my brain-activation levels straight to baseline. Those brightly colored pictures of brain activity are pretty though."

Both of *my* hemispheres say that understanding the brain bases of reading and every other behavior is important for what it tells us about who we are and why we are the way we are. A person has to be almost brain dead, so to speak, to lack any curiosity about how this protoplasmic lump—not one of the better-looking organs, to be frank—is able to accomplish all it does. Complexity is intimidating but also beautiful. The truly exciting prospect is that what is learned about the neural substrate will change how we think about behavior itself. With a better understanding of the brain bases of reading, we could devise behavioral activities that tune specific components of the neural substrate, helping more people to become better readers. Software purporting to do this is already being marketed, but its selling point is the promise of brain training, not demonstrated effectiveness.

Although people are fascinated by the brain, we are also easily bored by its complexity. You want the truth about the brain? You might not be ready for the truth. In 1992 David van Essen, a distinguished neuroscientist, constructed a now famous diagram of the visual system of the macaque monkey, which is very similar to the human system. It's worth a look; it resembles the circuit diagram for a microprocessor from that era. It includes over thirty distinct brain centers arranged in a ten-level cortical hierarchy, plus connections to systems for motor control, other senses, and cognition. We can assume that the system that underlies reading is as complex (for one thing, it includes a large chunk of the visual system). For anyone who attempts to write, teach, or talk about the brain, this conflict poses a challenge, often resolved by taking considerable scientific license. I just parodied one such simplification, the hoary stereotype of left- versus right-brain thinking.

It is very hard to write well about the brain without simplifying to the point of distortion. Many science writers think that such simplifications are justified because they aim to interest readers in a complex topic, not teach them about it. Simplification is inevitable, and the distortions can be severe, but the ends are thought to justify the means.

I'm not convinced. It's a problem if a reader cannot determine whether a description is true with many details omitted or a truthy version the writer has decided the audience can handle. Once registered in the public consciousness, the cartoonish rendering of complex science such as how the brain works is hard to displace. If it wasn't immediately obvious that I was *satirizing* left- and right-brain thinking . . . my point exactly.

Researchers are far from fully understanding the brain bases of reading or any other complex human behavior. A fair assessment is that much is known about the main components of the reading system and how the system develops and about some properties that vary across individuals and writing systems. I can therefore say quite a bit about how the brain solves the reading problem while abstracting away from neurobiological details that are an impediment for many readers (though of supreme interest to researchers). The result is not a cartoon version of what is known, just one that is neither overly broad nor numbingly specific. My aim is to achieve the level of description that the philosopher Hilary Putnam articulated in a famous analysis of the properties of a good explanation of complex phenomena. A good theory, he observed, explains the phenomena in terms of general principles that bring out their essential aspects rather than burying them in a mass of unessential detail.

Triangulating the Brain

The brain bases of reading cannot be discovered just by squinting very closely at the organ. Brains do not present themselves for study with the parts and their functions neatly labeled. It was an enormous advance when the German neurologist Korbinian Brodmann, working in the early 1900s, identified forty-three major anatomically distinguishable human brain regions, a taxonomic system still in use. In 2016, a team led by Van Essen used neuroimaging methods to create a modern Brodmann map that has 180 areas per hemisphere, including almost 100 that hadn't previously been detected. Yet looking at these diagrams alone, one wouldn't know what the thing can do, certainly not that it does what is called reading. It's necessary to have a theory of reading in order to know what questions to ask about the brain. But then our theories of reading are imperfect guides. They are incomplete and focus on behaviors, not properties of the neurobiological system that gives rise them. Since we do not have complete theories at either level, research involves working back and forth between them. It is another constraint satisfaction problem, like using letter and word knowledge to solve a Captcha. What is learned at each level constrains what can be correct at the other. The goal is to converge on a theory that is consistent with the facts at all levels and show how they are related.

The computational models I've described serve an important function, suggesting how something like a brain could give rise to something like reading a word or sentence. Initially the models were used to explore reading behavior with the expectation that they would be linked to the brain when more was known about it. The first step in that direction was using the models to make sense of the ways in which reading and language break down because of brain injury or disease. The pace of progress has increased since the advent of neuroimaging methods that allow observation of the intact brain in action. The computational modeling and imaging methods were both introduced in the 1980s and developed largely in parallel, but they are finally coming together in areas such as vision, memory, attention, and reading. It turns out that they capture phenomena at roughly the same level of specificity, or *grain*, a useful one for understanding cognition. Brain and behavior can then be linked via common computational concepts. Thus, even though the models I've described do not represent the brain, they are very helpful in understanding it.

Acquired Dyslexia

For many years the main source of evidence about the brain bases of reading and other higher functions came from studies of how they were affected by brain damage due to an injurious event (such as a stroke or a closed-head injury from, say, falling off a bike without wearing a helmet) or disease (such as Alzheimer's). The pioneering research was conducted by nineteenth-century neurologists including Paul Broca and Carl Wernicke, whose most famous work concerned spoken-language impairments (Freud was involved early on but developed a fixation on other organs). Evidence about the locus of the brain injury responsible for a behavioral impairment could only be obtained from postmortem inspection of the damaged tissue or, in later years, during neurosurgery. With the introduction of computerized tomography scanning in the 1960s, the locus of damage could be identified while the patient was alive.

Case studies of cognitively impaired patients are a staple of popular science writing because neuropathology can cause people to behave in extraordinary, unexpected ways, such as mistaking one's wife for an article of clothing or memories for new events that last only minutes. The plight of such patients has a tragic, morbid allure. The prominence of neuropathology in contemporary popular culture is similar to that of tuberculosis in the literature and educated culture of an earlier era. Tuberculosis is a painful, deadly disease, but treated with aesthetic distance, it was a rich source of metaphor in great works of literature by writers from John Keats to Thomas Mann. Rather than emphasizing the freakishness of a condition, the more sensitive modern writers use case studies as the starting point for ruminations on the nature of the self, consciousness, reality, and other issues at the outer limits of scientific inquiry. The late Oliver Sacks memorably focused on what it was like to be the patient with a condition such as agnosia rather than the malady's causes. It does not demean such patients to focus on the valuable evidence they provide about the neurobiological bases of human characteristics and how impairments affect behavior. What these injuries reveal about the brain is perhaps small recompense for their unfortunate existence.

The term "acquired dyslexia" refers to reading impairments that develop after a literate person experiences brain damage, rather than impairments in learning to read. Neuropathology causes several patterns of acquired dyslexia that reflect damage to different parts of the reading triangle. In all of these cases, the patient has experienced brain damage that affects reading and other behavior. The primary focus is again reading aloud, mainly because the behavior is observable; the patient's comprehension and production of spoken and written language, short- and long-term memory, and other capacities are also assessed.

The patient's task is simply to read aloud a word or nonword printed on an index card or displayed on a computer screen. Performance is impaired, but only in part. The critical data concern three familiar types of stimuli: "regular" words such as TAKE, MUST, and BLIMP; "exception" words such as HAVE, ONCE, and PINT; and nonwords such as NUST, FLANE, and DORST.

The several types of acquired dyslexia reflect the fact that in many cases, brain injury does not affect all components of the reading system equally. The researcher asks, How is the reading system organized such that neuropathology results in very impaired performance in reading one type of word but leaves another relatively intact? Each type of acquired dyslexia is defined by a characteristic pattern of errors. Performance varies within each type for several reasons: differences in the severity, locus, and cause of the brain injury; differences in reading skill that existed prior to the injury; time since injury; and type and extent of rehabilitation. The case reports in the literature emphasize patients for whom the damage was relatively selective because they are more informative than patients who are globally impaired.

I will describe the symptoms of four types of patients. I can tell you the usual locus of the brain lesion, but from this information you would not be able to predict the patient's reading errors. As the cognitive neuropsychologist, your task is to diagnose which part or parts of the reading triangle are impaired, which can then be related to the neuropathology.

Surface Dyslexia

The main symptoms are as follows:

- The patient can read many words and nonwords aloud fluently and accurately.
- Errors mainly occur with exception words, which are often "regularized." For example, BROAD is pronounced "brode," or PINT is pronounced to rhyme with MINT. The defining symptom is that damage affects irregularly pronounced words more than regulars, although the exact pattern varies for reasons I've noted. Some patients only mispronounce lower-frequency exception words; for others, the impairment extends to common ones such as SAID and GIVE.
- Comprehension of both spoken and written language, even single words, is poor. The patient might read the word BOOK aloud correctly but be unable to match the spoken or written word to a picture of the object. The patient's ability to speak and write coherent sentences is also severely impaired.
- Locus of the brain damage is in areas of the left temporal lobe known to encode semantic information or link to the regions where such information is represented.

The classic surface dyslexic case was a woman known as MP, whose head injury resulted from her being hit by a bus. MP recovered from the accident, but though she was alert and could engage in many activities, her language comprehension—both speech and reading—was severely impaired. She was something of a reading machine, however: she could read words aloud with high accuracy even though she did not understand them. She could also pronounce nonwords correctly, at speeds that were within the normal range for her age. Her errors were almost entirely limited to one type of word: lower-frequency exceptions such as BROAD and YACHT, which she regularized. Marlene Behrmann and Karalyn Patterson, distinguished neuropsychologists who tested MP extensively, found that she regularized every word in the phrase PUSH COMES TO SHOVE. Another surface dyslexic patient could correctly pronounce NASA, an acronym that is pronounced as if it were a bisyllabic word, but mispronounced IBM, an initialism whose pronunciation is the names of the letters, saying "ibm" (three phonemes), a regularization error, instead.

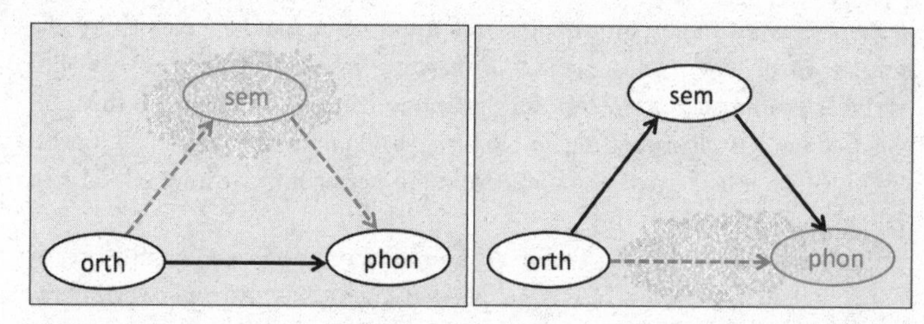

FIGURE 9.1. Surface dyslexia (left): Damage to semantics causes impairments in reading aloud irregular words such as MAUVE and SIEVE. For patient MP, the intact orthography → phonology pathway allowed most other words and nonwords to be pronounced correctly and fluently, but she did not comprehend what she said. Phonological dyslexia (right): Damage to the orthography → phonology pathway forces the patients to rely on orthography → semantics → phonology, which results in dysfluent speech and errors on nonwords.

Your diagnosis, please.

The fact that MP could not comprehend written or spoken words means that her semantic system was severely damaged (Figure 9.1, left). The fact that she could read many words and nonwords aloud suggests that her ability to map from orthography to phonology was intact. The errors in pronouncing lower-frequency irregular words resulted from her semantic impairment.

Why should the meaning of a word be relevant to saying it aloud? After all, MP could pronounce nonwords accurately, and they have *no* meaning. In Chapter 7 I described how the concept of "division of labor" applies to efficiently computing the meanings of words. Here the same principle applies to reading letter strings aloud. In reading aloud, the phonological code normally builds up as activation arrives from two parts of the triangle: orthography → phonology and orthography → semantics → phonology. The orthography → phonology pathway works more efficiently by letting the orthography → semantics → phonology pathway carry part of the burden for infrequent words with atypical pronunciations. We had to discover this; it was not predicted by any previous theory. The first hint came from our original model, which for practical reasons only included the orthography → phonology part. That model learned to pronounce both regular and exception words and also simple nonwords; the few words that it missed were lower-frequency exceptions such as SHOVE and YACHT—just like MP. Perhaps these words require input from the orthography → semantics → phonology side, which was missing from the model and severely damaged in the patient?

Subsequent studies of patients and unimpaired readers confirmed the greater involvement of semantics in reading lower-frequency words with atypical pronunciations, a very specific, nonobvious prediction. We then developed models incorporating both pathways that naturally developed this division of labor as a by-product of learning to perform the reading-aloud task quickly and accurately.

Although this division of labor is the most common one in skilled readers, the balance varies. Some people have very highly tuned orthography → phonology pathways, allowing even relatively uncommon spellings to be pronounced correctly, thus requiring little input from orthography → semantics → phonology. For other people, the orthography → phonology pathway is more coarsely tuned, representing only the most consistent spelling-sound patterns, requiring greater input from orthography → semantics → phonology to pronounce many words correctly. These folks may be familiar: they are very good at pronouncing the words they know but can be counted on to mangle the pronunciations of the words on "Hello! My name is" name tags.

In short, MP showed what one person's reading was like with the orthography → phonology component very well preserved but the orthography → semantics → phonology part completely disabled.

If this account is correct, two predictions follow. First, patients with impaired semantics will be surface dyslexics. Second, surface dyslexics will be impaired on other tasks that rely on semantics, such as speaking and object recognition. Nature, in the iniquitous guise of neuropathology, has provided tests of these predictions. The clearest evidence comes from patients with semantic dementia, a type of degenerative neuropathology that results in the gradual loss of conceptual knowledge. The impairment is due to progressive atrophy of the anterior temporal lobe, the brain area that acts as a hub linking the many types of information that constitute our knowledge of concepts such as cup or canary.

Unlike Alzheimer's disease, which has broad effects on cognitive function, semantic dementia only affects this type of knowledge until the disease is quite advanced. As the semantic system deteriorates, the patients get correspondingly worse at reading words aloud. As our models predicted, errors are initially concentrated on lower-frequency words with atypical pronunciations. As the condition worsens, other words are also affected. Remarkably, researchers were able to show that the status of a patient's semantic knowledge of individual words predicted whether they would pronounce them correctly or not. Also as predicted, these patients are impaired on other tasks that rely on semantics, such as hearing a word and then spelling it or generating the past tenses of irregular verbs such as TAKE-TOOK.

Phonological Dyslexia

The main symptoms:

- Many exception words and regular words can be pronounced correctly and at about equal accuracy levels.
- Nonword pronunciation is much more impaired. Patients can still correctly pronounce many words they know but are very poor at sounding out new ones. This discrepancy is the defining feature of the condition.
- The extent to which comprehension of spoken language is compromised varies; even when impaired, comprehension is much better than speech production, which is dysfluent. Patients usually understand the words they correctly read aloud.
- Lesions are found in parts of a temporoparietal circuit for generating phonological codes.

The classic case of acquired phonological dyslexia was known as WB. Over the course of extensive clinical assessment, WB was asked to read 712 words aloud and pronounced 85 percent of them correctly, which is remarkably high given that he was an elderly man who had experienced a major cerebrovascular accident (stroke) that left him with multiple impairments. His responses were not fluent, but they were accurate. Nonwords were a different story. Of a list of twenty simple nonwords such as COBE and DREED, he pronounced none correctly.

This degree of dissociation between word and nonword reading is rare. Acquired dyslexic patients are typically impaired at reading all types of letter strings but are somewhat more impaired on one type than others. WB, however, was very good at words and very poor at nonwords. Remarkably, WB could correctly pronounce a word such as MUST but failed when presented with a slight variant such as NUST. Most of the time he would not attempt a pronunciation; when he did, he usually gave the pronunciation for a visually similar word. For COBE, for example, he said "comb," the correct pronunciation of COMB; for MAVE, the patient might say "make." Responding with a word when the stimulus is a nonword is called a "lexicalization" error.

Your diagnosis?

The patient could still comprehend spoken language, indicating that semantics was largely intact. He could read familiar words aloud using the orthography → semantics → phonology pathway, though his speech was not fluent. Nonwords require the orthography → phonology part because they are meaningless; the severe nonword impairment indicates that this pathway was no longer functioning (Figure 9.1, right). The loss of this pathway includes some degradation of phonology itself, which caused his dysfluent speech production. The lexicalization errors result when the patient tries to pronounce nonwords using the orthography → semantics → phonology pathway, which only encodes known words. On most trials nonwords do not activate semantics enough to allow a response. However, sometimes a nonword such as COBE activates the meaning and pronunciation of a similarly spelled word such as COMB, which is then pronounced correctly.

Deep Dyslexia

The types of patients I've described so far have severe reading impairments, but errors such as regularizing an irregular word or lexicalizing a misspelled word also occasionally occur when people who aren't brain injured read aloud. Deep dyslexics misread words in several ways, some utterly unlike typical reading miscues. To illustrate, I will list a word that is presented for the patient to read aloud, a representative response, and a label for the error. The examples are all errors that patients actually produced.

Word (shown)	***Response (spoken)***	***Error type***
teacher	student	semantic
stable	staple	visual
sympathy	orchestra	visual then semantic
shirt	skirt	mixed visual and semantic

Deep dyslexics are also more impaired on abstract words (e.g., FATE, DESTINY) compared to concrete words (TABLE, BOOK) and on verbs more than nouns; they are poor at reading function words (e.g., conjunctions, articles, auxiliary verbs) compared to content words (nouns, adjectives, verbs), and they produce function word substitutions such as SHE → "is." One patient was reported to correctly read aloud the word CHRYSANTHEMUM but could not pronounce THE. Finally, the patients are very poor at reading nonwords, producing many lexicalizations, such as FLIG → "flag."

And the diagnosis is?

First let's pause to note how remarkable some of these errors are. The patient *sees* a word on an index card and yet makes a flagrant error: the word is TEACHER, but the patient says "student," which isn't spelled or pronounced anything like it. How can this be? There are two challenges: to identify the likely causes of these errors—how damage to the intact system could give rise to them—and to explain why these errors occur *together*.

The fact that nonword pronunciation is poor means that the orthography → phonology pathway is severely damaged. The patient is forced to rely on the orthography → semantics → phonology pathway, which is intact enough to allow correct pronunciation of many words but almost no nonwords. So far the patient looks like a phonological dyslexic. However, the errors suggest that the semantic side is also damaged. An error such as TEACHER → "student" would only occur if the patient had read TEACHER well enough to activate part of its meaning; the response was a semantically related word, not a random unrelated one.

How does the patient get from seeing TEACHER to saying "student"? The neuroanatomical evidence (about the location, extent, or fine structure of the brain injury) does not yet predict or explain such errors. This is a job for a computational theory that spells out the mechanisms involved in normal reading, which, if damaged in ways that are consistent with the neurobiological evidence, reproduces such errors. David Plaut, Tim Shallice, and Geoff Hinton explored deep dyslexia using models of the orthography → semantics pathway illustrated in Figure 9.2. The orthography → phonology pathway is omitted because it is severely damaged in these patients. As before, each oval represents a set of units, with the arrows indicating the weighted connections from units in one layer to another. As in our model of developmental dyslexia (which built on this earlier work), these models included a set of cleanup units, creating a feedback loop that turned the semantic part of the network into a dynamical (time-varying) system. The meanings of words are represented as attractors in this dynamical system. A spelling pattern activates semantic units, and the states of the semantic units change over time as activation passes to the cleanup units and back. The network eventually settles into a stable semantic pattern, the attractor for one of the meanings.

Plaut, Shallice, and Hinton's great advance was to show that distortion within this attractor network or in the input from orthography to semantics creates most of the errors in deep dyslexia, the others being due to the damaged orthography → phonology pathway. Consider semantic errors such as CAT → "dog." The spelling activates a part of semantic memory where cat, dog, and related concepts are represented. Normally the attractor system passes

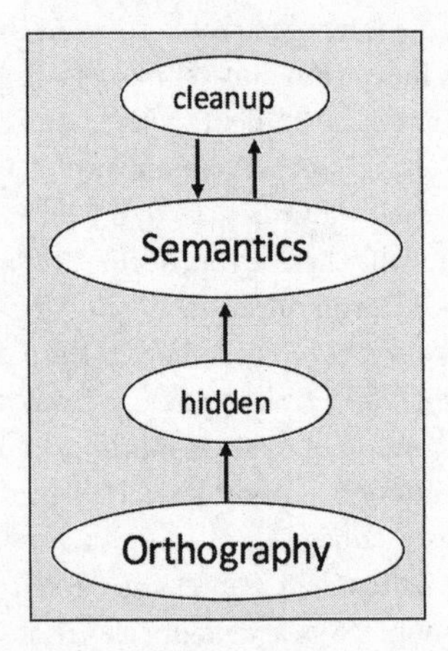

FIGURE 9.2. Hinton, Plaut, and Shallice (1993) showed that many of the puzzling and seemingly unrelated types of errors observed in these patients arose from the same underlying cause. The models learned to generate the meanings of words from print, using a semantic attractor network on the output side. Damage distorts processing within the semantic system, causing words to sometimes fall into the attractor for a nearby word.

activation around until it settles on the pattern for the correct meaning. However, if the system is damaged, it will sometimes get drawn into the pattern for a semantically related word: CAT gets drawn into the attractor for DOG which is then pronounced "dog."

The same mechanism yields others types of errors, even ones such as STABLE → "staple" that appear to be visual in origin. Here's the weird science explanation. The meanings of words are represented by bits of semantic information (features). There are a lot of them, far too many to represent in a manageable computer model (well, one that we can analyze very closely) but presumably not in a brain with billions of neurons. The meanings of words are learned attractors—stable patterns of activation. Each attractor can be thought of as a point in a high-dimensional semantic space determined by the number of semantic units. Plaut and Shallice (1993) used sixty-eight to represent forty words. Thus each word's meaning was located within a sixty-eight-dimensional semantic space (far smaller than the actual one). The distribution of points within this space was such that distortion sometimes caused the model to settle into the attractor for a semantically similar word, as when CAT falls into the attractor for DOG. Other times, however, it could be the attractor for a visually similar word. For example, CAT falls into the attractor for COT, which is then pronounced correctly, producing a "visual" error. The error is a quirk of how attractors are distributed within this high-dimensional space, not a visual impairment. A "visual then semantic" error such as CAT → "bed" results from combining these two events: CAT is drawn into the attractor for COT, which then resolves into the semantics of BED. Other deficits, such as poorer performance on abstract words compared to concrete ones, reflect properties of their semantic representations that affect the likelihood of such errors.

Deep dyslexia is a more extreme version of phonological dyslexia. Both types of patients are "semantic readers" who can only read aloud via the orthography → semantics → phonology pathway because the orthography → phonology pathway is broken. The deep dyslexics have more severe semantic impairments, however. These models addressed the two challenges I set out above: they showed how this unusual set of impaired behaviors can arise from damage to the normal system and also why the seemingly disparate set of impairments co-occur.

Alexia

This condition, also known as "word-blindness," was first described by Jules Dejerine, another pioneering nineteenth-century neurologist. The impairment involves the code that is unique to reading, orthography. Dejerine described two patients who were unable to recognize visually presented words from their spellings, but for different reasons.

One patient's brain injury had apparently decimated his orthographic knowledge. He could not read written words or spell spoken words that he comprehended normally. This condition is called alexia with agraphia because it affects both reading and writing of the orthographic code. The locus of the brain injury is the left hemisphere parietal lobe, including a structure called the angular gyrus, which Dejerine concluded was the spelling repository.

The other patient was also unable to read words, but his knowledge of spelling was intact. He could accurately spell a word when it was spoken. He could look at a spelling pattern and copy it accurately. Many such patients can read words aloud using a laborious compensatory strategy: they recognize the letters one by one, say them aloud as in oral spelling ("bee," "ay," "tee"), and then recognize the word by pronouncing it ("bat"). This strategy, called letter-by-letter reading, permits the word to be recognized through the auditory system, bypassing the damaged reading system. The patients are described as alexic but not agraphic because reading is impaired but knowledge of spelling is preserved. The condition is said to be "pure" because the deficit is confined to reading. As a British neurologist noted, "A patient with pure alexia who writes at a normal pace and then fails to read back to you what they have just written is one of the more striking signs in clinical neurology."

Diagnosis?

The key to understanding these patterns is the neuropathology. The first type of patient has experienced a brain injury that damages left hemisphere representations of spellings. In the second type, the spellings are intact, but the patient's ability to access them from print has been blocked by a brain injury (usually an infarct to the left posterior cerebral artery) that has two effects: (1) whereas visual information is normally processed in both hemispheres, in these patients the signals only reach the right hemisphere; and (2) connections between the hemispheres (via the splenium, a part of the corpus collosum) are also severely damaged. The processing of letter strings is thus marooned in the right hemisphere, unable to make contact with intact left hemisphere representations of spelling. The letter-by-letter strategy allows access to these representations via speech, which is not affected by the injury. A word can be copied using right hemisphere representations of the image, but only as arbitrary, meaningless squiggles. These patients show that orthography, the third major component of the triangle, can be compromised by brain injury in at least two ways.

Learning from Disorder

Research on acquired forms of dyslexia has provided important evidence about the neural bases of reading but also illustrates two general points about relating brain and behavior. First, understanding normal and disordered reading requires looking at behavior, computational mechanisms, and neurobiological bases *together*. The neural organization of the reading system and the bases of reading impairments cannot be reverse-engineered from behavioral evidence alone; nor can the properties of reading be deduced from the neurobiology.

Take the odd assortment of reading errors in deep dyslexia. Looking only at the behavioral symptoms, it appears that seven or eight different brain anomalies are involved, with no explanation for why they co-occur. Prior to the development of the computational models, it seemed intuitively obvious that a "visual" error had a different cause than a "semantic" error. The attractor mechanism introduced a new way to think about the genesis of such errors, whereby they arise from a common source. This mechanism is not an arbitrary computer hack; it is grounded in basic facts about how neural systems learn and function. Understanding the two types of alexia turned on neurobiological facts about the involvement of the two hemispheres in reading, properties of the visual system, the locations of the lesions, and how they disrupt the normal system.

A second lesson from the patient research is that reading cannot be understood just by studying reading. The impairments that underlie the acquired dyslexias also affect other kinds of behavior because reading makes use of capacities and neural structures that are employed for other purposes. Impairments to semantics and phonology affect performance on other kinds of tasks that also make use of them. Similarly, "pure alexia" is not pure: damage to the neural systems that support spelling also affects other kinds of behavior, such as face and object recognition. Sometimes a type of damage affects reading, the unnatural task, much more than anything else. Nonetheless, reading impairments are not reading specific.

Although close studies of certain types of patients continue to be a valuable source of evidence, most of what we are learning about the neural bases of reading comes from neuroimaging and related methods used with neurally intact people. Computational modeling and neuroimaging methods are now closely linked. The models are used in designing imaging experiments testing specific hypotheses and in analyzing neuroimaging data; the neuroimaging data feed the development of more sophisticated, neurobiologically realistic computational models, which in turn cause us to examine behavior in new ways. Tying neuroimaging to computational theories of complex skills such as reading is leading the research beyond the point-and-shoot studies of the "brain area for *x*" that characterized early research in this new field.

The "Simple View" of the Reading Brain

Research on the brain bases of reading has yielded broad agreement about the existence of the primary circuits and their general properties. The circuits are usually characterized in terms of the major brain areas involved, the probable functions of those areas, and how they are interconnected. fMRI studies with children (the method is safe, the main challenge being to keep the child from fidgeting) have yielded good evidence about how the circuits normally develop and anomalies that occur in developmental dyslexia. However, the brain is a tough nut. Although the major components have been identified, a detailed understanding of the system is not yet in hand. The fact that we still do not fully understand the functions of major brain structures identified in the nineteenth century is a sharp reminder of just how complex the brain is and how difficult to unpack.

With a nod toward Phil Gough (p. 118), here is a "simple view" of the reading brain (Figure 9.3).

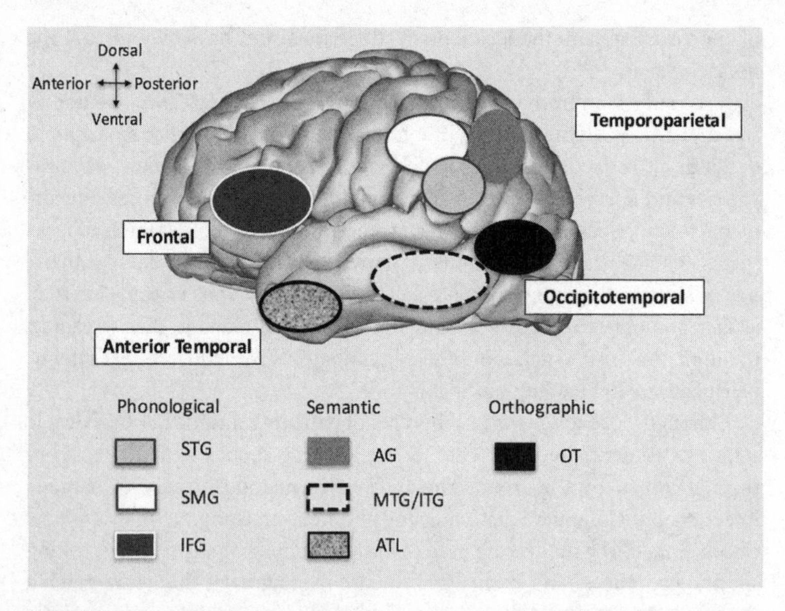

FIGURE 9.3. Some major components of the neural systems for reading. *OT*, occipitotemporal region: spelling hub, "skill zone". *AG*, angular gyrus: multimodal area, convergence of semantics, phonology, orthography. *MTG*, middle temporal gyrus, and *ITG*, inferior temporal gyrus: parts of the ventral orthography → semantics circuit. *ATL*, anterior temporal lobe: semantic hub. *STG*, superior temporal gyrus, and *SMG*, supramarginal gyrus: parts of the dorsal orthography → phonology → speech circuit. *IFG*, inferior frontal gyrus: multiple subareas, functions; involved in speech production, linked to semantic hub. Locations are approximate; designated areas are parts of larger regions (e.g., STG = a part of the STG). Connectivity between areas, not shown, shapes their functions, which are less narrowly tuned than the labels imply.

The *hub* of the reading system is an area at the junction of the occipital and temporal lobes, sometimes called the visual word form area (VWFA), or "letter box," which is the major locus of orthographic knowledge, the unique component of reading.

The *major reading circuits* result from learned connections between the left occipitotemporal hub and neural systems for language, which are already highly developed at the onset of reading and then modified by it.

- The dorsal pathway links spellings to temporoparietal brain areas known from the study of spoken-language comprehension and production. This is thought to be the primary circuit for computing phonological codes from print (orthography → phonology) and pronouncing them aloud.

- The ventral pathway links spellings to areas in the temporal lobe known for their role in determining the meanings of words, in reading but also in speech. It is a plausible candidate for part of the orthography → semantics side of the triangle.
- Connections from areas in the temporal lobe to the phonological pathway, known from speech production, fill out the orthography → semantics → phonology circuit.

Frontal areas focused on the inferior frontal gyrus (IFG) are implicated in several aspects of normal and disordered reading. This region, which includes the classical Broca's area, is activated in the performance of many tasks, with higher activation associated with greater task difficulty. One subcomponent, the pars opercularis, is involved in planning and producing speech and may be a part of the dorsal, phonological pathway in reading. A second component, the pars triangularis, is thought to have semantic functions and may be activated via the ventral orthography→semantics reading pathway. IFG links to adjacent frontal areas involved in "executive function," the planning and monitoring of goal-directed behavior. In reading, this includes establishing one's goals in reading a text, adjusting the depth with which a text is read in accordance with those goals, monitoring whether the goals are being met, and making adjustments as necessary.

Modules for Reading?

Although there is good neural evidence for the basic pathways in the triangle model, what is the status of the ovals labeled "orthography," "phonology," and "semantics"? The question has to be considered against the backdrop of current debates about how the brain works. One possibility is that each type of information is the purview of a corresponding brain area. The notion that human faculties are localized in distinct brain regions is familiar from the diagrams of the nineteenth-century phrenologists, with their creepily confident demarcations of areas associated with a screwball mix of "faculties" such as cautiousness, intuitiveness, urbanity, time, and beauty. ("Language" was located beneath the eyes, an apparent conflation of speech and reading.) Although the phrenologists were obviously wrong, the idea that brain regions might serve highly specialized functions has persisted. In the modern era, several such areas have been proposed. The visual word form area is one; the fusiform face area (FFA) exhibits specialization for faces; the parahippocampal place area, for visual scenes; the extrastiate body area, for body parts; and the lateral occipital complex, for objects. According to some researchers, categories such as foods, animals, and tools are also localized in distinct brain regions. The evolutionary story is that the brain evolved specialized systems for these types of information because of their survival value (distinguishing edible from inedible substances, distinguishing predators from nonpredators, devising weapons to fend off the predators). The brain has been likened to a Swiss army knife, a device consisting of many specialized tools.

A competing view holds that each major type of information is supported by a network of brain structures rather than localized to a single area and that each brain area is responsive to and participates in the processing of multiple types of information. Specialization exists—those forty-three Brodmann areas do not serve identical functions—but tuning is not as narrow as in the modular approach. With the availability of increasingly sophisticated methods for collecting and analyzing brain data and the use of computational models of how brain organization develops, the strongly modular view has been challenged, but the debate is not over. And the status of spelling is at the heart of it.

Studies of alexic patients and early neuroimaging research supported the view that spelling is represented in a neural module. Brain injury could strike the module itself or the pathways used in accessing it. Dejerine himself thought that the area was the angular gyrus, but modern research has yielded a different picture.

Experiments using fMRI have focused on a part of the fusiform gyrus in the left hemisphere (the putative VWFA, or "letter box") that is sensitive to various orthographic properties. It is involved in solving the letter-recognition problem, abstracting away from variations in form. Letter strings that are more word-like (e.g., STORT) elicit greater activation than ones that are less word-like (STCRT). Words elicit stronger activation in this area than random letter strings, checkboard patterns, geometric figures, or faces. The extent to which words elicit activation in the area is closely tied to progress in learning to read: activation is higher for young good readers than poor readers. With development, the pattern reverses: higher reading skill is associated with lower activation levels, with less-skilled readers producing higher activation levels.

Interestingly, the corresponding (homologous) region in the right hemisphere is the face area. It thus appears that the left and right fusiform regions are specialized for different types of visual expertise, spelling and faces, respectively. Damage on the left impairs the ability to recognize spellings (alexia), whereas damage on the right impairs the ability to recognize faces (prosopagnosia). Proponents of the modular view have argued that the capacity to recognize faces evolved in the species, resulting in a right hemisphere region that is innately sensitive to such stimuli. Spelling, the new technology, colonized (or "recycled") neural territory that would otherwise have been recruited for objects and other visual stimuli. Spelling gets routed to the left hemisphere region because of its proximity to language areas that are usually left lateralized and because the right hemisphere is preferentially tuned to the properties of faces.

At this point research and theorizing had yielded an elegant account of the origins and functions of two complementary modules of the brain. A beautiful theory, however, carries within it the seeds of its own destruction: rather than bringing research to a halt, it inspires even more research, which often yields findings that contradict what came before. The VWFA concept soon faced a number of challenges (as did the FFA and the other proposed modules).

1. If the VWFA is the brain's spelling module, damage to the area should cause alexia. The patient data are surprising, however. In a large study of eighty patients, Argye Hillis and colleagues found that the degree of damage in the ventral occipitotemporal region (the VWFA) was not associated with impairments in comprehending written words; patients could still perform tasks such as matching a printed word to a picture depicting its meaning. Spelling *is* represented in this area, as shown by the imaging studies of normal readers. The area does not have to be intact, however, because spelling is also represented in a network of brain areas, including the angular gyrus and the right hemisphere fusiform gyrus. Thus, the processing of visual word forms can be achieved without a visual word form area.

2. If specialized for spelling, the VWFA should have little involvement in processing other types of information. However, it is activated by objects, faces, Braille, and spoken words.

3. Activation in the VWFA is not just determined by the type of visual stimulus; it depends on the task the person is performing. In normal readers, the area is more highly activated when words are read aloud than when the task focuses on visual properties of the letters (e.g., whether they include ascenders as in L or descenders as in j). In the study by Hillis and colleagues, damage in the area affected performance on generating spoken responses for several kinds of stimuli: reading words and nonwords aloud, but also naming pictured objects and objects identified by touch rather than sight. Matching words to

pictures, which does not require a spoken response, was unimpaired. Similar findings from a study of nonimpaired readers also led to the conclusion that this area is closely involved in using visual information, mainly spelling, to gain access to phonology.

4. The modular view of the VWFA was supported by studies in which greater activation was elicited by spelling patterns than other types of stimuli. The standard methods used in analyzing fMRI data until recently contributed to the impression that an area was narrowly tuned to the type of information that elicited the strongest responses by filtering out the somewhat weaker activation elicited by other types of stimuli, which could be spurious. More sensitive methods are now able to pick up the whole range of responses, and analyses can be conducted over smaller subareas that can be tuned differently. Moreover, other brain areas also respond more strongly to spelling than to other types of stimuli, and orthography elicits activation in areas such as the FFA. Thus, the processing of spelling is more widely distributed than previously thought.

5. The neuronal recycling account of the VWFA has taken an interesting twist. It was thought that the spelling-specific area repurposed territory in the left hemisphere because the corresponding right hemisphere region had evolved sensitivity to faces, which have markedly different visual properties. More recently Behrmann and Plaut have plausibly argued that the sequence is the opposite: faces are initially processed bilaterally, with right hemisphere specialization occurring as a result of the acquisition of orthographic knowledge, which is on the left side because of its proximity to language-related areas.

To summarize, rather than acting as the brain's orthographic oval, the VWFA is the hub of a system that supports orthographic processing and linking the spellings of words to their sounds and meanings. The term "visual word form area" is inapt because it incorporates claims about its functions that turned out to be too strong: the area is not restricted to visual information or to words, and the task of recognizing "visual word forms" is accomplished by a distributed network, not a single module. The practice of naming an area based on its hypothesized function creates confusion because the name can persist long after the original hypothesis has been abandoned. The term "visual word form area" survives because it is not wholly incorrect—the area *is* critically important for spelling—and because the label is less of a mouthful than "ventrolateral occipitotemporal cortex" or "posterior region of the left hemisphere fusiform gyrus."

The fact that views about cerebral specialization as exemplified by the VWFA have changed radically since the early 2000s shows that the science is progressing rapidly, not that the earlier studies were terribly flawed or the conclusions unjustified given what was known at the time. fMRI is a relatively new methodology and the techniques used to acquire and analyze the data have been developed along the way. Groundbreaking studies that used the best available methods in innovative ways might nevertheless have led to conclusions that do not hold up for long, as seems true for much of the research about neurocognitive modules. The value of the research is not simply whether, with hindsight, the conclusions were entirely correct. Rather, the research was a necessary intermediate step that enabled further advances. The same is surely true of current theories, which are also fated to be replaced by ones that build on current insights but move beyond them.

The story is very similar for semantics: information is distributed across numerous brain regions and linked via an integrative hub. In the simple view of the reading brain, the ventrolateral pathway links the spellings of words, integrated through the occipito-temporal hub, to semantics in parts of the temporal lobe. However, we know that the meaning of a word is more than a simple definition (see discussion, pp. 109–111). The information that constitutes the meaning of CANARY is encoded by many brain structures. Properties such as color, sound, and characteristic movement are thought to be represented in or near subsystems responsible for processing such information, which are distinct from ones that encode encyclopedic information, such as where canaries are found or episodic memories of particular experiences with canaries.

How is all this widely distributed information linked to the word CANARY? The best-supported current theory holds that the anterior temporal lobe acts as a hub linking amodal representations of words to the satellite areas in which associated information is stored. The "semantics" in the ventral "orthography→ semantics" pathway is this hub and its projections. Semantics is not "in" the temporal lobe, just as orthography is not "in" the visual word form area, but each area is crucially involved in *using* these types of information.

Finally, where's phonology? Recall that phonological representations have two important properties. First, phonology is an abstract, intermediate code that links speech comprehension and production. The need for this level of representation arises from the fact that the acoustic pattern that we recognize as a word is different from the articulatory pattern used in saying the word. Second, phonological representations are shaped by exposure to spelling (pp. 27–29). We are looking for a brain area (or areas) consistent with these clues.

The first property suggests that phonological representations should be found somewhere within the circuits involved in comprehending and producing speech. I've already mentioned some of the main areas. The short (but not exhaustive) list includes

Primary auditory cortex
Superior temporal gyrus (including Wernicke's area)
Middle temporal gyrus
Supramarginal gyrus
Inferior frontal gyrus (including Broca's area)
Motor areas relevant to producing speech

On the hearing side, all types of sounds are routed through the primary auditory cortex. An adjacent area, the superior temporal gyrus, responds to properties of the signal specifically relevant to speech. On the output side, the formulation of a plan for saying a word and the execution of that plan are accomplished by a network including parts of the IFG, premotor and motor areas for speech, and other structures. That leaves the supramarginal gyrus as a candidate for representing phonology—the intermediate linguistic code.

As Holmes nearly said to Watson, "You could not possibly have come—to the brain evidence—at a better time."

Research on spoken-word recognition has shown that brain regions in and around the SMG are sensitive to phonological properties of words. Here are two representative findings. First, the SMG distinguishes between spoken words and artificial stimuli that have the same acoustic properties but lack phonological structure. Second, activation in the SMG (but not other areas) is affected by neighborhood size—the extent to which a spoken word overlaps with other words. Overlap is measured in terms of phonological properties: syllables, onsets, rimes, phoneme-like segments. These results suggest that the SMG is involved in representing spoken words and their phonological constituents.

Is the SMG also shaped by orthography? An experiment: the stimuli consist of words such as "dust" and pseudowords such as "frab"; on each trial the subject hears one of the stimuli and presses a button to indicate whether it is a word or pseudoword. This task, an auditory version of the lexical decision task, is trivially simple. The kicker is that there are two types of spoken words. D-U-S-T is the only legal way to spell "dust" in English. There just aren't any good alternatives. In contrast, "seat" could be spelled several ways: S-E-A-T (the actual spelling), S-E-E-T, or S-E-T-E, to name a few. The behavioral finding is that subjects are quicker in identifying the word they hear when it can only be spelled one way: they respond faster to "dust"-type words than to "seat"-type words. The effect has been observed in several languages.

This is another surprising example of the impact of orthography on listening. The relevant factor must be spelling because the words are similar in other respects. As in the rhyming studies (pp. 28–29), spelling affects spoken-word recognition.

The experiment was then repeated using the method called transcranial magnetic stimulation (TMS). As before, the subject is in the lab, listening to the stimuli, pressing buttons, yes or no. The experimenter uses a wand-like tool in the shape of a figure eight to apply brief bursts of electromagnetic stimulation over specific parts of the subject's scalp. This stimulation technique is being used in many research labs and in clinical settings (because stimulation of this type may be effective in treating mood disorders and other conditions). The stimulation affects activity in a targeted region of the brain. For example, stimulating a part of the motor area involved in hand movements produces involuntary finger tapping. TMS to other regions can temporarily interfere with routine behaviors such as talking. The technique may seem odious—voluntary transient brain damage?—but the effects are mild and brief, the method is safe, and it permits tests of the functions of parts of the brain using a very simple logic: disrupt the area and observe the effects.

Researchers found that applying TMS to the putative phonological area, the SMG, eliminated the effect of spelling on spoken-word recognition: words such as "dust" lost their advantage over words like "seat." Why? Spelling influences spoken-word processing because it restructures phonological representations in the SMG (and possibly other areas). TMS to the area therefore removes the orthographic effect. TMS to the VWFA interfered with overall performance but did not eliminate the "dust" advantage. This particular effect of orthography on phonology therefore seems centered on the SMG.

Phonology is not located "in" the SMG any more than spelling is "in" the VWFA. The brain is filtering and transforming an acoustic signal across a highly interconnected set of structures (and now we know there are more of them; Glasser et al. 2016). Phonological sensitivity seems to emerge from a conjunction of properties of acoustic signals, the demands of tasks such as reading and speech comprehension and production, and neuroanatomy, with SMG as a critical hub.

Brain Bases of Dyslexia

Signatures of developmental reading impairments are found in brain structure, function, and development. The cutting edge of research focuses on their genetic and neurochemical causes. This work focuses on the usual suspect areas for language and reading, including temporoparietal areas (audition, phonology), the occipitotemporal area (the spelling hub), inferior frontal areas (speech planning and production, top-down processing), and subcortical structures (basal ganglia, thalamus, cerebellum) implicated in learning.

fMRI studies show that acquisition of reading skill is associated with a transition from bilateral involvement to left hemisphere dominance for language and then reading. The right hemisphere does not go silent, but it is the supportive partner. In dyslexia the leftward shift is delayed and not fully realized, as Samuel Orton, an early-twentieth-century physician who conducted pioneering research on learning disabilities, had proposed. The effect is to interfere with the acquisition of orthographic knowledge and its integration with language.

In comparison to peers who are good readers, dyslexics show underactivation in areas critical for orthography and phonology while reading. That result confirms that they are indeed behind in learning to read. In a few studies dyslexics also exhibited underactivation in comparison to younger children who read at the same level, evidence that the activation level is due to the reading impairment rather than simply reflecting that level of reading proficiency.

The evidence that phonological impairments interfere with orthographic learning (which feeds back on phonology) is strong. The intuition that dyslexia results from visual processing impairments is also strong, but despite years of research, the evidence is not. Some studies have found that dyslexics performed more poorly (as a group) than nondyslexics on tasks assessing an aspect of visual processing (e.g., motion detection, persistence of visual images). Some such effects have failed to replicate with new subjects or testing methods or in other research laboratories. Sometimes the effect is only seen in a subset of dyslexic subjects. Some nondyslexics also perform poorly on such tasks. The effect may be due to a property such as neural noise that has multiple effects. Often the putative visual processing impairment (e.g., motion detection) is unrelated to the reading problem (e.g., pronouncing individual words). Finally, poorer performance on visual tasks could be a consequence, rather than a cause, of poor reading.

Although the phonological deficit is clearly primary, it may be that subtle visual processing impairments are a lesser but relevant risk factor for some individuals. Suggestive evidence is emerging from studies of pathways in the visual system leading into orthographic areas, using methods that were not available until recently. Visual impairments may be like phonological ones: they affect the idiosyncratic aspect of reading, the integration of written and spoken codes, without having broad effects on vision or language.

Dyslexics also exhibit atypically high levels of activation (hyperactivation) in left and right hemisphere inferior frontal areas compared to better readers. This activation has several causes. Recall that IFG activation is associated with task difficulty. Hyperactivation occurs because reading is indeed difficult for dyslexics. Overactivation in the right hemisphere is due in part to the relative underdevelopment of the left hemisphere circuit and also results from a compensatory circuit that can develop in reaction to the left hemisphere reading circuit's failure to thrive. Activation in the left hemisphere IFG normally decreases as word-recognition skills (and activation in posterior spelling areas) increase, which is not happening for dyslexics. The initial left hemisphere frontal involvement may also reflect reliance on less efficient top-down guessing strategies, which also decline in skilled readers but not dyslexics.

The global impact of these conditions is on the integration of reading and speech. Studies from an international research team led by Kenneth Pugh have assessed the extent to which the patterns of brain activity for reading and spoken-language comprehension are similar. The degree of overlap in the usual suspect areas was closely related to reading ability in four languages with different types of writing. For skilled readers, the two uses of language become neurobiologically alike despite the obvious modality differences. Less-skilled readers show less overlap. The degree of integration of reading and language at the neural level is thus a marker of reading development.

These fMRI studies, like many others, measure a by-product of neural processes to quantify brain activity by region. Typical and atypical activation patterns in reading are now being linked to properties of the corresponding brain structures. Analyzed in a different manner—like tuning a radio to another station—signals obtained using MRI yield detailed evidence about neuroanatomy. Although structural imaging has been around for decades, a recent explosion of innovative methods for collecting and analyzing such data has transformed it into a hot area. There is further excitement because of emerging links to the genetic mechanisms that underlie brain development.

The product of the newer structural imaging research has been documentation of multiple anomalies in dyslexia. Morphometric studies examine the volume, density, and surface area of neural tissue, mainly the gray matter (neurons and local connections among them), in a region. Connectivity studies measure properties of white matter tracts, which contain the axons that propagate neural signals over the longer distances between cortical regions. The integrity and functionality of the tracts depend on the myelin that sheaths the axons.

Gray matter is reduced in the major reading/language areas in dyslexics compared to nondyslexics. Gray matter differences have also been observed in comparisons of literate and nonliterate adults. Such findings are highly suggestive, of course, but not yet fully interpretable. How large a difference is important? How much do such measures vary across individuals, ages, reading levels, and brain areas? Two studies have reported increases in gray matter volumes in relevant areas following a reading intervention, a very promising result. How large an increase can be attained is not yet known. An excess of gray matter could also be problematic at some points in development or in some parts of the system.

Connectivity between regions is assessed using methods for measuring or interpolating properties of white matter tracts. These methods are rapidly gaining in precision and resolution. The most widely used method to date, diffusion tensor imaging, may be familiar from beautiful visualizations of the findings. Anomalies in white matter tracts are seen in dyslexic children and adults. The affected tracts connect regions in the reading/language circuit. An important tract, the arcuate fasciculus, links the left posterior superior temporal cortex (part of Wernicke's area) and the left IFG (the pars opercularis, part of Broca's area). This is a major chunk of the dorsal orthography → phonology reading pathway. Myers and colleagues recently reported that increases in white matter volume in the arcuate fasciculus and in a nearby region between kindergarten and grade three were strongly related to increases in reading ability. Another major tract (the left inferior fronto-occipital fasciculus) is a plausible basis for the ventral orthography → semantics reading pathway. It is less well studied because of technical difficulties in obtaining MRI signal from the anterior temporal lobe, the hypothesized semantic hub, but some evidence indicates it is also affected. Other tracts that are part of the reading system are being identified for the first time. These tracts plausibly contribute to the integration of speech and language and are likely to vary with reading skill, and we will soon know for sure.

These anatomical properties are likely to have several effects on the neural computations underlying reading. At this point, however, the best documented is that in dyslexia, signal propagation between and within regions seems to be—wait for it—*noisier*. The brain research is converging on the actual mechanisms underlying the simulated neural noise in our models.

In the brain, "noise" involves at least three phenomena. First, neural responses to stimuli are more variable, as in the studies of dyslexics' neural responses to repetitions of a stimulus. Second, neural activity along the white matter tracts is more variable and less reliable. This reduces the effective (that is, functional) connectivity between regions, in addition to any reductions in anatomical connectivity. Third, learning is adversely affected. Dyslexics are less able to benefit from learning experiences because their responses are more variable but also because noise affects the modification of neuronal responses and their retention (consolidation). The net result is a failure to develop coherent, reliable circuits that support reading skill.

Noisy processing has been tied to susceptibility genes for dyslexia that modulate brain development in relevant ways. Some affect myelinization, decreasing the speed and reliability of signal transmission within and between reading/language areas. Some produce neuronal hyperexcitability: neurons that fire randomly or at excessive rates in areas such as the auditory cortex. Some affect neural migration, an important process that affects the brain's basic wiring. Early in brain development neurons migrate to positions in the cortex. Migration errors produce ectopias, clumps of neurons that have dead-ended in the wrong positions. All of these anomalies disrupt the functionality of neural networks.

Oh, there's one final bit. Neuronal activity is modulated by neurotransmitters such as GABA, glutamate, dopamine, and choline. Atypically high concentrations of choline and glutamate have been observed in studies of individuals with ADHD and autism. In animal studies high choline is associated with white matter anomalies. Glutamate modulates neuronal excitation, with high glutamate producing hyperexcitability.

Atypically high levels of glutamate and choline have now been observed in a small number of studies of children and adults with reading impairments. This research brings us very close to the source of neural noise. Finally: two of the candidate genes implicated in reading disability, DCDC2 and KIAA0319, have been tied to hyperexcitability via spontaneous firing in the auditory cortex. Now we just need to know whether this hyperexcitability is caused by abnormal glutamate or choline production.

And what is the executive summary of this research? I introduced dyslexia by observing that it is complicated because it can be analyzed at multiple levels: behavior, proximal causes, computational mechanisms, neural systems, genetic influences on development. I also said that dyslexia had been confusing to some onlookers because our explanations kept changing as the research advanced. It is clear that what is known about normal and disordered reading has increased markedly. The field is a great illustration of what power science—multilevel, multidisciplinary, high technology, international—can achieve.

What's surprising to me is how rapidly the field is converging on a unified theory, the one that connects the levels. That is something—along with a telephone that recognizes my speech—that I did not expect to happen this soon. We do not have this unified theory in hand (and the speech recognition apps still need work), but the form of the answers to questions about dyslexia has come into focus. The biggest need, in my view, is for large-scale epidemiological studies in which many readers at many skill levels are assessed on many measures—behavioral, neural, genetic—in many languages in order to determine the frequencies of occurrence and co-occurrence of various impaired behaviors and underlying causes in the population. We don't know who gets what how often in what combinations and at which severities. That kind of study is only justifiable because of the progress that has been made.

Will these scientific advances lead to earlier and more precise identification of reading difficulties, more effective interventions, and better outcomes for more people? Certainly: they already are, though on a limited basis. Will the day arrive when MRI and fMRI screening tests for dyslexia and other developmental disorders are required for children to attend school, like vaccinations for mumps and rubella? Not any time soon—or perhaps ever. Leave aside the high likelihood that MRI will be replaced by tools that collect brain measures more easily, cheaply, and reliably, with less inconvenience and discomfort for the subject. It isn't necessary to scan everyone to benefit from the research. Here's why.

The field is clearly on the path to understanding the neural systems for reading and the causes of reading impairment. We need a much better grip on how individuals vary, but that requires further studies of hundreds or perhaps a thousand research participants, not millions. We need the readomes for these people. With sufficient data, we could ask, Which behavioral measures, or combinations of measures, are sensitive to important neural properties, ones that facilitate or interfere with performance as children develop? Instead of scanning every child, behavioral measures could provide a basis for inferring the status of the components of the brain's reading system, similar to the way a neurological exam—a series of simple, noninvasive tests—is used to assess the status of the nervous system. A person could always go through the imaging protocol if circumstances called for it, but it would not be required for all.

At the outset I said that we conduct research by using our understanding of behavior to frame questions about the brain, but also that as more was learned about the brain, it could change how we view behavior. Viewed the right way, behavioral measures may provide more direct evidence about brain structure and function than was possible prior to the advent of neuroimaging and other noninvasive methods, back when a brain-injured patient's lexicalizations of nonwords were the coin of the realm.

THE EDUCATIONAL CHALLENGES

CHAPTER 10

How Well Does America Read?

IN 1843, HORACE MANN, THE secretary of the newly created Board of Education of Massachusetts, journeyed to Europe "to observe the schools in several countries, feeling an intense desire of knowing whether, in any respect, those institutions were superior to our own; and, if anything were found in them worthy of adoption, of transferring it for our own." Mann was deeply impressed by schools in Scotland and Prussia, finding "many things which we, at home, should do well to imitate."

Education is an international endeavor, and curiosity about how our system compares to those in other countries is inevitable, an expression of our deep-seated interest in the varieties of human culture. The increasing urgency of such comparisons reflects a sense that it is in the national interest to take the literal measure of the other players in global politics and economics. Many citizens and their elected representatives believe that America is not doing well enough at educating children in reading or in math and science, but is this true? Is the conviction that Finland and Shanghai do a better job just a modern version of Horace Mann's veneration of what he saw in Scotland and Prussia?

Concerns about how well America reads have accelerated in response to data from two large-scale programs, the Programme for International Student Assessment (PISA) and the National Assessment of Educational Progress (NAEP) or Nation's Report Card. Each release of the latest round of data from these assessments elicits much handwringing, with the president or the secretary of education usually compelled to express concern. The validity of the results and their implications about the state of education depend on how the studies were conducted and the data analyzed, technical issues that aren't suitable for soundbites. My message is that although the story is more complex than the way it is told in press releases, the main narrative is correct: Both assessments indicate that the US has literacy issues that have immediate and longer-term consequences. Reading education works very well for some, OK for many, and not well at all for others. That is not good for the US and contrasts with other countries of interest. This narrative has been challenged, however, by educators who think the data misrepresent the true state of affairs. Because the tests have been drawn into ongoing policy arguments that greatly affect what happens in schools, they demand a closer look.

It is important to recognize that these assessments are not connected to the "high-stakes testing" that was a major component of the No Child Left Behind (NCLB) legislation. The NAEP and PISA results do not trigger consequences such as closure of low-performing schools or reallocations of federal funding. Antipathy to high-stakes testing has carried over to these assessments, but they are conducted for wholly different purposes.

The PISA conducts assessments of reading, math, and science performance that are familiar from the league tables, the ranking of countries by scores. The assessments are conducted by the Organisation for Economic Co-operation and Development (OECD), as one of the activities it undertakes to promote the economic goals of its members, a core group of thirty-four market economy countries. The PISA exercises are conducted in these countries, in many partner countries, and in regions such as Shanghai, Macao, and Hong Kong; I will refer to them all as "countries" for brevity. The PISA is the largest and most rigorous of several cross-national assessment programs.

Conducted every three years since 2000, the PISA is a massive undertaking, with 510,000 students participating in 2012, the most recent cycle as of this writing. Fifteen-year-olds are tested because the goal is to assess the knowledge and skills of young people as they near the end of compulsory schooling. The assessments are impressive in regard to what is tested, how it is tested, how the results are analyzed and presented, and the massive scale of the undertaking. It is challenging to develop an instrument that assesses reading skills intelligently, but doing so in a manner that allows meaningful comparisons between countries is remarkable. The reading assessment is based on a thoughtful, inclusive notion of literacy, focusing on a detailed set of component skills and the ability to apply them in real-world situations.

The PISA is also an open-book operation: the methods are described in technical reports covering every angle in fine detail, and the organization provides online tools that allow anyone to analyze the data in simple ways. Although the league tables attract the most attention, the program's stated focus is on identifying the factors that underlie the results, not declaring winners and losers. There are results to engage even the most zealous dataphobe. For example, girls outperformed boys in *every* participating country in the 2012 assessments (as in the previous round). The smallest gap was in Albania, the largest in Jordan; as on many other measures, the US gap was close to the average for the OECD member countries. Surprisingly, high-performing Finland had one of the biggest gaps.

The PISA is a serious endeavor, the product of a dream team of researchers from around the world, and it has yielded an unprecedented amount of valuable data. If the participating countries could cooperate this successfully on political, environmental, and economic policies, the world would be a far better place. It is not, however, the embodiment of perfection. No project of this type (statistical estimation of a cognitive ability) and scope (conducted on an international scale in multiple languages) could be. No one who understands the nature of the project would claim otherwise.

The PISA exercise has been conducted six times since 2000. For the US, the reading results can be described in the terms that might be used for an underachieving child: passing grade, should be doing better, little change from previous years. On every round, the US has scored close to the average for the core OECD countries, lower than countries including Canada, South Korea, Finland, Japan, and Germany. The results gained much wider attention following the release of the 2009 assessment, the first to include Shanghai, which scored highest by large margins in reading, math, and science, the first time any participant had won the triple crown. These results, taken with the high scores for Hong Kong, Singapore, Korea, Taiwan, and Japan, generated a stronger reaction than when the US was only being lapped by countries such as Finland, Canada, and Australia. The concern, of course, is that, by analogy to British football, the US is a Premier League competitor in danger of relegation to a lower division, global economic competition–wise (baseball version: sent to the minors).

The NAEP is an assessment of American students' performance in math, reading, science, and other subjects. It is a congressionally mandated program run by the Department of Education. The test has been given approximately every two years since 1992. The main NAEP data concern the performance of fourth- and eighth-grade children; twelfth graders are also assessed using a somewhat different approach. The test is taken by children from the fifty states, District of Columbia, and Department of Defense–run schools, totaling approximately 139,100 fourth graders and 136,500 eighth graders in 2015. Participating schools and children are chosen to be representative of schools at national, state, and district levels.

Like the PISA, the NAEP assessment is based on a sophisticated theory of reading that specifies the skills that children at each grade level should be able to demonstrate. The assessment yields data about groups, not individuals. Results are scored on a scale from one to five hundred and also presented as the percentage of children whose scores fall at achievement levels termed Below Basic, Basic, Proficient, and Advanced. Like the PISA, the NAEP is an open-book operation: the methods, data sets, and results are described in numerous documents that can be downloaded for free, your tax dollar at work. The NAEP website also provides tools for conducting simple analyses to look at how scores relate to variables such as race/ethnicity, income, public versus private schooling, and gender, among others.

The main NAEP findings discussed extensively in the media include the fact that overall scores have changed very little over the more than twenty years of testing, moving slightly higher (Figure 10.1) within the Basic range. Each time the test results are released, commentaries focus on whether there has been a statistically significant change rather than whether the change is meaningful. Differences of one or two points are often statistically reliable. The fact that such small differences can be detected is a testament to the skill of the test's designers. The fact that only such small differences are observed indicates how little change has occurred. Figure 10.2 shows the percentages of children scoring at the four levels of reading skill. In every year of testing, about two-thirds of the children have scored at the two lowest levels.

These statistics on the state of US children's reading seem pretty grim, but are they accurate? Both the PISA and the NAEP have been the focus of persistent criticism calling into question the validity, interpretation, and relevance of the findings. Concerns have been raised by academics, politicians, educators, and practitioners of new and old media.

Because the NAEP and PISA data are freely available, we can investigate the controversies by exploring the results ourselves. Examining the data also advances a goal that I've emphasized throughout the book: helping readers to hone their own abilities to evaluate claims over disputed territory. What can be done when experts disagree about an important issue? Delving into the technical literature to dissect conflicting claims and counterclaims is reasonable only if that is what one does for a living. A person could, however, increase the strength of their *Really?* reflex. Someone makes a definitive, surprising assertion about what is true. *Really?* Is the assertion likely to be true, given other things we know? Can it be confirmed by another source of evidence? What kinds of evidence are cited in support of the claim and what other kinds must be assessed in order to judge its validity? Claims that depend on the assurances of an authority without supporting evidence warrant an automatic *Really?*, as do claims that are clearly in the service of a political or ideological agenda or commercial enterprise, which are not necessarily false but require checking. Really.

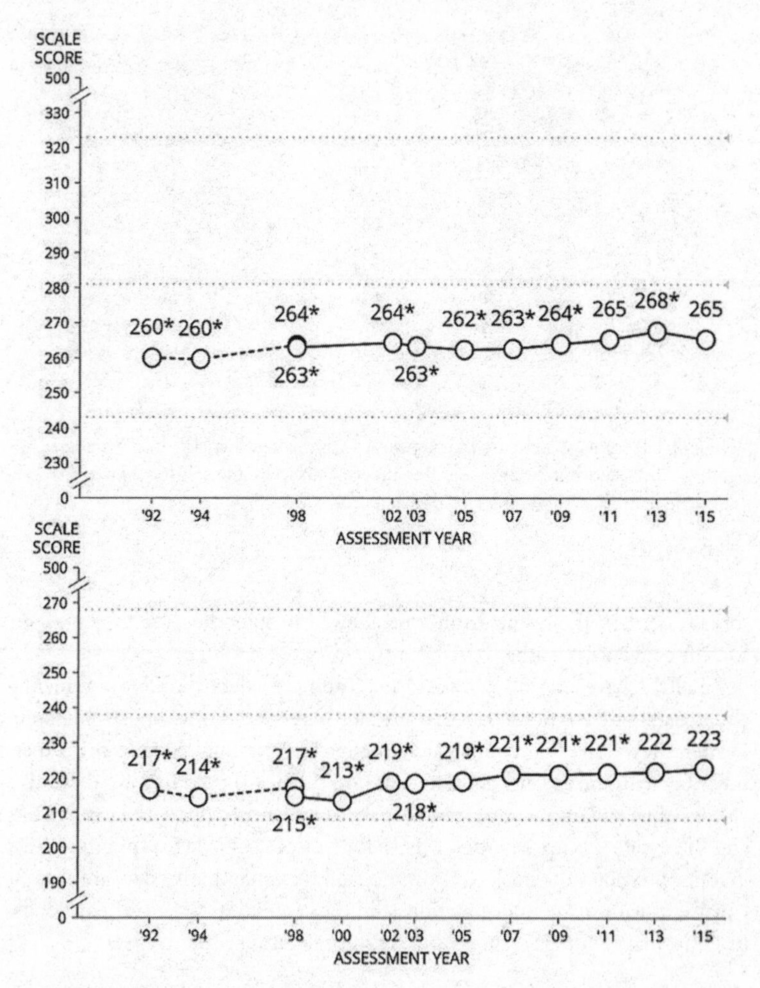

FIGURE 10.1. Overall scores ("trend lines") on the NAEP; top, eighth grade; bottom, fourth grade. *Source:* "National Average Scores," Nation's Report Card, http://bit.ly/2jcpSsY. Asterisk indicates score that differs statistically from 2015.

I will also offer a perspective on the pushback against the tests that places the concerns in the relevant context, the ongoing battles over educational reform and the future of public education. The tests are not a lightning rod because someone thinks the way that item response theory was used to calculate scale scores in the NAEP led to misleading results. It is more likely that, for

Year	below *Basic*	*Basic*	*Proficient*	*Advanced*
2015	31	33	27	9
2013	32	33	27	8
2011	33*	34*	26*	8*
2009	33*	34*	25*	8*
2007	33*	34*	25*	8*
2005	36*	33	24*	8*
2003	37*	32*	24*	8*
2002	36*	32	24*	7*
2000	41*	30*	23*	7*
1998	40*	30*	22*	7*
1998[1]	38*	32	24*	7*
1994[1]	40*	31*	22*	7
1992[1]	38*	34	22*	6*

Year	below *Basic*	*Basic*	*Proficient*	*Advanced*
2015	24	42	31	4
2013	22*	42	32*	4*
2011	24	42	30	3
2009	25	43*	30*	3*
2007	26*	43*	28*	3*
2005	27*	42	28*	3*
2003	26*	42	29*	3*
2002	25	43*	30	3*
1998	27*	41	30	3*
1998[1]	26*	41	31	3*
1994[1]	30*	40*	27*	3*
1992[1]	31*	40	26*	3

100 90 80 70 60 50 40 30 20 10 0 10 20 30 40 50 60
PERCENT

FIGURE 10.2. Percentage of children scoring at each achievement level; left, fourth grade; right, eighth grade. Asterisk indicates score that differs statistically from 2015. *Source:* "Overall Achievement Levels," Nation's Report Card, http://bit.ly/2AL8JAV.

some observers, the results cannot be allowed to stand because they present some inconvenient truths.

Finally, I have one other excuse for leading readers through arguments about data and methodology that are not nearly as fun as boustrophedon. Reader, they aren't that complicated, though they can be made to seem that way. Stay with me to get a good look at how science is used in ways that subvert its value and create unwarranted distrust in the enterprise. The vast NAEP and PISA data sets are easy to scavenge for factoids suitable for spinning. This practice promotes the misleading impression that quantitative data are always consistent with many interpretations and thus equivocal. Scientists can't control how their findings are used, but we can call out flagrant misuses.

Move Along, Nothing to See Here

The NAEP and PISA results are not directly comparable because the tests and target age groups are different. Many questions can be addressed using both assessments, however, allowing each to serve as a cross-check of the other. The evidence from the two sources converges on the same conclusions: overall US reading scores have been flat or nearly flat since testing began. On the PISA, the US performs more poorly than many peer countries, but some subgroups (e.g., higher-income students in Massachusetts) score in the range of the top-tier countries. Disparities related to gender, race/ethnicity, and income are seen in both data sets. Both assessments indicate that a substantial proportion of the students in the tested age ranges have limited reading skills, and few are highly advanced.

Opinions vary, however. The most prominent critic of the tests is Diane Ravitch, the author, blogger, and policy advocate who has dissected them in her recent books, articles, talks, interviews, and social media bulletins. I mainly focus on her influential arguments, which have been carried forward by many others.

The debate over these tests is ultimately about perceptions of the quality of public education in the US and control over educational policy. It is taking place in the context of the educational reforms that were introduced at the national level with the passage of the 2001 NCLB legislation, some of which are being rescinded under the 2015 Every Student Succeeds Act (ESSA). NCLB was a response to concerns about education that began to snowball after the publication of a 1983 report, *A Nation at Risk: The Imperative for Educational Reform*, which warned, "The educational foundations of our society are presently being eroded by a rising tide of mediocrity that threatens our very future as a Nation and a people." Concerns about reading were magnified by two reports from the National Endowment for the Arts (NEA), *Reading at Risk* (2004) and *To Read or Not to Read: A Question of National Consequence* (2007), which offered dire warnings about the future of reading and high culture, supported by numerous graphs and tables said to document decline in interest and engagement in the activity.

NCLB was the leading edge of standards-based education reform, with its now familiar elements, including annual assessments in math and reading, holding schools and teachers accountable for student progress, tying federal contributions to educational funding to quantitative measures of performance, promoting charter schools and vouchers to expand school choice, and allowing commercial enterprises to operate public schools.

Ravitch's own story is well known: a strong supporter of these initiatives who had a hand in implementing them as an education official in the George H. W. Bush and Bill Clinton administrations, she eventually repudiated them as ineffective and part of a conspiracy to undermine public education, becoming a vocal advocate for tossing them out. For Ravitch, the reforms resulted from a toxic combination of misinformation about the state of US education and the opening of the massive but underdeveloped educational market to commercial interests. Parts of her argument are compelling and important. It is true, as she has argued, that concerns about the quality of American education were intensified by high-profile "white paper" reports of dubious quality. The reports on the state of reading by the National Endowment for the Arts were particularly weak. Ravitch is also correct that high-stakes testing has been a disaster, and her critiques helped bring about the apparent elimination of the most onerous requirements under ESSA, although they were retained for the lowest-performing schools. Good riddance, if that is indeed ESSA's effect.

Antipathy to high-stakes testing does not explain the attacks on NAEP and PISA, which have no connection to that testing policy. They are informational exercises, like a census report for reading, math, and science achievement. Moreover, these in-depth assessments are a much-needed corrective to misinformation from other "authoritative" sources such as the NEA. The results, however, get in the way of Ravitch and colleagues' main claim, that the US education system is doing very well, better than ever; educational reforms such as NCLB were therefore both ill-conceived and unnecessary. On this view, the deficiencies that such laws were intended to address were exaggerated to serve political and corporate interests. NAEP and PISA enter the debate because they provide substantive evidence of deficiencies in our educational system. Whereas the high-stakes tests were subject to, indeed created out of, political interests, PISA and NAEP are independent operations not tied to specific policies, political parties, or presidents. If the Ravitch view is correct, something must be wrong with the NAEP and PISA. Hence the many attempts to explain what they "really" show. Ravitch argues as follows:

A. The American educational system is successful because children learn far more than in earlier eras. Students are studying and mastering far more difficult topics in science and math than did their peers forty or fifty years ago. Expectations are higher, and curricula are more challenging. A student who excelled at middle school math and science in the 1970s would be a struggling student in the modern era.

B. The system is working well for most students. High school graduation rates are at their highest level ever, and more people are entering college.

C. The assessments have little real-world relevance. They are not strongly predictive of economic growth; the US has always produced mediocre scores but is still the world's largest economy.

D. On the NAEP, reading scores in both fourth and eighth grade have improved slowly, steadily, and significantly since 1992 for almost every group of students.

E. Both the NAEP and PISA underestimate true reading levels because they are low-stakes tests that do not trigger immediate consequences. Because there is no incentive to do well, motivation to excel is low for the child, teacher, and school. Hence the scores are not as high as they would be if the tests mattered more.

F. The overall trend lines (Figure 10.1) look almost flat, but this is an artifact. The increase in the group scores but not in the overall scores is an instance of Simpson's paradox, a statistical curiosity explained below.

G. The NAEP achievement-level data (Figure 10.2) look bad, with many more children scoring at the two lower levels than at the two higher ones, but that is because the cutoffs were set too strictly. Ravitch offers her own rescaling of the data, according to which most children score B or C or better.

NAEP	Ravitch
Advanced	A+
Proficient	A/B+
Basic	B or C
Below Basic	D or below

Ravitch concludes, "Don't believe anyone who claims that reading has not improved over the past 20 years. It isn't true."

These arguments bear careful scrutiny. Some claims are correct; some are technically correct but do not support the conclusions; some are misleading because relevant information is omitted; some are contradicted by other research that isn't mentioned. And the implications of Ravitch's analysis of whose scores "count" are rebarbative.

Point A about children learning far more than in previous eras is correct; it is a reflection of the well-known Flynn effect, the steady increase in IQ since the 1930s. People do need to know more, and schools need to teach it. The question is how well they are meeting these demands. The fact that schools would be doing superbly if judged by fifty-year-old standards is irrelevant.

Point B, that high school graduation rates are increasing, is contested because it depends on how general equivalency diploma (GED) holders are counted. The US graduation rate was once the highest among the OECD countries, but it is now below the average. Ravitch's focus on graduation and college-entry numbers ignores more important questions about *why* rates change. Graduation rates are driven by social and economic factors as well as educational ones.

Point C: It is true, unfortunately, that US scores on the PISA and similar assessments have never been in the top tier. Ravitch bases the claim that the results are unrelated to real-world outcomes on results from the First International Mathematics Study (FIMS), a cross-national comparison conducted in 1964. As she wrote, "In my recent book, *Reign of Error*, I quote extensively from a brilliant article by Keith Baker, called 'Are International Tests Worth Anything?,' published in 2007." Baker stated that math scores on that early cross-national assessment were not related to subsequent economic development in the twelve participating countries. Ravitch concludes, with Baker, that the data from international comparisons are not relevant to real-world outcomes and that results from the PISA can also be discounted.

Here are two criteria that can be used to judge whether an expert is a trustworthy informant. First, does the authority apply the same critical standards in assessing evidence whether it favors their view or not? Second, does the authority consider all of the relevant evidence or only evidence consistent with their view?

The Keith article is a four-page document that describes results from a fifty-year-old study for which few details are available in English, in contrast to the massive amounts of publicly accessible data from the PISA and NAEP. It also includes short, dismissive comments about later cross-national studies. Ravitch accepts this author's characterization of his study uncritically, even though it cannot be independently assessed, but then takes issue with detailed aspects of the NAEP and PISA methods and results.

She also ignores the many other studies of the same question. Hanushek and Woessmann analyzed the relationship between academic achievement and economic outcomes using the results from twelve rounds of international tests conducted between 1964 (the FIMS) and the 2003 PISA (the analysis also included data from the FISS, FIRS, SIMS, SISS, SIRS, TIMSS, and PIRLS assessments). The analyses indicated "a close relationship between educational achievement and GDP growth that is remarkably stable across extensive sensitivity analyses of specification, time period, and country samples." Hanushek and others have pursued these issues in depth, and the positive relation has been established beyond reasonable doubt.

Point D: Like the NEA, Ravitch focuses on numerical differences but not on whether they are meaningful. As I've noted, the differences in reading since 1992 are tiny. A five- or eight-point change on a five-hundred-point scale within the Basic range over a twenty-three-year period is small. In discussing the increases reported in 2007, Hanushek (who is a senior fellow at the Hoover Institution at Stanford) noted that the reading results couldn't be taken as an upward trend at that point. "If we could get two points every year for the next 10 years, that would be a lot," he said. "Two points by themselves don't mean anything if that's all that happens." From 2007 to 2013, the fourth-grade scores increased by one point (0.17/year); the eighth-grade scores increased by five (0.8/year).

Point E, about these being low-stakes tests, should trigger a strong *Really?* response. Do students perform *less* well on the NAEP or PISA because they are low-stakes tests? Or do they perform *better* because there is less pressure? Performance on these tests obviously matters a great deal to many people—presidents, education officials, commentators like Ravitch—an attitude that may filter down to the test takers. The stakes are the reputations of their home states and countries. They matter enough to raise concerns that some states have gamed the NAEP by categorizing low-performing students as having disabilities in order to exclude them from testing, and that the same may have occurred in some high-scoring countries in the PISA. The claim that the scores on these assessments are depressed because students have little incentive to do well is hand-waving.

The remaining arguments, F and G, involve two issues. The less important one concerns the settings of the cutoffs for achievement levels. The more important one is inflammatory. It concerns the performance of various groups of children and its impact on scores.

Is the NAEP Graded Too Harshly?

The concern is that the scores needed to be classified as Advanced or Proficient were set too high. The designers of the NAEP established the range of scores that would be categorized as Basic, Advanced, or Proficient. The cutoffs reflect the consensus of an advisory board of experts (on which Ravitch served from 1998 to 2004). Could it really be true that only 3 to 4 percent of US eighth graders are advanced readers? Ravitch thinks not, and so she regraded the categories more leniently. Teachers know that assigning higher grades on a softer curve does not mean that students learned more, but it may be good for self-esteem.

Deciding how to curve these tests should be a trivial issue because it does not change how well children read. Still, the achievement levels are a useful type of summary data, and it could be asked which cutoffs better reflect children's skills. Here the PISA data are helpful. The PISA also groups scores into proficiency levels (six levels rather than four for the NAEP). The tests are different, as are the ages of the test takers, the expectations about reading performance, and the placement of the cutoffs. Two questions can be asked nonetheless. First, how does the US compare to other countries using the PISA categories? The US results were again close to the average of the main OECD countries. A

TABLE 10.1. PISA 2012 Reading

	Below 2	*5–6*
OECD	18	8
US	17	8
Finland	1	13
Canada	11	13
Shanghai	3	25
MA	11	16
CT	13	15
FL	17	6

Note: Percentage of students who scored below achievement level 2 (lowest) and at levels 5–6 (highest). *OECD:* average for OECD countries; *MA:* Massachusetts, *CT:* Connecticut, *FL:* Florida

smaller percentage of US students scored in the upper, 5–6 range and a larger percentage in the lowest, < 2 range than in the higher-achieving countries (Table 10.1). Thus, when compared to other countries using the same test and cutoffs, the US yields fewer high scorers and more low scorers than many peer countries. Education-positive states Massachusetts and Connecticut run with Finland and Canada.

Second, we can use the PISA cutoffs to ask whether the NAEP was curved too strictly. A comparison of the results shows clearly that the NAEP achievement cutoffs are stricter, but not by much:

Eighth graders (NAEP)	4% Advanced	22% Below Basic
Fifteen-year-olds (PISA)	8% advanced levels (5–6)	17% lowest levels (1–2)

Do these assessments greatly distort how well children read, emphasizing the bad over the good? Does the picture change greatly if the cutoffs are jittered slightly? I do not think so but will leave the last word to Diane Ravitch. In a 2005 *New York Times* column titled "Every State Left Behind," she noted that there were large discrepancies between the results of the standardized tests the states were using to meet the NCLB requirement and the NAEP results. The states ensured that their children's scores would be high by dumbing down the tests and fiddling with the cutoffs. Ravitch objected, "It is fair to say that we will not reach [the goal of having the best-educated, hardest-working, best-trained, and most productive workers in the world] if we accept mediocre performance and label it 'proficient.'"

What Are the Causes of Reading Failure?

Where to set the cutoffs for high scorers is a quibble compared to what is happening at the other end of the scale: on every NAEP assessment, about a third of the fourth graders and a quarter of the eighth graders score at the Below Basic level, and about two-thirds scored at Basic or Below Basic. Garrison Keillor was referring to the 27 percent of Minnesota fourth graders who scored Below Basic in 2007 when, having reaffirmed the moral obligation to teach children to read, he added, "That 27 percent are at serious risk of crippling illiteracy is an outrageous scandal." We are not in Lake Wobegon anymore.

Ravitch and colleagues have an explanation for why there are so many low achievers: poverty. The system is working well, except for children from low-income backgrounds.

I recommend following closely here.

On each round the NAEP tests a nationally representative sample of children, meaning that the characteristics of the test takers closely approximate the characteristics of the entire population. So, for example, if *n* percent of the fourth graders in Wisconsin are African American, the sample of African American fourth graders who take the test there should come as close as possible to that value, and similarly for other states and the country as a whole. On each test cycle, the composition of the sample changes in response to demographic changes in the US population.

In step with these demographic shifts, the main changes in NAEP participation have been (1) an increase in the proportion of Hispanic students and a decrease in the proportion of whites (Table 10.2), and (2) an increase in the proportion of children who are eligible for free lunch (Table 10.3). Participation in the National School Lunch Program is used as a proxy for socioeconomic status (SES) (free, reduced-price, ineligible = low, medium, higher SES). It is only a rough index; sociologists evaluate SES using combinations of measures that yield more accurate estimates, as in the PISA.

Here is the crucial bit of accounting: white students score higher overall than Hispanic and black students. Higher percentages of blacks and Hispanics score in the lower achievement range than whites. The scores for lower, medium, and higher SES groups also differ in the corresponding way.

Ravitch correctly notes that scores for all these groups have increased gradually at similar rates. If scores have increased, why is the overall trend line flat? Because the increases are offset by the fact that the proportion of test takers from lower-scoring groups has also increased. That is Simpson's paradox.

For Ravitch, the fact that scores are increasing incrementally shows the educational system is making progress. The fact most students get a C or higher

TABLE 10.2. Percentage of Participants from Three Largest Racial/Ethnic Groups on Each NAEP Testing Round

Year	*White*	*Black*	*Hispanic*
2015	51	15	25
2013	52	15	24
2011	54	15	22
2009	56	16	20
2007	58	16	19
2005	59	16	18
2003	61	17	17
2002	61	17	16
2000	63	17	14
1998	66	15	14
1994	72	17	7
1992	73	17	7

on her grading scale indicates that educators know how to teach children to read. Very high-achieving groups can be found within the large, heterogeneous NAEP sample, such as the 20 percent of Massachusetts fourth graders not eligible for subsidized lunch who score at the Advanced level. These educational successes are taken to show that the low scores must be due to other factors. Ravitch focuses on poverty, which disproportionately affects blacks and Hispanics. The way to solve the literacy problem in the US is by eliminating poverty. Low achievement in reading (or math or science) is therefore an economic and social justice issue, not an educational one.

These arguments raise fundamental questions about the functions of public education.

Gaps in educational opportunity are real. Poverty exists in this country and elsewhere, and eliminating poverty would have positive effects of monumental scope, including better reading outcomes for children. Debate focuses on the causes of and remedies for poverty, not its existence or whether lower poverty levels would be beneficial. In the absence of agreement about how poverty could be reduced to whatever its bare minimum might be, many people engage in efforts large and small to reduce poverty or ameliorate its effects. Ravitch contributes to this effort by reminding her enormous audience of the devastating effects of poverty on education, helping to keep attention focused on the issue and perhaps intensify efforts to find solutions.

TABLE 10.3. Percentage of NAEP Participants Eligible for Free or Reduced-Price Lunch or Ineligible

Year	*Not Eligible*	*Reduced-Price Lunch*	*Free Lunch*
2015	42	5	47
2013	44	5	45
2011	46	5	43
2009	50	6	38
2007	52	6	35
2005	51	7	34
2003	51	8	33
2002	49	8	33
2000	49	7	31
1998	51	7	31

Note: Data not available prior to 1998.

I want to firmly distinguish this issue, the very real economic one, from the ways that poverty enters into arguments about educational achievement and the goals and obligations of public education.

Poverty does not explain the US's mediocre showing in cross-national comparisons. Every PISA assessment includes extensive analyses of the impact of socioeconomic status and its various components on the scores in every country. These analyses show that the US does not perform more poorly than other nations because we have a higher proportion of participants from low-income backgrounds. (The major difference between the US and other OECD countries is that we have a higher proportion of socioeconomically *advantaged* children.) US scores lag behind those of countries such as Canada and Finland at all SES levels, not just at the low end (Table 10.4). Statistically adjusting the PISA 2012 results to remove the effects of SES had little impact on the rankings or the US score. Finally, the impact of SES on reading achievement is not the same in every country. To take an example from close to home, SES varies widely in both the US and Canada, but it is a stronger predictor of educational outcomes in the US. Then secretary of education Arne Duncan drew attention to these contrasting situations:

> The chief reason that U.S. students lag behind their peers in high-performing countries is not their diversity, or the fact that a significant number of public school students come from disadvantaged backgrounds. The problem,

> OECD concludes, is that "socioeconomic disadvantage leads more directly to poor educational performance in the United States than is the case in many other countries." . . . Our schools, in other words, are not doing nearly as much as they could to close achievement gaps. As schoolchildren age in America, they "make less progress each year than children in the best-performing countries," according to the OECD. By contrast, high-performing countries not only dramatically boost student achievement, they do so while closing achievement gaps at the same time. In Finland, there are consistent, strong, and predictable educational outcomes for children regardless of where they go to school.

The poverty hypothesis—that US education is fine but dragged down by high poverty—is not supported by the PISA data. However, it can be kept alive by redefining poverty, as in a 2013 report from the Economic Policy Institute (EPI), a Washington think tank. PISA's measure of SES (the ESCS in Table 10.4) is a composite based on indicators such as "parental education and occupation, the number and type of home possessions that are considered proxies for wealth, and the educational resources available at home." The EPI researchers estimated SES using one measure: number of books in the home. The effect was to categorize a larger proportion of US test takers as "low SES." The authors also made another change: they compared the US to six countries (Canada, Finland, South Korea, France, Germany, United Kingdom). With fewer countries in the comparison and a more favorable, if arbitrary, index of SES, the authors could show that US performance is more heavily affected by low SES, and scores improve if that is taken into account.

Is this analysis convincing? The relevant comparison is to the far more extensive analyses conducted over a many-year period by an international consortium of researchers and by others who study these data. Unfortunately, critics do not have to be correct to create distrust of established results. The Internet is now filled with chatter about the PISA's validity, and an international group of one hundred educators has called on the OECD to halt the PISA exercise because of its "damaging world-wide effects."

A much easier way to improve US performance has been discovered, as reported by the *Charlotte Observer*'s Kay McSpadden:

> [A newly released] NCES [National Center for Education Statistics] report shows that in schools with less than 25 percent poverty rates, American children scored higher in reading than any other children in the world. In. The. World.

TABLE 10.4. PISA 2012 Reading Scores, by Quartiles of the PISA Index of Economic, Social, and Cultural Status (ESCS)

	Lowest			*Highest*
OECD average	**456**	**486**	**509**	**542**
Australia	471	501	530	537
Canada	494	517	534	562
Finland	494	515	535	559
Germany	467	504	535	556
Hong Kong/China	521	542	550	571
Italy	452	480	504	528
Japan	505	532	552	575
South Korea	506	525	546	568
Netherlands	472	503	527	551
New Zealand	458	507	527	571
Shanghai-China	528	561	583	608
Singapore	494	528	554	596
United Kingdom	463	485	514	550
United States	**461**	**480**	**510**	**546**
Connecticut	467	502	546	579
Florida	459	483	489	543
Massachusetts	480	503	545	583

The takeaway is simple. Our middle-class and wealthy public school children are thriving. Poor children are struggling, not because their schools are failing but because they come to school with all the well-documented handicaps that poverty imposes—poor prenatal care, developmental delays, hunger, illness, homelessness, emotional and mental illnesses, and so on.

Really? American children score higher in reading than the rest of the world only if scores from low-income schools are removed from the US data *but not from any other country's.* That works. "Less than 25 percent poverty rates" refers to schools where fewer than 25 percent of the students are eligible for free or reduced price lunch—18 percent of US public schools in the 2012–2013 academic year. The recommended comparison, then, is the wealthier 18 percent of American schools against 100 percent of everyone else's.

Here we reach the crux of the matter. Focusing on the role of poverty in US educational failures draws attention away from other factors that affect outcomes, making it seem as though poverty is the only determinant and that outcomes can only be improved by addressing it. I question the validity of the claims and reject the argument's unstated corollary, which is to release the educational system from responsibility for lower-performing children from low-income strata.

Focusing policy discussions on whether low achievement is due to "poverty" rather than "education" encourages the simplistic view that they are distinct factors. Sociologists and economists attempt to quantify the effects of economic well-being using SES statistics in order to address certain kinds of research questions, such as, What was the impact of 1990s welfare reform on people's economic well-being? SES affects people's lives via various associated conditions that differ in susceptibility to change, many of which are education related: access to quality schools, public libraries, computers, and the Internet, among other things. Educational practices can magnify differences in SES, as when they are predicated on the availability of resources such as computers, Internet access, and supplementary instruction. Education can also act as a "protective" factor against disfavorable circumstances. However, it functions this way less often in the US, which has a low percentage of what the PISA calls "resilient" children, high-achieving readers from low-income backgrounds, compared to other OECD countries. In short, education is deeply implicated in differences in SES, not a neutral bystander.

The focus on SES also creates a "Waiting for Poverty" attitude that outcomes can't change until poverty is eliminated by superhuman effort. I do not advocate focusing on educational policy at the expense of efforts to address poverty. Rather, a narrow focus on poverty draws attention away from effective steps that could be taken in the meantime. Education may not be "the great equalizer," but it can ameliorate the effects of disparities for many people. For example, children are born into impoverished homes, but they are not born into impoverished schools. Yet the schools they attend are worse than those of children who are better off with respect to personnel, physical plant, safety, teaching materials and resources, and much else. Poverty being so difficult to eliminate, we could take a cue from the Finns: "One of the basic principles of Finnish education is that all people must have equal access to high-quality education and training. The same opportunities to education should be available to all citizens irrespective of their ethnic origin, age, wealth or where they live."

I am greatly concerned that, whatever the best intentions of advocates such as Ravitch, however sincere their commitment to alleviating poverty, the emphasis on poverty also serves to relieve the educational establishment of responsibility for educational failures, offering a context for plausible denial. Looking at the achievement-level data (Figure 10.2), can it be said that the system is working well for most people? Who are those people? Going to great lengths to demonstrate that NAEP scores would look much better were it not for the growing number of low scorers, who are mainly lower-income minority children: What does that say about who this educational system is for? "Our middle-class and wealthy children"? The message is particularly disturbing given that as of 2015 more than half of all public school children qualified for free or subsidized lunch. Ravitch and colleagues' arguments seem to provide a rationale for educational redlining: it would be unwise to invest in this area, educationally speaking, because the people are too poor to benefit—which, as in the case of mortgage redlining, makes it more difficult to break the poverty cycle.

More Than Poverty: The Black-White Achievement Gap

The PISA data show that the United States has an "achievement gap" compared to other countries of interest. In the US, as in other countries, SES matters, but so do other factors that SES does not capture. At some point, showing that performance is correlated with a measure of SES yields diminishing returns because we need to get at the underlying causes. These same issues arise in connection with the gaps between groups within the US.

There can't be a serious discussion of how well America reads without a consideration of achievement gaps. The term refers to disparities in performance between groups of individuals in areas such as reading, math, and science. It is mainly used with reference to minority groups—African Americans, Hispanics, Native Americans—compared to whites, but gaps are everywhere. There are gaps associated with income, gaps for first-generation children of immigrants to the US compared to later-generation children, and gaps between geographical regions. Males score below females, and whites score below Asian Americans. My interest is whether what we have learned about reading can help in determining why such gaps exist and what we could do to reduce them.

Although many gaps exist, this discussion will focus on the reading gap for African Americans. Why single out this group? Because this gap has serious negative consequences for a large group of American citizens. Because it is part of a long history of racial inequality in the US. Because it has been the focus of attention from politicians, educators, and economists for many years. Because it has existed for as long as relevant data have been collected, with little change despite efforts from the War on Poverty through No Child Left Behind to Race to the Top. The most important research consideration is that achievement gaps arise from constellations of conditions and circumstances that are not the same for every group. Focusing on a group such as African Americans is warranted and necessary if greater progress is to be made; however, it is essential to recognize that rather than being unique to this population, this gap is part of a much broader phenomenon, documented in the NAEP, PISA, and other studies, that the US education system is working much better for some children than others.

Black-white achievement gaps are seen in reading, math, and science, as indicated by grades, standardized test scores, the kinds of classes students take, high school and college completion rates, and other measures. The reading gap has been thoroughly documented in numerous studies. Figures 10.3 and 10.4 show how the gap looks on the NAEP. A significant disparity has been observed in every year of testing in both grades, with scores for black children lagging about 10 to 20 percent. It is encouraging that the size of the gap has been decreasing slowly in both grades since 1992. However, the severity of the gap is conveyed by the achievement-level data, which show that black children are overrepresented at lower levels and underrepresented at higher ones.

Many factors contribute to this achievement gap, and I cannot consider them all here. My principal goal is to bring reading science into a continuing discussion. The topic is also sensitive for many reasons. One is that "gaps" are generalizations about overall differences between groups. For skills such as reading, the individuals within each group vary a great deal, and the distributions of scores for the groups overlap (as in Figure 10.4). What is said about statistically reliable differences between groups therefore does not apply to all members of the groups and cannot be assumed to apply to a given individual. The same is true of other attention-grabbing group differences, such as those between males and females or people from Eastern and Western cultures. Such groups overlap, but journalists and sometimes scientists write about them as though they do not (e.g., "East Versus West: One Sees Big Picture, Other Is Focused").

I also know that anything that is said about this issue, however well reasoned, backed by evidence, and carefully stated with necessary qualifications attached, can be distorted for political purposes that researchers cannot control. I nonetheless think it's important to pursue the issue here because the achievement gap is real, the differences between groups are substantial, they affect many individuals, and their consequences are severe. By examining the group data appropriately, we can hope to identify factors that place individuals at serious risk for failure or serve a protective function. People—parents, community leaders, educators, politicians, philanthropists—could then make better-informed decisions about how to promote desired outcomes.

It will not be surprising to learn that large-scale socioeconomic studies of the achievement gap point to SES as an important factor but only part of the

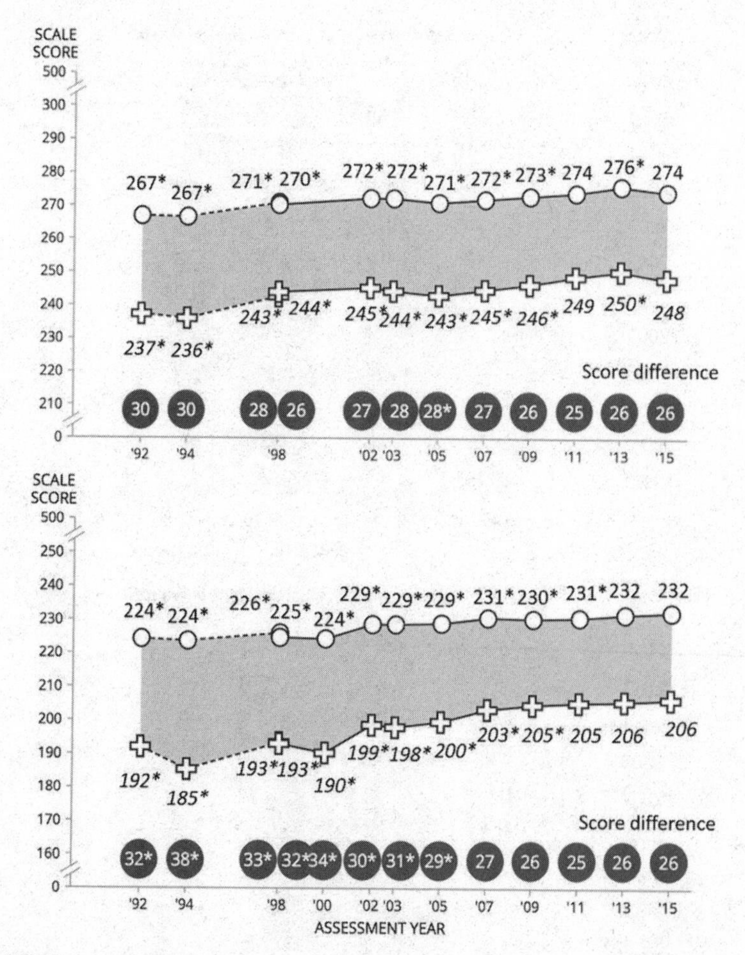

FIGURE 10.3. Achievement gap as measured on the NAEP. Asterisks: scores differ statistically from 2015. Circles, white; crosses, black. Above, eighth-grade; below, fourth-grade. *Source:* "National Score Gaps," Nation's Report Card, http://www.nationsreportcard.gov/reading_math_2015/#reading/gaps?grade=8.

explanation. A good place to look is the reading data for blacks, whites, and Hispanics by income level from Stanford educational sociologist Sean Reardon and his colleagues that appeared in the *New York Times*, a superb example of how complex data can be visualized and explored. The results show that reading performance is related to income and that incomes for whites are generally higher than for blacks or Hispanics. It follows that scores for blacks and

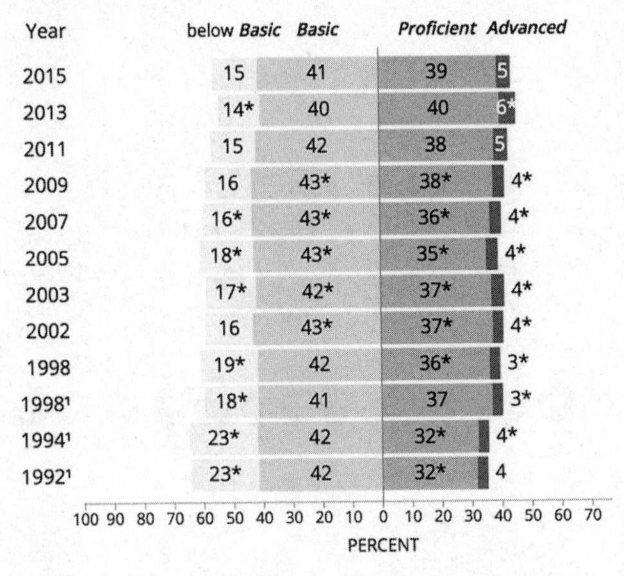

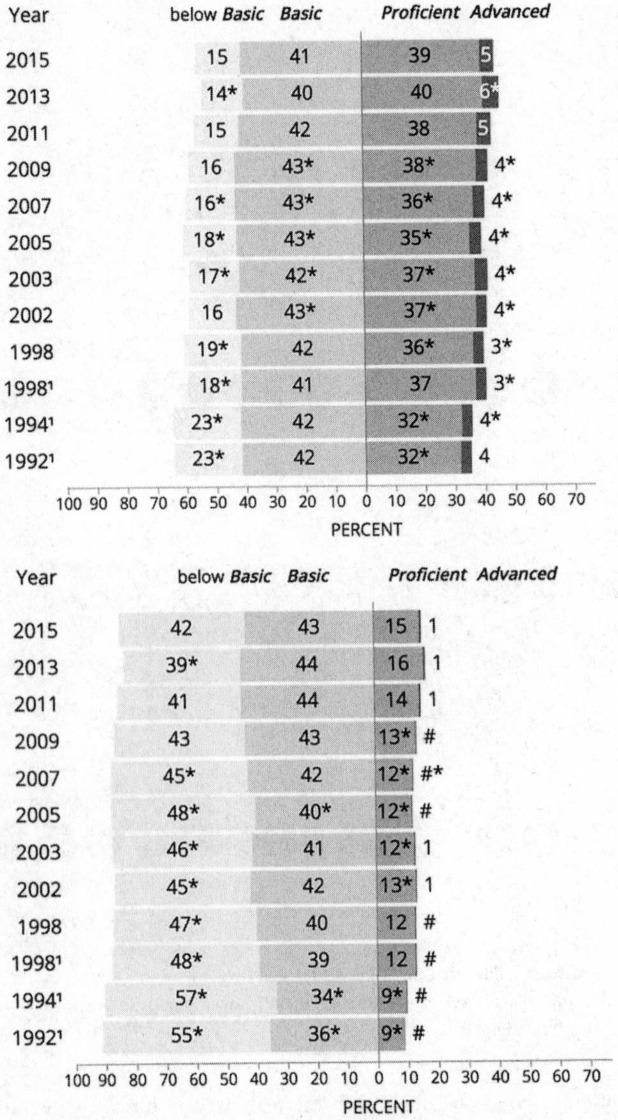

Year	below Basic	Basic	Proficient	Advanced
2015	42	43	15	1
2013	39*	44	16	1
2011	41	44	14	1
2009	43	43	13*	#
2007	45*	42	12*	#*
2005	48*	40*	12*	#
2003	46*	41	12*	1
2002	45*	42	13*	1
1998	47*	40	12	#
1998¹	48*	39	12	#
1994¹	57*	34*	9*	#
1992¹	55*	36*	9*	#

100 90 80 70 60 50 40 30 20 10 0 10 20 30 40 50 60 70

PERCENT

FIGURE 10.4. Eighth-grade achievement levels; top, white; bottom, black. *Score differs significantly from 2015. [1]Years before accommodations for students with disabilities and English language learners were permitted. #Score rounds to zero. The pattern is similar for fourth-graders. *Source:* "National Achievement Level Results," Nation's Report Card, http://www.nationsreportcard.gov/reading_math_2015/#reading/acl?grade=8.

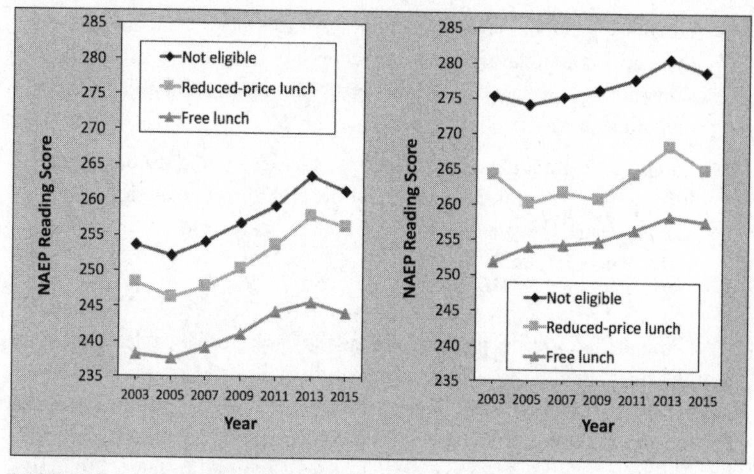

FIGURE 10.5. NAEP scores, eighth grade, by eligibility for free or reduced-price lunch, a proxy for SES. *Left,* black; *right,* white.

Hispanics will tend to be lower than for whites, which is true, even when they attend the same schools. SES matters.

It is a little harder to see that the gaps are also observed in comparisons of blacks, whites, and Hispanics whose SES levels are the same (the *Times* feature also highlights examples of these results). Moreover, as in the comparisons between the US and other countries, the gaps are not limited to the lowest-income individuals. In Figure 10.5 I've plotted NAEP reading results grouped by children's eligibility for free or reduced-price school lunch, the SES proxy. There is a clear SES effect in both groups, but there is also a black-white gap at each level. Studies of African American youth from higher-income households and/or with more highly educated parents show that they do much better in school than African American peers who lack these advantages but not as well as whites from similar backgrounds. These kinds of data suggest that some factor or factors not captured by measures of SES contribute to the group differences.

One possibility that should be obvious from the rest of this book is language. At this point you can probably supply the reasons:

- Reading is closely linked to speech. Children's knowledge of spoken language is the most important determinant of early reading success.
- Children's verbal skills, including vocabulary size and lexical quality, vary greatly.

- Language plays an increasingly important role as the child progresses through the first several years of schooling. Differences in language skill and their impact on reading become more apparent as curricular demands increase.
- Language is not a major focus of large socioeconomic studies, which include very little information about the children's knowledge of spoken language or properties of the speech they are exposed to at home, daycare, or preschool.

We know that vocabulary size and quality are strongly related to early reading achievement. A groundbreaking study by Betty Hart and Todd Risley documented large differences in the amount and variety of adult speech to children and in the types of verbal interactions that took place associated with SES. Children from lower-income families were exposed to fewer utterances and acquired fewer vocabulary words than higher-SES children, and they later showed poorer progress in learning to read. Variability in spoken-language acquisition associated with SES is apparent as early as eighteen months of age.

Like poverty, early spoken-language experience is unquestionably an important factor that demands close attention. Recognition of the importance of the child's spoken-language environment has led to the introduction of language-enrichment programs, which may have beneficial effects. However, the extent to which this factor contributes to the black-white gap is unclear. The main facts to have emerged since Hart and Risley's research are that spoken-language environments are highly variable at all SES levels and across racial/ethnic groups. For example, a study of fifty-three low-income mothers from diverse backgrounds found wide variability in the amount of speech to children, vocabulary, and verbal interaction. Similar variation has been observed in studies of low-income white and Latino families and in middle- and higher-SES families. As Anne Fernald and Adriana Weisleder have noted, "Parents' verbal engagement with their infant is often a better predictor of that child's developing language proficiency than is family SES." Moreover, variability in adult speech to children isn't limited to the home environment. A longitudinal study of eighty-five children from low-income families found that teachers' verbal engagement with preschool children was also a significant predictor of reading development.

These findings indicate that children's spoken-language environments differ substantially, with direct effects on reading. However, this variability is a general phenomenon rather than limited to lower SES minorities. The black-white achievement gap seems to involve one or more additional factors that clearly differ between these groups. Here another aspect of the spoken-language environment, dialect, may be relevant.

Most African American children and their families routinely speak some amount of African American English (AAE), a minority dialect that overlaps with the mainstream dialect (called mainstream American English [MAE] or "standard" American English) but also differs from it in many ways. Children who speak AAE in their homes and communities also have to accommodate the mainstream dialect used in school. The situation is different for white children, who are highly likely to use the mainstream dialect at both home and school.

The question then is whether these differences in language experience contribute to the achievement gap in reading and the ability to benefit from schooling in general. The question is not about the linguistic validity of AAE, which has been established beyond doubt. Nor is it about the characteristics of AAE speakers or their culture. Rather, the question arises because of a specific sociological circumstance: a child's home dialect can be the same as the school dialect or differ. Leaving other factors aside, children who speak a minority dialect such as AAE in the home are at risk for performing more poorly when compared to children who speak the mainstream dialect. The minority dialect speaker faces the task of gaining greater facility with the mainstream dialect while also learning school skills such as reading, writing, and arithmetic. These additional linguistic demands do not arise for monolingual speakers of the mainstream dialect. Although task requirements are intrinsically more difficult for one group compared to the other, both are given the same amount of time and assessed against the same standards. Part of the "achievement gap" is therefore *built in*, a product of these circumstances.

It is easy to see that a child who speaks a language other than English will have difficulty in a classroom where English is the language of both instruction and the books they are learning to read. Speaking a minority dialect of English poses some of the same difficulties. As a reading researcher, I have focused on how differences between the dialects alter learning to read, making it more complex for bidialectal children than for speakers who only know MAE. To take one example, the task of learning how the written code relates to speech is more difficult because the pronunciations of many words differ across dialects. This occurs for several reasons; one is that AAE speakers can optionally drop final consonants under some conditions, such as GOLD (rhymes with BOWL in AAE), END (rhymes with PEN), and DRIFT (rhymes with CLIFF). Say a teacher explains that the word "gold" is spelled "gee-oh-ell-dee," writing out the letters. For an MAE speaking child, this is a lesson about the alphabetic principle and the correspondences between four graphemes and four phonemes.

The message for an AAE-speaking child who pronounces the word "gole" is more complex. In the mainstream dialect, deleting a phoneme from a word usually leaves behind either a completely different word (e.g., "bust"→"bus") or a nonword (e.g., "dust" → "dus"). For an AAE speaker, language works differently because deleting the final phoneme more often yields an alternative pronunciation of the same word. The child also has to learn that the final letter in words such as GOLD either maps onto a phoneme or is silent, depending on the dialect. The impact of such pronunciation differences is to take a spelling system that already lacks transparency and make it even more opaque for African American readers.

Like children who speak a different language in the home, the speakers of a minority dialect must learn alternatives to words and expressions they already know. For example, both MAE- and AAE-speaking five-year-olds use inflectional morphology to mark tense and noun-verb agreement, but the systems are somewhat different. AAE speakers need to learn the MAE versions because they will be encountering sentences such as "There is no lake at Camp Green Lake" and "Grace was a girl who loved stories," which employ grammatical constructions that are obligatory in MAE but usually expressed by other means in AAE. Both bilingual and bidialectal speakers also have to learn when, where, and with whom they should speak each code and gain facility in switching between them.

Acquiring this additional linguistic expertise has advantages but also costs. Bilingual children in the early grades generally perform more poorly in each language in areas such as vocabulary and reading compared to monolingual speakers because their experience is split between the two. When an English-language learner falls behind, it is attributed to the fact that they are still learning school English. A speaker of a minority dialect such as AAE may fall behind because they too are still learning school English, but that is rarely how they are perceived.

Like low SES, speaking a minority dialect is a risk factor whose impact depends on other conditions, yielding a range of outcomes. African American children's progress in learning to read is strongly related to their knowledge of MAE because books are written in it. Greater usage of AAE is usually negatively related to reading progress because the children have weaker knowledge of MAE. However, that isn't always the case. For verbally proficient children, speaking AAE is a nonissue because they are also proficient in MAE. Moreover, under favorable conditions children can rapidly accommodate the second dialect. When other conditions are less favorable (e.g., because of limited learning opportunities), the additional demands associated with using two dialects have greater impact. Low SES is a highly visible risk factor and the focus of much attention, as is being an English-language learner. Dialect is a subtler concept susceptible to misinterpretation, and minority dialects have been the focus of political and cultural controversy. The impact of dialect differences on children's opportunities to benefit from schooling is underrecognized, and little discussion has focused on ameliorating or obviating the effects.

The evidence that speaking a minority dialect is a risk factor for low reading achievement has been accumulating over the last decade. The findings are particularly sensitive because of the history of deep prejudice and divisive controversies about dialect. I must assert again as strongly as possible that this issue is not about the linguistic validity of the dialect, other cultural differences, or the capacities of the child. It is about circumstances. I have used computational models of word reading to demonstrate the point. The models can be used to examine how language background affects learning to read, independently of factors such as SES and vocabulary. The models I described earlier were analogous to MAE speakers who use the same phonological codes in speaking and in reading words aloud. My colleagues and I later explored a hypothetical condition in which AAE phonology was used in both speaking and reading. No problem: performance was like the MAE models, which shows that properties of AAE are not an obstacle. We then looked at the bidialectal case: the model was taught AAE phonology for speech (the home dialect) but had to generate MAE phonology for reading (the school dialect). About half the words had different pronunciations in the two dialects. Having two pronunciations made it harder to learn the spelling-sound mappings for these words. The same thing happened when MAE was treated as the home dialect and AAE the school one, something that does not occur in the real world but indicates that MAE speakers would inherit the "gap" if it did. In short, learning the spelling-sound mappings was harder when the home and school pronunciations differed. We also found that these effects could be greatly reduced by providing contextual cues that reliably indicated whether AAE or MAE was required.

Dialect differences complicate learning to read, but the effects can be modulated by SES, parental education, classroom experiences, personal characteristics, and other factors. Language background is important, but outcomes are determined by multiple factors.

If familiarity with the school dialect is important, what are the educational and sociocultural implications? Here is one view. Knowledge of MAE is strongly related to early reading achievement. Given the opportunity, children who speak AAE in the home could certainly gain familiarity with the mainstream dialect before they start school. Young children are remarkable language learners and could absorb both dialects with sufficient experience, as in preschool settings with staff who are native speakers of the different dialects. Gaining additional familiarity with MAE does not require extinguishing AAE; children can become bidialectal. We could also look at how similar issues are addressed in other countries. Both Australia and Canada have Second Dialect Acquisition programs that are designed to increase knowledge of the mainstream English dialect among low-SES speakers of minority dialects. Finally, it would be important to increase teachers' awareness that lack of fluency in the mainstream dialect is a significant risk factor for poor reading. Preservice teacher education rarely includes instruction about dialect and other basic linguistic concepts. This *knowledge gap* could be closed, which could also rectify negative perceptions of AAE and AAE speakers that contribute to lower achievement expectations.

Here is a second view. The differences between home and school dialects place AAE speakers at an intrinsic disadvantage that is overlooked and unaddressed. Having to learn two dialects in order to succeed in school is another example of having to be twice as good to succeed. Compelling children to learn MAE—or even pointing out that differences between AAE and MAE affect learning—can be taken as pathologizing AAE as "bad" English, intrinsically inferior to MAE, linguistic evidence notwithstanding. The concern is certainly valid given the history of prejudiced attitudes toward black English.

Schools of education already emphasize the importance of recognizing, validating, and accommodating cultural variability, as David Steiner and Susan Rozen found in their study of teacher education. After listing several of the books used in courses (e.g., *Dream Keepers*, by Gloria Ladson-Billings; *White Teacher*, by Vivian Paley), they observed, "Often powerful, provocative, and rightly disturbing, these texts largely share and promote a particular argument about education: teachers should champion the particular voices and experiences of repressed minorities, engaging in what Ladson-Billings calls culturally relevant teaching. The alternative viewpoint, namely that teachers should focus on assuring that all students master universally valuable knowledge, is underrepresented. We found only one course syllabus, in one program, that offered readings presenting this countervailing view."

Greater understanding of language variation is badly needed, including the fact that which dialect functions as the "standard" is determined by social, historical, and economic factors, not linguistic superiority. Once the linguistic integrity and cultural importance of a minority dialect such as AAE are recognized, the question is what to teach. AAE is an oral dialect. Texts are written in MAE. The first view was predicated on the observation that being able to read is crucial for mental, physical, and economic well-being. For AAE speakers under existing conditions, that requires gaining facility with MAE. From this perspective, the goal should be to enable all children to gain these skills, and I would start by changing teacher training, focusing preschools on language exposure, and modifying curricula and assessment practices to reflect differences in language background with particular attention to removing casual but pernicious biases and unfair comparisons. Much attention has recently focused on the role of personal traits such as self-motivation, tenacity, perseverance, and "grit" in achievement. I suggest instead that engagement in a task such as learning to read depends on success, and that failure is a powerful disincentive, especially when it is in part built in.

The alternative view is that the disadvantages associated with using AAE are due to social and cultural conventions that are unjust—dialect privilege, if you will. As self-identified agents of social justice and cultural change, educators' goal should be to create alternatives to a biased system rather than perpetuate it. Such alternatives might build on the additional ways of acquiring information made possible by digital technologies. Watching a video or listening to a recording can lessen dependence on reading. As pedagogical theorist Dawnene Hassett has written, "Using a theoretical approach that combines poststructural theories of power and knowledge with sociocultural theories of literacy, [my] analysis challenges traditional alphabetic print literacy as a political and historical sign of the times, and reframes educational reasoning about 'appropriate' early reading instruction in terms of new technologies, changing texts, and sociocultural forms of literacies." In other words, if traditional reading is an obstacle, consider other types of "literacy," including ones that do not involve print. Rather than reduce the achievement gap, work around it.

The evidence that the gap is to some extent "built in" is certainly consistent with some of these observations. However, advocating alternative forms of literacy seems like poor advice given that print remains an essential medium, and reading skill (the traditional kind) continues to be a prerequisite for engagement with major institutions that greatly affect quality of life. Are alternative literacies a means to empower a minority population or to ensure their disenfranchisement? A bleeding-edge theory that creates additional barriers to print literacy or encourages opting out does not look like progress to me, a higher-SES white person who has benefitted from the traditional concept of literacy. Perhaps these options would be judged differently by parents for whom it has been an obstacle. More likely, they would not have a say in the matter because they would be unaware that this educational theory was being tried out on their children.

Where Does Education Matter Most?

Although education matters for all, it especially matters for children subject to risk factors such as low SES or speaking a minority dialect. A higher-SES background, for example—truly higher, not merely above the subsidized-lunch cutoff—comes with numerous conditions that are conducive to success in school and in life. Quality of education matters, but less so than in unfavorable environments. I think this point is obvious, but now it is supported by interesting new evidence.

I mentioned before that genetic influences on reading are larger for high-IQ children than for low-IQ children, as indexed by the heritability statistic derived from twin studies. SES has analogous effects. For higher-SES individuals, heritability of reading-related skills is high, whereas environmental influences are low. When environmental influences are consistently positive, the genetic component accounts for much of the observed variability in children's reading. For lower-SES individuals, the effects are reversed: the environment accounts for much more of the variation in reading and school success than the genetic component.

This evidence puts Ravitch and colleagues' arguments about the health of US education in an even worse light. The US education system works well, just not for lower-SES students, it's said. Ravitch and friends would take credit for successes in the segment of the population where educational quality has less impact and relinquish responsibility for the children for whom educational quality matters most. Rather than focusing narrowly on the undeniably large role of poverty in poor achievement, we might also focus on the undeniably large role that education could play in improving outcomes.

I will close with Diane Ravitch's admonition: "Don't believe anyone who claims that reading has not improved over the past 20 years. It isn't true." She is correct. Reading *has* improved. Just not enough for enough people.

CHAPTER 11

The Two Cultures of Science and Education

AMERICAN EDUCATORS HAVE NEVER BEEN able to settle on how to teach children to read. Horace Mann and his schoolmasters disagreed about it back in the 1840s. Mann thought it wrong to focus on the alphabet, describing letters as "skeleton-shaped, bloodless, ghostly apparitions" because they lack intrinsic meaning. He encouraged teaching children to "read whole words at a time—a lesson that would be like an excursion to the fields of Elysium" compared to the "phonic" practices that his teachers favored. The pendulum has swung back and forth ever since. Mann's tone—authoritative assertion coupled with contempt for other views—is characteristic of much of the subsequent 150 years of debate.

Like the rest of human behavior, reading abilities are shaped by forces that come at us from several directions: cultural (such as the value placed on reading in comparison to other activities), economic (the impact of socioeconomic status on access to books, libraries, the Internet), personal and familial (children's abilities and personalities, parents' education levels), and educational (school funding, quality). Knowing that so many other things matter, I insist on asking another question: How much does the way that reading is taught affect reading achievement? If education matters, we are obligated to examine whether and how much accepted practices could be improved. Better practices that led to more successful outcomes for a greater number of children could also modulate the impact of some of those other factors.

The chronic disagreements about how reading should be taught suggest that we should look closely. Most people know about the reading wars of the past thirty years—the polemical debate about teaching methods that centered on phonics and whole-language approaches. The first decade of the 2000s saw the emergence of a compromise called balanced literacy, said to incorporate the best aspects of both. Balanced literacy allowed educators to declare an end to the increasingly troublesome "wars" without resolving the underlying issues. The solution was to encourage teachers to combine whatever worked best for them and for individual children. In practice, that usually means the whole language approach (for those who came into the field in its heyday), or its methods and principles under other labels.

The unresolved issues about reading education matter because instructional practices make a difference, affecting children's proficiency but also whether they engage in the activity, enjoy it, or assign value to it. The reading wars and their aftermath also brought to light broader issues about education itself: the philosophical and empirical bases of educational theories and practices; relations between education and cognate disciplines such as psychology and neuroscience; the functions of elementary education and how well they are being fulfilled; where change is needed, what to do, and who should decide; who goes into the profession and how they are trained. Disagreements about these issues are why debates about reading continue, despite exhaustion with the topic and pressures to move on.

In this chapter I argue that a major factor contributing to our national underachievement in reading is the culture of education, by which I mean the beliefs and attitudes about how children learn, the role of the teacher, and the educational mission that dominate the schools of education, which are the main pathway into the profession, and the advocacy organizations closely tied to them, such as the National Council of Teachers of English. This culture is an obstacle to improving educational outcomes. What has happened in reading is an especially prominent example of a more general phenomenon.

I want to be clear about the scope and limits of my concerns. The field of education is housed under many institutional structures in academia; for simplicity I will refer to schools of education, which are common on large campuses. Such schools encompass a diverse group of disciplines such as educational policy analysis, educational psychology, rehabilitative psychology, special education, and others. My concerns are specific to the divisions of this academic conglomerate concerned with what used to be called pedagogy: the preparation of educators for positions in K–12 settings, continuing education for in-service teachers, and the development of curricula and instructional methods, tools, and practices. My references to the "culture of education" are shorthand for "the culture of education related to the teaching vocation and its practice." K–12 educators have a variety of roles and titles: they include teachers in regular classrooms, special education teachers, resource teachers, learning coordinators, and others. The jobs differ, but for brevity I will refer to "teachers" or "educational practitioners." Finally, in this context, the term "educator" refers to the people who prepare students for these positions and devise the curricula, materials, and methods used on the job.

My comments about the culture of education may seem harsh to readers who are unfamiliar with the landscape. I know that exceptions to these generalizations exist among the several thousand programs that offer coursework leading to teacher certification. I also know that many of the same concerns are being aired by educators and professional organizations that represent them, as they have been for decades. In the case of reading, however, the extent to which the dominant beliefs and practices continue to be misaligned with children's needs is not adequately recognized, in my view. I must also emphasize that my concerns focus not on teachers—their integrity, commitment, motivation, abilities, effort, sincerity, or intelligence—but rather on what they are taught about child development in general and reading in particular and about the teacher's role. Responsibility rests with the educators who teach the teachers, shaping their expectations about the profession and curating the ideas and methods to which they are exposed. The people who enter the field of education are being underserved by the authorities they have entrusted with their careers.

I also offer this analysis knowing full well that educational outcomes are affected by factors that do not originate in education and are not within its control. However, that is also reason to look closely at whether improving teacher training and classroom practices could make a meaningful difference. Whereas poverty will not be eliminated any time soon, the culture of education could be changed more easily, in principle. Such changes have not occurred because pedagogy is an academic island, cut off from other disciplines and well defended against incursions. The barriers between educational training programs and the related sciences—psychology, neuroscience, cognitive science—are especially well entrenched. Those barriers could be overcome, and, as I will argue, the benefits could be substantial. What is required is ending Education's hegemony over education.

What Are Teachers Taught?

Parents who deliver their children to school on that momentous first day of kindergarten, proudly starting them on the venerable path to education, make a big mistake: they assume that their child's teacher has been taught how to teach reading. They haven't. Thankfully, many teachers are excellent anyway, being dedicated people who worked hard to learn what is effective and why. In America, prospective teachers aren't trained to teach; they are socialized into a view of the teacher's role. Teaching children to read has been superseded by an emphasis on developing literacy, which includes text but also using sound, pictures, video, and other media. Reading as I've construed it is hardly a part of literacy in this sense, its role being comparable to spelling or typing. Yet the reading education part—helping children to become skilled readers, to be motivated to read, and to be able to comprehend and interpret texts for varied purposes—is not easy. Teachers need to know at least as much about how reading works and how children learn as you do from this book, but they are rarely exposed to this information.

The path to becoming a teacher is not like the ones for other professions. As Jal Mehta, a professor at the Harvard Graduate School of Education, has observed,

> Teaching requires a professional model, like we have in medicine, law, engineering, accounting, architecture and many other fields. In these professions, consistency of quality is created less by holding individual practitioners accountable and more by building a body of knowledge, carefully training people in that knowledge, requiring them to show expertise before they become licensed, and then using their professions' standards to guide their work.
>
> By these criteria, American education is a failed profession. There is no widely agreed-upon knowledge base, training is brief or nonexistent, the criteria for passing licensing exams are much lower than in other fields, and there is little continuous professional guidance. It is not surprising, then, that researchers find wide variation in teaching skills across classrooms; in the absence of a system devoted to developing consistent expertise, we have teachers essentially winging it as they go along, with predictably uneven results.

Not only does the culture of education lack a professional model: it has constructed an ideology around the idea that that adopting such a model is inappropriate. The rationale is that education cannot be codified like law or medicine because it is a culturally determined construct that is always in flux (unlike, say, tax law!). What children need to know changes, in accordance with the increases in the complexity of the information and skills relevant to everyday life underlying the Flynn effect. Educational programs and practices fall into obsolescence. Vocational schools once trained students for trades that no longer exist. Children used slide rules and learned to write checks. Their computers ran DOS, and screens didn't respond to being touched or talked at. Being raised with television/cable TV/video games/computers/DVRs/the Internet/laptops-tablets-smartphones has changed how children acquire information. Teachers whose training narrowly focused on what works now will be lost when the next change inevitably occurs. Just as today's students do not learn what was relevant thirty years ago, teachers today cannot teach the way their counterparts did then. That is what is said.

Concerns about low standards for educating teachers led the American Educational Research Association (AERA) to commission a task force to survey the research literature on teacher preparation. AERA is a large organization that brings together educators and researchers representing many viewpoints and disciplines. Regarding the status of courses on teaching methods, the task force's 2005 report noted,

> Across the studies it is clear that the term methods course has evolved from a training environment in which specific strategies are transmitted and practiced or from a text and lecture based environment. The researchers see teaching and learning as very interactive and at times even collaborative. Methods courses are seen as complex and unique sites in which instructors work simultaneously with prospective teachers' beliefs, teaching practices, and creation of identities. A methods course is seldom defined as a class that transmits information about particular methods of instruction and stops with a final exam.

If teachers are not trained to teach, how are they prepared for the profession? The principal function of schools of education is to socialize prospective teachers into an ideology—a set of beliefs and attitudes about children, the nature of education, and the teacher's role. Prospective teachers are exposed to the ideas of a select group of theorists who provide the intellectual foundations for this ideology.

The ideology is derived from a theoretical framework known by labels such as constructivism, sociocultural theory, and the social construction of reality, among others. I will not go into this framework in detail; that is what the Internet is for. To my nonexpert eye it is a pastiche of some prominent but dated psychological research and the postmodern literary criticism satirized by Frederick Crews and pranked by Alan Sokal. Many aspects of the culture of education, including approaches to teacher preparation, curricula, and classroom practices, have developed in accordance with a central precept of this framework: that knowledge—from perception of the physical world to scientific theories to the interpretation of a text—is constructed, by individuals and groups, under the influence of beliefs, experiences, and cultural practices. As one observer put it, "If we accept constructivist theory . . . we have to recognize that there is no such thing as knowledge out there independent of the knower, but only knowledge we construct for ourselves as we learn. Learning is not understanding the true nature of things, nor is it (as Plato suggested) remembering dimly perceived perfect ideas, but rather a personal and social construction of meaning out of the bewildering array of sensations which have no order or structure besides the explanations (and I stress the plural) which we fabricate for them."

This is a deeply skeptical, relativistic stance toward the world—and super popular too. That there are thirteen ways of looking at a blackbird is seen as somehow vitiating questions about whether the blackbird is a Eurasian or New World species, its evolutionary history, territorial range, diet, or breeding habits. The processes by which people interpret visual illusions (see discussion on p. 22) are of less interest than the fact that their interpretations depend on cultural knowledge that is not universally shared.

Much of the air may have gone out of the Theory balloon at the Modern Language Association, but it remains influential in education because it complements existing beliefs about learning. The idea that children learn by being taught came to be seen as fundamentally flawed because learning is the process of constructing knowledge. Cognitive constructivist theories emphasize the individual's role in this process; sociocultural theories emphasize the role of the group.

The emphasis on the individual originated with psychological research on what we now call the interaction of top-down and bottom-up processing. Some classic research by Jerome Bruner dating from the late 1940s showed that subjects' expectations (the top-down part) affected their perception of simple stimuli (from the bottom up). In an early, influential experiment, ten-year-old children were asked to indicate the size of coins and cardboard disks by adjusting a lighted circle. The sizes of the cardboard disk were judged accurately. The sizes of the coins were overestimated, with larger monetary values producing larger overestimates. The overestimates were larger for poor than for middle-class children. The study was picked up by educational theorists as an illustration of the culturally and experientially dependent construction of knowledge. Since each person's experiences differ from everyone else's, so do the meanings they construct. Since learning is the process of constructing meaning, it follows that individuals learn in different ways and thus that no single pedagogical approach can work with all learners.

The sociocultural approach originated with Lev Vygotsky, the Russian developmental psychologist whose 1920s research was rediscovered some fifty years later. His theory of cognitive development held that higher mental functions were social in origin. As applied to education,

> Sociocultural theory describes learning as not simply what happens in the brain of an individual but what happens to the individual in relation to a social context and the multiple forms of interactions with others. This theory has had an impact on teacher education because it helps educators to better understand the possibilities for teaching culturally and linguistically diverse students. This perspective prioritizes helping prospective teachers understand their own cultural practices and those of others, the impact of cultural experiences on teaching and learning, and the value of implementing culturally supportive reading instruction. (Risko et al. 2008)

The impact on teacher education has been enormous. Scholar David Steiner, later New York State commissioner of education and now at Johns Hopkins University, analyzed the syllabi and program descriptions from sixteen schools of education to determine what prospective teachers were being taught. Regarding the topic "foundations of education," Steiner and his coauthor, Susan Rozen, found that the schools offered classes on multiculturalism or cultural diversity but almost no coursework focusing on major debates in contemporary education (such as high-stakes testing, achievement gaps, charter schools, and vouchers). Developing awareness of cultural differences and their impact in the classroom is a central theme in teacher education programs. Summarizing the AERA report, Kate Walsh wrote,

> Teacher educators now view their job as forming the *professional identities* of teachers. They aim to confront and expunge the prejudices of teacher candidates, particularly those related to race, class, language and culture. This improbable feat, not unlike the transformation of Pinocchio from puppet to real boy, is attempted as candidates reveal their feelings and attitudes through abundant in-class dialogue and regular journal writing. Once freed of their errant assumptions, teachers can embark on a lifelong journey of *learning*, distinct from *knowing*, as actual knowledge is perceived by teacher educators as too fluid to be achievable and may even harden into bias.

I have no direct experience with these activities, but as described they are uncomfortably reminiscent of methods used to restructure people's beliefs to bring them into alignment with a dominant political ideology. The assumption that the individual's beliefs are insufficiently advanced, the engagement in extended self-criticism, the public confession and repudiation of errors in thinking: the goals may be more enlightened and the methods kinder, but they share characteristics with other forms of indoctrination. Perhaps the more benign comparison is to 1970s-era sensitivity training and consciousness raising.

I am not taking issue with the need for prospective teachers to broaden their awareness of cultural variation among children attending public schools and in the communities where they are likely to work. This is especially important because the backgrounds of teachers (who are more likely to be white and middle-class) and of the students (who are more likely to be nonwhite and less well-off) will often differ. Sociocultural matters matter, but there are two concerns. One is about the need to dig deeper into how such differences affect the child's ability to benefit from school and how they can be systematically addressed rather than merely acknowledged. The other is that attention to sociocultural issues does not obviate the need to teach reading but is often taken that way.

As preparation for developing children's literacy, candidates are exposed to an eclectic assortment of approaches on the view that no single approach works for every child or every teacher. The goal is for the future teacher to synthesize a personal educational philosophy in which literacy exists in multiple culturally specific forms.

To illustrate, here is a description of the process taken from the syllabus for a class on reading education offered not long ago at a major university:

> Our class readings represent a variety of perspectives on language and literacy development. Many readings will contradict each other: our job is to find places where different perspectives might meet in the best interest of our students.

The requirements for the final paper are described as follows:

> Review current theories, perspectives and research on reading and writing acquisition, language development, cognition, and literacy learning. Consider the impact that cognitive, social, and cultural factors have on literacy learning. Write a short (3–8 page) position statement about literacy education in relation to the major theories of language and literacy development—*and* in relation to your own role as an educator and scholar. This is *your* position on literacy learning which you can take with you anywhere.

According to this view, becoming a teacher is *itself* a constructive process. Modern-day teacher education is the application of the constructivist philosophy to teaching. Children are not taught by instruction; neither are prospective teachers. Children learn via discovery in naturalistic problem-solving contexts; so do teachers.

Given these assumptions, what is the role of the classroom teacher? Since knowledge is actively constructed rather than passively received, the teacher is not seen as an expert who imparts knowledge to a novice. Educational practice has evolved from "the sage on the stage" to "the guide on the side" to "the peer at the rear." Learning occurs via transaction, not transmission. The prototypic learning activity is therefore discovery through activity. Here the classic figure is John Dewey, the philosopher who founded an experimental school at the University of Chicago in 1894: "Education through experience formed the foundation of the Laboratory School curriculum. Students learned practical skills from weaving to woodworking to sculpting. Science was mastered in the garden as well as in the classroom, where sandboxes offered opportunities for individual experiments in landforms and erosion." The teacher's role in this scenario is to guide the discovery/construction of knowledge as a participant in an environment that engages the child's innate curiosity and desire to learn through active engagement in meaningful activities.

Reality Bites

That is the theory. How much of it survives the transition from student of teaching to teacher of students? The realistic part is that teachers are left to discover effective classroom practices because they haven't been taught them. One of the first discoveries is the irrelevance of most of the theory they learned. Some of the concepts are impractical, or don't work, or don't work as well as something else, like instruction. Some of the concepts are so abstract as to resist translation into something a teacher could do with a class of thirty six-year-olds.

Take discovery learning. It is an appealing idea: learning by exploring, knowledge as the final frontier, a many-year voyage to discover strange new worlds and civilizations. What is the evidence that it works, or about the conditions under which it does or does not work, or whether it works better or worse than some other method, or works better in combination with other methods, or for some topics more than others? In 2004, the cognitive psychologist Richard E. Mayer published an article titled "Should There Be a Three-Strikes Rule Against Pure Discovery Learning? The Case for Guided Methods of Instruction," which argued that there was sufficient empirical evidence to reject discovery learning as the preferred instructional method. The research suggested that to be workable, discovery had to be closely coupled to explicit guidance and instruction. Such findings have low status in education, however, and are easily discounted if they conflict with personal experience.

The feasibility of discovery learning is also at issue. Creating the mise-en-scène under which it can occur is a tough proposition. Under favorable conditions, it might be very effective for some types of subject matter (the research literature also includes such demonstrations). But as the principal means to cover, as an example, the laundry list of topics included in the Common Core State Standards for third-grade math or reading—hopeless. The Philosophical Lexicon, a satiric take on philosophers and their work, includes an entry for *deweyite*: "adj. Full of vague and impractical but well-intentioned ideas." The approach probably worked better for Dewey's students, who were weaving, woodworking, gardening, and cooking their meals. Some vestiges of the Dewey ideal can be observed at the present-day Laboratory Schools, where children still learn by doing marvelous things, but not to the exclusion of other types of learning and instruction. That is what a wealthy, modern private school serving a small, elite population of students can achieve. I say this as an alumnus of the Laboratory Schools. I took photographs in the garden for my "unified arts" class with Mr. Erickson, and one summer I was surprised to find some tasty lettuce planted by another class. It is a great school but, like the Dewey original, not a model that is easily replicable.

The fact that new teachers are left to discover what works means that their students are participants in the teacher's personal learning voyage. How much will a child learn about reading or arithmetic in a classroom where the teacher is also a learner? Well, that will depend. Perhaps the teacher is a quick study in a school that provides quality in-service mentoring. Perhaps the child will learn to read regardless of what happens in school because they are being taught at home. Or perhaps the child would benefit from more expert instruction. I question the ethics of a teacher education curriculum predicated on learning on the job. It's an important job. It's an important time in the child's development. The fact that a substantial proportion of fourth graders have already fallen behind, as indicated by the NAEP and other assessments, may be related to the fact that many of them will have had K–3 teachers who were learning on the job. That would be one noteworthy difference between the US and Finland, where longer internships under experienced teachers and continued mentoring are the norm. It may also be related to the high percentage of teachers who leave the field within the first five years.

We know that teachers are underprepared because they say so. As Geraldine Clifford and James Guthrie, two former deans of the Graduate School of Education at the University of California, Berkeley, noted, "Educational courses have always been fair game for skeptics, with practicing teachers among their most intense critics. Everywhere there is anecdotal evidence about heroic instructors who were thrown, 'sink or swim,' into the turbulence of teaching without benefit of formal training or practice teaching, and who survived to brag about it. That society condones such 'practicing on the patients' may be testimony of how little, in fact, it regards children—at least other people's children."

The American approach leaves teachers vulnerable to failure because they are not given tools that would make it easier to succeed at their jobs. Much of the discussion about improving teacher quality focuses on the need to attract "talented" people into the profession. If the teaching profession attracted some of the Ivy League graduates who currently find employment on Wall Street, that would certainly add many bright, ambitious, well-educated people to the profession. But how much better would the people who *already* go into the field perform if their training were adequate? Is there something deficient about people who want to be teachers rather than hedge fund managers or with a system that deposits them at the classroom door underprepared for a difficult job? What matters is how well children are taught, which depends on the teachers' talent but also their teaching ability, a type of expertise that requires adequate training.

Here too Diane Ravitch dissents. For Ravitch it is especially important to eradicate poverty because the quality of teachers cannot be relied upon.

> Poverty, writes Ravitch, is the most important factor contributing to low academic achievement. As for the claim . . . that improving the quality of teachers will lift educational outcomes, Ravitch is skeptical. She raises an eyebrow at the oft-cited research that seems to show that great teachers can produce 18 months of test gains in a year while bad teachers produce only six. "Perhaps such 'great' teachers exist," she writes, "but there is no evidence they exist in great numbers or that they can produce the same feats year after year for every student."

That is the sound of American schoolteachers being thrown under the ideological bus. For Ravitch, the current system establishes the upper limits on what can be expected from teachers. The alternative hypothesis is that the way teachers are educated limits how effective they can be and that the upper limits on what could be accomplished if teacher education were better are not known.

The Cross-Cultural Divide

The culture of education that has rationalized why teachers can't be taught how to teach has also rationalized why science of the type I've described is of marginal interest. Education as a discipline values observation and hard-earned classroom experience, setting up a conflict with science's emphasis on understanding that supersedes personal experience. The deal breaker for educators, however, is that reading science assumes that complex phenomena can be understood by examining their parts. Vygotsky thought that "higher mental functions" such as reading needed to be studied holistically, in all their complexity; studying components of the system (how people read words, say) is seen as artificial and hopelessly reductive, though it is standard practice in mature sciences.

Ensuring basic scientific literacy and familiarity with modern research in relevant areas (reading science, child development, cognition, language, and learning) is therefore not a priority in educating teachers or in designing curricula or innovative uses of technology. The science then has little impact on educational practice. This disconnection has been harmful. Teachers' personal philosophies about reading and learning come to include assumptions that are not consistent with other things we know, resulting in practices that make learning to read harder than it should be. Translating research into practice requires close collaboration between the areas, but the two cultures are so disconnected as to prevent this conversation from taking place.

The literacy problems in this country are not simply a consequence of failing to pay attention to relevant science, and the science cannot solve all of them. Making good use of good science is a fine principle to live by, but how much it can affect outcomes depends on other conditions. It matters more in a country like the US, where many important factors outside the school are not very malleable, than it does in South Korea, where the challenges are of a different sort: educational reforms since the end of the Korean War have resulted in high achievement but also the unhappiest students in the entire 2012 PISA cohort. It also may matter more in the US because ideological battles have had a greater impact on reading education than elsewhere. I am not advocating a mindless scientism here, the attitude that science provides the best solution to every problem. However, in cases such as reading, where there is such a large body of relevant but underutilized evidence, it is realistic to think the science could be extremely helpful.

Science and education occupy different territories in the intellectual world (literally so on many university campuses). The result is that people who are studying the same thing, such as how children learn to read, can nonetheless have little contact. The cultures of education and science are radically different: they have different goals and values, ways of teaching new practitioners, and criteria for evaluating progress. They gather at separate conferences sponsored by parallel professional organizations attended by largely nonoverlapping audiences and publish their scholarly articles in different journals. There are commercial publishing houses that service one audience but not the other. These cross-cultural differences, like many others, are difficult to bridge.

These differences are more than a petty academic turf war: they affect the quality of children's educational experiences via their impact on the setting of policies at the school, district, state, and national levels. The nature of the cultural conflicts and their impact on children are manifestly apparent from what has happened to reading. This history makes it clear that the differences are too deep to be overcome by simply getting well-intentioned people into the same room—once again—to hash out their disputes. In my view, the depth of this divide needs to be more widely recognized in order to open new paths to change.

Pressing the issue also seems necessary because the culture of education is well protected against outside influence. As we've seen, prominent figures in the field do not acknowledge the existence of any problems that could not be solved by eliminating poverty, getting the government out of the classroom, reducing the influence of private enterprise, and letting teachers teach. Nor is there acknowledgment of the connection between the educational establishment's inability to recognize or address the problems in its own house and legislative intervention, including the retrogressive No Child Left Behind (NCLB) Act. Instructional practices might be improved by making better use of existing science, but that requires changes in educational culture and some on the science side as well. At present we have the worst of both worlds: potential benefits of good science that have been left on the table combined with casual misuse of science in support of current practices.

How Science Functions in the Educational Context

Teaching children to read is traditionally the purview of an educational establishment that includes the schools of education that turn out most of the personnel and the government agencies that run school systems, with close linkages to the corporations that sell textbooks, teaching materials, standardized assessments, and, lately, school and classroom management services. Efforts to improve how children are educated naturally focus on working within this educational-industrial complex. One has to ask how well this strategy is working. We keep going back to institutions to solve problems they have helped to create and maintain.

Whereas teaching children to read is the purview of education, we have been able to learn more about how reading works and children learn from science of the type I've described in this book—research on cognition, development, language, and learning, central topics in modern psychology and cognitive science. Research on the neurobiological and genetic bases of behavior is flourishing within newer hybrid fields such as cognitive neuroscience, behavioral and molecular genetics, and developmental neurobiology. Treating this science as one among many sources of insights about practices does not require a high level of scientific literacy, and so teacher education programs do not focus on it. The field instead coalesced around the work of a few major historical figures: John Dewey, Lev Vygotsky, Jerome Bruner, Jean Piaget, and Maria Montessori. Allegiance to great theorists of the past obviates the burden of engaging newer research. These individuals were noted figures in modern psychology, but with fifty to one hundred years of additional research and thought, their main insights have been assimilated, and the field has moved on. In education, however, their work is treated as the source of axiomatic truths.

Whereas science is adversarial and theories compete, on the education side greater effort goes into maintaining a shared belief system. The field has converged on a framework that has considerable intuitive appeal, can be explained fairly easily and thus taught to students who have little background, and gains scientific credibility from the fact that the main ideas originated with important historical figures. Research focuses on expanding the application of these ideas (e.g., educational computer games based on Vygotskyan principles) and on studies and practices that assume their essential correctness. Had this framework been more successful, these concerns might be moot. But the schools are not working well enough, especially in areas such as reading, and these concerns are relevant to understanding why.

This framework/ideology/worldview is so restrictive it even excludes important modern work that originated within the field of education itself. John Bransford is a researcher who has spent most of his career studying how people learn. He trained as a cognitive psychologist and conducted several classic laboratory experiments about fundamental characteristics of human cognition. He soon turned his attention to questions about learning in the broadest sense: What is knowledge? How is it acquired? What is expertise? How is it achieved? Bransford is a remarkably acute observer and creative thinker. He is a scientist who has used a variety of tools in developing, testing, and refining his ideas: laboratory studies, school-based studies, and computer-based learning environments. Bransford recognizes three forms of learning: the implicit type (which I've mainly discussed), learning as it occurs in informal settings such as finding one's way around a new city or talking to a doctor, and learning in formal settings such as schools. In academia he is an esteemed figure. It in no way maligns the work to say that it has had far less impact on education than the dead white guys (and one woman) I've mentioned. Why? As with most of the reading research I have discussed, comprehending Bransford's work requires a substantial amount of background in cognitive science, which makes it difficult to integrate into a teacher education program. His ideas cannot be reduced to aphorisms (like "learning by doing" or Vygotsky's "zone of proximal development").

If it is a mistake to rely on teaching methods from forty or fifty years ago, why doesn't the same reasoning apply to the relevant *science*?

This reliance on an intellectual canon coexists with a readiness to appropriate new research findings that are consistent with existing beliefs and practices. The special role of science—to find out, to the best of our ability, what is true, letting the implications fall where they may—is subverted if it is mined for nuggets, transforming research findings into another form of anecdote. Educators also use our research as a source of novel findings that feed the relentless demand for educational innovation. Educators are incorrigible early adopters. Often this means getting far too carried away far too rapidly with a finding that is interesting but not well established or understood and possibly wrong. Findings from neuroeducation—the offshoot of cognitive neuroscience concerned with reading, math, attention, learning, reasoning, and other topics close to education—are particularly susceptible to misuse, given their novelty and the seductive appeal of brain evidence.

From the perspective of modern studies of cognition, educators' confidence in the reliability of their classroom observations and experiences is baffling. A good teacher has to be a good observer, to be sure. What is learned about each child over the many hours of a school year is used in planning, evaluating, and adjusting teaching activities. Close observation is important to being an effective teacher and yet not very informative about basic questions about how reading works and children learn. What people observe depends on what they already believe. Inferences based on observation are subject to deep-seated biases that required Nobel Prize–caliber research (by Daniel Kahneman, conducted with the late Amos Tversky) to uncover. If what happens in classrooms is too complex to capture using scientific methods, how can it be grasped by the individual at the center of those activities? How much observation and trial and error involving how many students over how many years are required to attain the skills of a master teacher?

The educational worldview takes subjectivity as an existential condition that extends to science itself. But the limitations of subjectivity are among the reasons why we conduct this other, scientific kind of research: to understand the elements of reading and much else that would otherwise be hidden from view and to do so in ways that allow findings to be independently verified.

Science is not infallible or bias-free, but that makes it difficult, not equivalent to personal observation. Elevating a folk psychology about how we read based on intuition and observation to the status of educational dogma does not make it valid.

The Pseudoscience of Reading

These features of the culture of education help to explain what happened in reading. Most of the research I've discussed is discounted as irrelevant. The clearest illustration is the fate of the report of the National Reading Panel (NRP), the most prominent of several reports commissioned by various agencies to review the scientific literature relevant to learning to read and effective pedagogical practices. The NRP began by establishing criteria for including studies in the review. Their criteria excluded research that many educators value: observational studies of individual schools, teachers, classrooms, and children that do not attempt to conform to basic principles of experimental design or data analysis. The report was therefore of little interest to reading educators except as evidence for a scientistic bias at odds with their core values. That justified ignoring its conclusions, which, like the other reports, called for practices they oppose.

Reading scientists think that their job is done when the data have been gathered, analyzed, reported, replicated, extended, meta-analyzed, summarized in reports like the NRP's, and communicated to a broader audience in accessible terms. At each step, however, the research is shadowed by educational experts who reinterpret the work for their audience, teachers and other personnel, mitigating whatever impact it might have had. These lobbyists target the parties that determine what happens in classrooms: teachers, administrators, politicians, and publishers.

Reading educators rely on authorities whose names are not as well known as Dewey or Montessori but who play a similar role. They formulated the strong, confident perspective on how reading works that became the whole-language approach and continues to pervade reading education. The claims were plausible and easily explained to preservice teachers without requiring specialized knowledge. The claims were not updated as evidence accumulated. They were compatible with the educational zeitgeist I've described and contributed to its ascendance. The key figures made a point of communicating directly with teachers, through lectures, workshops, talks at educational conventions, books, and materials direct-marketed to them. In the contemporary era, social media provide a powerful platform. As their ideas gained wide acceptance and their personal stature increased, they acted as the experts to whom educators could turn when documents such as the NRP report challenged their core beliefs.

I will present three case studies about the misuse of science in reading education, chosen because they involve beliefs that have had enormous influence on classroom practice but were backward: what was said to be true of good readers is actually true of poor readers, what was said to be easy to learn turns out to be hard, and what was seen as hard is easier. The point is not to use the hindsight provided by thirty to forty years of additional research to demonstrate how little was known in the old days. People cannot be faulted for having been wrong; in fact, these erroneous claims were good for reading science because they stimulated research that greatly extended what is known. People *should* be faulted, however, for having made definitive claims based on weak evidence, for sticking with them long after they had been contradicted beyond reasonable doubt, and for continuing to market their stories to a trusting but scientifically naive audience. These ideas are now deeply embedded in an educational culture from which they cannot easily be resectioned.

The modern era in research on reading began with the emergence in the late 1960s of a new field, psycholinguistics, that focused on the study of language acquisition and use, drawing on the advances in cognitive psychology, linguistics, and artificial intelligence that underlay the turn away from behaviorism now known as the cognitive revolution. From 1960 to 1972 the locus of the revolution was the Center for Cognitive Studies at Harvard. The center's many affiliates over the years included nearly all of the scientists who were to create the modern study of cognition and then dominate the field for several decades.

The educational implications of the new field were taken up by Frank Smith (who received one of the first PhDs in the area, at Harvard) and Kenneth Goodman, a professor in the College of Education at Wayne State University. They are prominent figures in these case histories, along with Marie Clay, an educator from New Zealand.

Phonics

Opposition to phonics has been a central theme in both Smith and Goodman's work; Table 11.1 provides a sample of relevant quotes. The main arguments are as follows:

1. Phonics is irrelevant because people read for meaning, whereas phonics emphasizes connections between print and sound.
2. Phonics is unworkable because of the properties of written English: too many irregularities.
3. Even the patterns that are consistent are too complicated to teach.

4. Children who learn to read using phonics become poor readers.
5. The drill-and-kill methods used to teach phonics are soul-draining exercises that stifle children's interest in reading (and tedious to teach).

If you've followed the book this far, you know that Smith and Goodman were wrong.

1*. People unquestionably read for meaning; the question is how. Whether they rely on phonology is a question of fact, which was answered using appropriate empirical methods. The answer is, they do. The reason is, they have to, given the deep integration of orthography and phonology in writing systems, in behavior, and in the brain. Many of the major findings are listed in Table 11.2.

2*. The properties of written English do not preclude phonologically based reading. Early theories emphasized that children learn pronunciation rules for most words and memorize pronunciations of the exceptions. Our later models showed that the problem is even simpler once it is seen as an example of statistical learning.

3*. The statistical learning theory shows how the correspondences between spelling and sound can be learned, via a combination of implicit learning and timely instruction.

4*. This is the part they got backward. It is the good readers who make more rapid progress in mastering the mappings between spelling and sound. Children who are able to use this information can recognize words fluently and automatically, allowing them to focus on comprehension. Children who struggle with these mappings must continue laboring at the word level rather than developing comprehension skills and learning from texts.

5*. It is time to divest the phrase "drill and kill" (or "drill and skills" as it is also known) of its magical power to make phonics instruction disappear. The empirical question is, What do good readers know and do? The pedagogical question is, What are the effective, engaging methods for helping children achieve that skill? Teachers (I am one) do not have the option of skipping a topic because we find it hard to teach. Moreover, phonics can be taught without inducing severe print aversion, and the amount of time and effort involved isn't excessive unless a child has a developmental impairment requiring focused intervention.

TABLE 11.1. Smith and Goodman on Phonics

Reading by "phonics" is demonstrably impossible. Ask any computer. (Smith 1973)
To the fluent reader, the alphabetic principle is completely irrelevant. He identifies every word (if he identifies words at all) as an ideogram. (Smith 1973)
Phonics, which means teaching a set of spelling to sound correspondence rules that permit the decoding of written language into speech, just does not work. (Smith 1985)
Phonics is a flat-earth view of the world since it rejects modern science about reading and writing and how they develop. (Goodman 1986)
The worst readers are those who try to sound out unfamiliar words according to the rules of phonics. (Smith 1992)
It has become crystal clear to me—and it has taken about ten years to come to this understanding—that children learn phonics best after they can already read. I am convinced that the reason our good readers are good at phonics is that in their being able to read they can intuitively make sense of phonics. (Smith 1994)
Some studies in reading development are being centered around a narrow and sterile concept of phonemic awareness. All children who learn to understand oral language must be aware of the phonemes (significant perceptual sound units of language) or they could not comprehend speech. (Goodman 1994)

Some questions remain.

Does anyone still believe that phonics is the route to poor reading? Of course. Several generations of teachers learned that this is so. The science is still discounted and can be excluded from the curriculum. Or, more likely, the pro and con sides can be presented as options for the prospective teacher to pick from. Truly studying the conflict—reviewing the opposing claims in depth, guided by instructors with relevant expertise—would be a great learning experience. Presenting the alternatives as equally valid but different realities from which to choose is like providing equal time for the pro- and anti-vaccination sides on cable news.

Isn't balanced literacy the obvious solution? It might be, if it were more than a slogan. By the close of the twentieth century, the amount and variety of research on phonology and reading, combined with dissatisfaction over

TABLE 11.2. Evidence Concerning Phonology and Reading

Dependence of written code on speech
Writing systems evolved to represent sound and meaning
Role of phonological, phonemic awareness in beginning reading, impact on reading skill
Positive impact of explicit phonics instruction on reading outcomes
Impaired representation of phonology associated with poor reading, reading disability; behavioral studies and computational models showing causal connection
Remediation of phonological impairments associated with increases in reading skill
Dyslexia observed in children with phonological impairments due to atypical language development (e.g., speech sound disorder)
Neural integration of orthographic and phonological codes
Automatic activation of phonology in skilled silent reading
Use of phonology in integrating information across saccades
Similar use of phonological information across different types of writing systems
Use of phonology in resolving lexical, syntactic, and semantic ambiguities in sentences
Use of phonology in maintaining information while sentence continues; role in resolving long-distance dependencies between parts
Impact of phonological deficits on reading in brain-injured patients

achievement levels and some excellent coverage of the reading wars in the mass media, made it impossible to maintain the "phonics = reading death" position. Balanced literacy allowed the educational establishment to diffuse the controversy by acknowledging phonics without specifying how practices should change.

Incorporating phonics in a serious way requires addressing some tough questions: how to teach it, how much is enough, how much is too much, how to integrate it with reading and literacy activities, how much to individualize instruction, and so on. Balanced literacy provided little guidance for teachers who thought that phonics was a cause of poor reading and did not know how to teach it. Documents such as the report of the National Reading Panel weren't much help. The panel did a fine job identifying major components of early reading: phonemic awareness, phonics, fluency, vocabulary, and comprehension. The components can always be carved slightly differently, but they covered the major skills and potential obstacles. However, the panel was not charged with making specific curricular recommendations. Translating the panel's findings into effective practices required additional expertise, effort, and evaluation, which were in short supply. That led to blunders.

For educators opposed to phonics, only a nod in that direction was required for curricula and practices to be marketable as a "balanced" program consistent with the recommendations of the NRP. For educators seeking curricular guidance, the NRP report could be read as describing a set of independent building blocks. Of course the components aren't independent; they bootstrap each other. Phonics is necessary but not a skill to be acquired before the child can be exposed to something else. Vocabulary affects comprehension but also the discovery of the components of words, which gains support from phonics instruction, which contributes to fluency, and so on and so on. It's a system of interacting components. Worse, under this mistranslation of the panel report, mastering the component skills can take precedence over reading, the actual goal. A teacher needs considerable expertise to integrate necessary instruction in the components with reading itself, but how to acquire it?

What about Smith and Goodman's own evidence in support of their positions? It did not exist. Smith's influential books were extended arguments about characteristics of reading and learning to read that the author believed must be true, given his analysis. No need to consider the stack of relevant research findings because, as Smith assures readers in the preface to the current edition of *Understanding Reading*, not much has changed, people still disagree as they always will, *c'est la vie*. Smith recycled the same arguments in a book for teachers, *Reading Without Nonsense*, which should have been called *Reading Without Evidence*. Here is his recommendation for how to help a struggling reader: "The first alternative and preference is to skip over the puzzling word. The second alternative is to guess what the unknown word might be. And the final and least preferred alternative is to sound the word out. Phonics, in other words, comes last." The reader is expected take him at his word because he mentions no relevant research.

Goodman took the same approach. What a person learns from the book *Ken Goodman on Reading* is the views of Ken Goodman, trusted authority. Goodman tells teachers what they need to know, in his expert view. He mentions that some studies have contradicted his claims and good-naturedly pooh-poohs them.

As the evidence accumulated, did the scholars adjust their views? In his excellent history of the reading wars, James J. Kim writes, "By the end of the 1970s, Smith's writings reflected the actions of a policy entrepreneur seeking to persuade teachers to reject phonics instruction and to encourage children to use context clues to identify words."

Goodman also appealed directly to teachers, aligning the whole-language approach with teacher empowerment. Kim notes, "Kenneth Goodman asserted that teachers are not relying on gurus and experts to tell them what to do. The whole-language movement, according to Goodman, was generating a knowledge base passed from teacher to teacher in personal contacts, in teacher support groups, and in local conferences. Rather than following the findings of experimental research published in academic journals, Goodman urged scholars to do research that was useful for teachers, and predicted that practitioners will move ahead, with or without this support."

Goodman's prediction that practitioners would adopt whole language regardless of the findings was correct. He was speaking as one of the gurus who encouraged them to do so. The impact was enormous and continues to be felt.

The Psycholinguistic Guessing Game

Goodman also contributed the idea that reading is a "psycholinguistic guessing game," one of the most famous concepts in the study of reading. The guessing game was an extreme take on the roles of top-down and bottom-up processes in reading. Goodman pointed out that people can identify a word such as R_BBIT even with a letter missing. From this demonstration he concluded that skilled reading does not require close attention to the letters. Rather, the reader uses knowledge of the language and the topic to guess the upcoming word. They only have to read enough letters to confirm or disconfirm the guess. Poor readers are poor guessers and thus stuck with identifying letters and words from the bottom-up, which is slow and prone to error. The guessing game idea was adopted as a tenet of whole language (Table 11.3). For teachers, the psycholinguistic guessing game meant that they did not need to develop beginning readers' word-recognition skills, allowing them to focus on "literacy."

Goodman's guessing game theory was grievously wrong.

Beginning readers can often predict the words in texts that have pictures. They can also predict the words in books they have read enough times to memorize, as often happened in whole-language classrooms in which a book was read many times as part of literacy activities tied to it (e.g., reading the book to children, group reading aloud, making a copy of the book, writing about the book using invented spelling, talking about the book, using the book

TABLE 11.3. The Psycholinguistic Guessing Game

The more difficulty a reader has with reading, the more he relies on the visual information; this statement applies to both the fluent reader and the beginner. In each case, the cause of the difficulty is inability to make full use of syntactic and semantic redundancy, of nonvisual sources of information. (Smith 1971)

Accuracy, correctly naming or identifying each word or word part in a graphic sequence, is not necessary for effective reading since the reader can get the meaning without accurate word identification. Furthermore, readers who strive for accuracy are likely to be inefficient. (Goodman 1974)

The art of becoming a fluent reader lies in learning to rely less on information from the eyes. (Smith 1975)

Skill in reading involves not greater precision but more accurate first guesses based on better sampling techniques, greater control over language structure, broadened experiences, and increased conceptual development. (Goodman 1976)

Guessing in the way I have described it is not just a preferred strategy for beginner and fluent readers alike; it is the most efficient manner in which to read and learn to read. (Smith 1979)

Early in our miscue research we concluded . . . that a story is easier to read than a page, a page is easier to read than a paragraph, a paragraph is easier than a sentence, a sentence is easier than a word, and a word is easier than a letter. Our research continues to support this conclusion and we believe it to be true. (Goodman & Goodman 1981)

In efficient rapid word perception, the reader relies mostly on the sentence and its meaning and some selected features of the forms of words. (Clay 1991)

[Beginning readers] need to use their knowledge of how the world works; the possible meanings of the text; the sentence structure; the importance of order of ideas, or words, or letters; the size of words or letters; special features of sound, shape, and layout; and special knowledge from past literary experiences before they resort to left to right sounding out of chunks or letter clusters, or in the last resort, single letters. (Clay 1998)

to teach concepts such as opposites or colors). This only created the illusion of reading, however, because what was learned from memorizing one book did not transfer to the next.

Goodman's theory focused on the predictability of words, which is assessed using a technique called the Cloze procedure. In one variant, subjects are asked to guess the last word in a sentence like "He mailed the letter without a _______" and "Jill looked back through the open ______." In another variant they are asked to guess every *n*th word in a text, say the fifth or sixth. Measured in this way, the predictability of adult-level texts is low. Philip Gough (the "simple view") found that adult skilled readers could correctly predict about 40 percent of the function words but only 10 percent of the content words in undemanding articles from *Reader's Digest*. He also reported the results from an informal experiment conducted with a generous colleague who was asked to guess, one by one, the first one hundred words from nine texts from several genres (he guessed a word, was told the correct answer, and went on to the next one). Accuracy ranged from 20 to 39 percent, in line with other research. The subject managed to correctly predict 74 percent of the words in one text—a textbook he had coauthored. A 1995 review concluded, "It is often incorrectly assumed that predicting upcoming words in sentences is a relatively easy and highly accurate activity. Actually, many different empirical studies have indicated that naturalistic text is not that predictable."

The coup de grâce, however, was the simple finding that it is poor readers who rely more on context to predict the words in texts, not good readers (see "Evidence from Children," pp. 129–131). Here again the educational theorists' claim was backward. Guessing provides no advantage if the reader is already good at recognizing four or five words per second. People are forced to guess if their word-recognition skills are poor.

The conclusions from this body of research can be summarized by rephrasing Smith's statement from Table 11.3: "The more difficulty a reader has with reading, the more he relies on the *context*; this statement applies to both the fluent reader and the beginner. In each case, the cause of the difficulty is inability to *read words accurately, rapidly and automatically*."

Teachers who jumped on the whole-language bandwagon, believing the guessing game story and not having any reason to doubt it, proceeded to engage in literacy activities that inadvertently encouraged their students to read the way that poor readers do. The product of being taught to guess and sample rather than decode and integrate is familiar: skimming. Children might manage the literacy demands of K–3 classrooms reading this way, but the text hits the fan in fourth grade, when they are expected to begin reading materials of greater complexity and topical breadth.

In short, Goodman's psycholinguistic guessing game was a good theory—of poor reading. Several questions remain.

Does anyone still believe it? For reading scientists, the answer has been no for about thirty years. On the education side: of course. It is part of the knowledge base passed from teacher to teacher in support groups and "communities of practice." It is still presented in textbooks as one of the approaches from which teachers synthesize their personal reading philosophies. It's right there in the National Council of Teachers of English's position statement on what is known about reading, with citations of the scientifically but not educationally discredited Goodman and Smith claims.

What about Goodman's own evidence supporting the theory? There wasn't any. The original guessing-game article presented a theory said to be based on a suggestion by Noam Chomsky. Goodman later presented evidence from children's errors in reading aloud (which he called miscue analysis). Say the upcoming word is DOWN. When a good reader makes an incorrect prediction, he said, the erroneous word is usually very close to the meaning of the actual word (e.g., BELOW), which can be easily corrected or ignored. Poor readers were said to produce errors that sound somewhat like the correct word but are semantically unrelated, as in misreading DOWN as DOG, which disrupts comprehension. Neither claim held up. By 1980 numerous studies had shown that even good readers' accuracy in predicting upcoming words is low and that good and poor readers make similar types of errors.

As the evidence accumulated, did views change? The theory was one of the foundations of the new whole-language method, which was adopted remarkably quickly. As the evidence contradicting the claim mounted, Goodman and his followers doubled down, offering more extreme versions. The 1981 statement (Table 11.3) is hyperbole—a story is easier to read than a letter?—but his audience took him seriously.

Little of what I have said about these first two cases is new; I've mainly updated observations from the 1980s and 1990s. The persistence of the ideas despite the mass of evidence against them is most striking at this point. In normal science, a theory whose assumptions and predictions have been repeatedly contradicted by data will be discarded. That is what happened to the Smith and Goodman theories within reading science, but in education they are theoretical zombies that cannot be stopped by conventional weapons such as empirical disconfirmation, leaving them free to roam the educational landscape.

Basic Skills and Comprehension

Everyone agrees that the goal of reading is comprehension, in the broad sense of understanding and interpreting written language for varied purposes. Theorizing about alternative literacies and the social construction of meaning notwithstanding, a primary goal is for people to be able to understand the kinds of texts they will encounter in school, on a college admissions exam, in college-level reading assignments, on a job, or on their nightstand. In the reading wars period, the contentious issue concerned the balance between basic skills and comprehension. It's a telling reflection that whole-language proponents viewed "skills" and "comprehension" as competing approaches. Ignoring the extreme rhetoric, it seems clear that basic skills are an important ingredient in comprehending texts. Beyond this observation, there are two contradictory views.

Reading educators assumed that basic skills are relatively easy to acquire, but comprehension is hard. Teaching basic skills is mostly a matter of providing a literacy-rich environment with activities that engage and stimulate the child. Learning to read is like learning a spoken language, neither of which depends on explicit instruction, unless there is a problem. In a highly motivating environment, full of authentic literature and literacy activities focused on extended, multisensory engagement with a book, the child will discover the mechanics. Discovering how reading works was assumed to have greater value than being taught. The teacher's role was to facilitate literacy, not teach reading. Skills instruction was thought to be toxic in large doses, and so children should be exposed to as little of it as possible.

Comprehension, in contrast, was thought to be hard. Imagine setting yourself the task of identifying all of the types of knowledge and mental operations involved in comprehending a text—a short story, say, or *Best Friends for Frances*, a book for second- and third-grade children. Several reading researchers took up this challenge, yielding intricately detailed theories. Table 11.4 is a representative list of major components. It is long and could easily be expanded. A comprehension theory would then show how these elements function from moment to moment as a text is read. Only bits of this process have been spelled out because of its complexity. Reading is one of highest expressions of human intelligence, and the theories of this advanced skill were extremely complicated.

These theories nurtured the sense that comprehension is an enormously complex process; word recognition looks simple in comparison. They also suggested new pedagogical possibilities: perhaps learners could benefit from instruction targeting these components, for example, teaching children about "inferencing" or building a "story grammar." Teaching reading comprehension became a focus of research, pedagogical innovation, and classroom activity.

On the science side, the story is *the exact opposite*. Basic skills are difficult to acquire (mainly because of the partial and abstract way that writing systems represent spoken language) and thus the area where instruction matters most. For the beginning reader, comprehension does not require instruction

TABLE 11.4. Components of Text Comprehension

Theories of text comprehension attempt to specify all the types of knowledge and cognitive processes involved. The theories divide things differently but usually include the following:

Lexical decoding: components of words (graphemes, phonemes, syllables, morphemes); relations between print and speech.

Words: vocabulary, grammatical categories (noun, verb, adverb, etc.); syntactic patterns associated with particular words.

Syntax: constituents (noun, verb, and prepositional phrases; clauses); word and constituent order.

Textbase: propositional structure; referring expressions such as anaphora; bridging inferences that connect explicit propositions.

Situation model: situation conveyed in a text; agents, objects, locations, instruments, etc.; temporal, spatial, causal, intentional dimensions of the situation; inferences that elaborate a text and link to the reader's background knowledge; images and mental simulations of events.

Genre, rhetorical structure: type of genre (e.g., narrative, expository, descriptive); speech act categories (e.g., assertion, question, command, request); reliability, validity of assertions; theme, moral, or point of the text/discourse.

Pragmatics: goals of the author and reader; attitudes and beliefs (humor, sarcasm, irony, etc.).

Adapted from Lesgold & Welch-Ross (2012).

because they already understand speech. Bringing reading comprehension up to that level turns on gaining facility with print: basic skills.

Children whose spoken language is age and school appropriate can acquire basic skills rapidly if given relevant instruction and support. A child who can comprehend a story when it is read to him but has difficulty reading it has a basic skills problem due to inadequate instruction or a learning impairment. Such children may indeed be poor at parsing sentences, understanding the situation being described, or drawing linking inferences because they cannot read the words well enough to get that far, not because they lack an appropriate "text comprehension model."

For children who have acquired good basic skills, comprehension depends on two main factors: knowledge of language (especially "academic" language) and content knowledge. Both types of expertise are acquired by reading what Marilyn Adams has called challenging texts—ones that include vocabulary and grammatical structures that are relevant to a topic but occur infrequently in casual speech or reading. In reading about dinosaurs, for example, the child learns about the topic but also encounters uncommon words such as PALEONTOLOGIST and EXTINCTION and the concepts and expressions associated with them. Reading skills depend on the sheer amount and variety of reading experience because it is how these types of knowledge are acquired.

Children develop comprehension skills in the course of reading varied texts for varied purposes guided by a teacher whose activities, assignments, and feedback are the principal source of learning (group work in which children provide such feedback for each other is a noisier version of this process). Writing plays an important complementary role. Promoting the development of comprehension through engagement and feedback—traditional teaching activities—is different from teaching the child a theory of text comprehension. Philosophers distinguish between "knowing how" (e.g., to ride a bike) and "knowing that" (e.g., the theory of bicycle motion and balance dynamics). Comprehension is a matter of knowing how, not that.

In short, theorists on the education side also had the instructional demands of acquiring basic skills and comprehension backward. Generations of teachers were then taught that skills come naturally and that comprehension requires extended instruction. That inversion made learning to read more difficult for many children.

Repercussions

That's my account, but how is a reader to know if it is accurate? Well, you could look it up, as Casey Stengel, the manager of the Amazin' Mets, used to say. The copious notes for this chapter are meant to enable that. Here's an additional test. If reading education has been as flawed as I have suggested, there would have been other repercussions, wouldn't there? Here are four.

The Phonics Industry

Phonics never went away; it was outsourced. If the schools were not providing adequate basic skills instruction, concerned parents could try to fill the gap by other means. Supplemental instruction could take place in the home, using phonics workbooks or downloadable worksheets. Software for learning prereading skills, vocabulary, and phonics proliferated, followed by web-based applications and eventually apps. Reading tutors could be hired, or the child could attend a commercial after-school program such as Kumon, which employs an explicit-instruction-with-practice method that American educators abhor.

Outsourcing phonics is a poor alternative to effective classroom instruction. A person can legally raise a child without any demonstrable ability to teach. Some parents will succeed at schooling their children, but it is a hard job, and many will not. Although whole language nourished a large phonics aftermarket, the software has been terrible, covering only what is simplest to teach and not sufficiently engaging to maintain children's interest for long. Phonics programs of marginal educational value were sold on television like Thighmasters. Children could get their phonics from toys that spoke microchip English. A home curriculum can be assembled from materials on the Internet, but it takes considerable expertise. Given the difficulties involved, parents might gladly hand off the task to tutors and learning centers.

All of these options require resources: a parent or other adult who is available to teach the child, speaks the language, understands what needs to be taught, and is able to teach it, as well as the financial resources to pay for the teach-at-home materials, computer, software, Internet connection, private tutor, or learning center. Transferring responsibility for phonics from the school to the home magnifies the impact of socioeconomic differences.

Educators emphasize the importance of the partnership between the school and the home; although it may not be obvious to all parents, that partnership now includes their supplementing classroom instruction. Phonics is on its way to becoming one of the skills, like knowing the multiplication tables or cursive writing, that schools relinquish responsibility for teaching.

Why "What Works"?

I've stated that NCLB was a deeply flawed response to real problems. Let's look at one of the lesser-known provisions: the establishment of the What Works Clearinghouse (WWC). The Department of Education was directed to identify educational programs and practices in reading, math, and other areas whose efficacy had been demonstrated in appropriately designed field tests. The effort was loosely modeled on the movement toward evidence-based practices in medicine. This provision of NCLB was also a response to a real problem: educational programs and practices can be adopted on a massive scale without having been shown to be effective. Whole language was the quintessential example.

Although educational theory changes very slowly, the culture places high value on innovations in educational practice. There is a sense that great advances, not merely incremental changes, can be achieved by developing novel learning activities, environments, and technologies that are better attuned to children's interests and abilities. Rolling out a new educational approach is therefore not like marketing a new drug. A drug will have gone through several stages of clinical trials to determine its efficacy and safety. A drug is not marketed based on the theory about why it should be safe and effective or on the drug developer's assurances; rather, it is subject to empirical tests under controlled conditions using appropriate experimental designs.

Whole language, in contrast, *was* brought to market based on the developers' theory about why it should be safe and effective. The "clinical trials" took place in the schools, as an unregulated experiment on millions of children who did not know they were participants. A small behavioral study of children's reading with forty six- to seven-year-olds is subject to far greater institutional oversight than a new educational method used with millions of children. In the current climate, the pressures for school systems to improve are so intense that new, untested approaches are rolled out very rapidly in the hope that something will stick.

The What Works Clearinghouse was supposed to correct this, but it has not worked well. As with the National Reading Panel report, the sticking point is what should count as evidence. The WWC narrowly focused on randomized control trials (RCTs), the "gold standard" method used in medical research. This excluded not only the qualitative research that the NRP had rejected but almost all research on reading because there are so few RCTs. In pharmaceutical medicine RCTs are funded by the National Institutes of Health and the pharmaceutical industry. Nothing of comparable scope exists on the education side. Moreover, the RCTs that have been conducted are not necessarily convincing because applying the method in educational settings is even more complicated than a drug trial. Effects of instructional practices depend on characteristics of the child but are mediated by classroom, school, community, and home factors that are exceedingly difficult to measure or control on the large scale demanded by an RCT. Such studies tend to favor methodological rigor at the expense of the significance of the research question. The WWC has also run afoul of a problem that occurs in medical trials: RCTs conducted by researchers who are not at arm's length from the products (commercial reading curricula or software, a drug or medical device) they are evaluating.

Reading Recovery

If the whole-language/balanced-literacy approach is as flawed as described, many children will struggle to learn. What happens to them? In thousands of schools in America and other English-speaking countries, those children participate in Reading Recovery (RR), a remedial program devised by Marie Clay, a New Zealand educator. Reading Recovery is a short-term program focused on first graders who have made little progress. Clay, a whole-language popularizer, focused on the guessing idea and the use of many types of information—semantics, grammar, background knowledge, and so on—to identify words and understand texts. Children who have not learned to do this fall behind. Reading Recovery provides tutoring in the use of these strategies.

Reading Recovery is as controversial as whole language because it is more of the same thing. Advocates can point to numerous studies showing that RR is both beneficial and cost-effective. RR is included in the What Works Clearinghouse as effective for teaching many skills. Serious concerns have arisen nonetheless about its assumptions, its effectiveness, how it compares to other types of interventions, and its high cost because it is delivered one-on-one by instructors with specialized training. The Reading Recovery organization responds forcefully when such concerns are raised.

My perspective is that, having popularized a reading theory that hinders many children's progress because its core assumptions are mistaken, proponents of the approach developed an expensive remediation program based on the same principles. Fewer children would need Reading Recovery if they had received appropriate instruction in the first place. For these children, the easy cases, many types of intervention are successful. Reading Recovery does not address the hard cases, children whose difficulties persist beyond first grade despite intervention. Effective interventions with these children focus on their main problem areas, basic skills and spoken language and on comorbid conditions such as attention deficit hyperactivity disorder.

Can't We Talk About Something More Pleasant?

The reading wars were about whether instruction should emphasize "skills" or "literacy." It was a false dichotomy because literacy requires basic skills, but educators equated "skills" with drudgery. Balanced literacy diffused this debate by encouraging teachers to use whichever combination of methods worked best for them and for a given child. That is what teachers thought they were already doing, and so little changed.

Balanced literacy also consolidated the focus on "literacy" rather than "reading." The change might seem inconsequential because the word "literacy" has meant "ability to read and write" since it was coined in the 1880s, but it actually mattered a great deal. Criticisms of how reading is taught were suddenly moot because the goal was developing literacy, which was treated as a different enterprise. Faced with a mountain of research contradicting basic assumptions about how to teach reading, educators didn't alter their practices; they changed the topic.

Having distinguished Literacy (I'll capitalize the educational term) from reading, educational theorists pushed further, decoupling Literacy from written language. Literacy has become the umbrella term for communication involving print but also other types of information (pictures, sounds, video, and combinations thereof). As the International Literacy (formerly Reading) Association has put it, "Literacy is the ability to identify, understand, interpret, create, compute, and communicate using visual, audible, and digital materials across disciplines and in any context."

So defined, Literacy is the exchange of information by linguistic and nonlinguistic means. The educational challenge is seen as familiarizing children with the "multiple literacies" that technology now affords. The lists of literacies vary but this one from the American Association of School Librarians is representative: visual ("the ability to think, learn, and express oneself in terms of images"), digital ("The ability to understand, evaluate, create, and integrate information in multiple digital formats"), and technological (ability to use the technologies of today and the future). Also literacy literacy (reading and writing), now called textual literacy.

The "multiple literacies" concept has taken off in education the way that whole language did in the 1990s. Some interpretations of it are benign; others seem likely to maintain, even rationalize, disparities between readers and nonreaders. Alternative "literacies" have existed since the 1940s, when "literacy" became a term for specialized expertise in expressions such as "cross-cultural literacy," "financial literacy," and "scientific literacy." The new screen-based technology of the 1950s required "television literacy." These expressions are metaphoric extensions of the original word, just as "bee language" and "body language" are for "language." However, educators who specialize in reading began taking some of these figures of speech way too literally.

Technology has unquestionably changed how we acquire and exchange information; text is no longer the only option, and it's great. Compare the experience of reading the front page of the *New York Times* from a century ago with using the nytimes.com website. The website integrates other media with text, but the format also makes the text itself easier and more inviting to read. It's win-win. Or this book: a traditional text with URLs for videos, images, and activities and for sites that combine text and other media. (Now we just need easy ways to integrate these elements in a convenient format and perhaps a new business model for marketing the product.) Using the Internet for informational purposes involves teachable skills: how to ask the right Google questions and how to evaluate the answers that come back, for example. Aside from visiting social media sites, many of my college students seem to use the Internet mainly to look up facts, like who was Dr. Seuss and what is the definition of "mutual information"? As David Kellner has observed, "In addition to reading, writing, and traditional print literacies, one could argue that in an era of technological revolution, we need to develop robust forms of media literacy, computer literacy, and multimedia literacies, thus cultivating 'multiple literacies' in the restructuring of education. Computer and multimedia technologies demand novel skills and competencies, and if education is to be relevant to the problems and challenges of contemporary life it must expand the concept of literacy and develop new curricula and pedagogies."

Yes, being able to effectively use these technologies is important, but a person still has to be able to read. Written language has unique expressive capacities. *Moby Dick* can be written with emojis, but not very well. On multimedia websites from the *Wall Street Journal* to People.com, print dominates, directing users to videos, interactive graphics, and slide shows. (Imagine trying to do it the opposite way.) A person can access hundreds of pictures of oranges in a few seconds and dozens of videos featuring Annoying Orange, an Internet character with 5 million followers, but that doesn't obviate the need to be able to read the word ORANGE. The availability of so much information in so many forms is a spectacular resource for those who have access to it, but writing retains advantages over pictographs, even animated ones.

My concern about the emphasis on multiple literacies is that it devalues the importance of reading and teaching reading at a time when they need more attention, not less. Educators, the early adopters, seem to be overanticipating a future in which written language is a legacy technology, replaced by other means of communication. It's a familiar argument. Teacher education needn't focus on methods because they will eventually be obsolete, and the same goes for teaching traditional reading and writing in the classroom. I am waiting for someone to claim that children only have to sample the letters and words in texts because they can be guessed from the pictures and videos: voilà, multimedia skimming. Multimedia software may be quite helpful for struggling readers or dyslexics. Are "multiple literacies" another case of treating what is beneficial for poor readers as the norm for skilled reading?

Information technologies will continue to change and to change us, but we are poor at predicting how. In the meantime, an approach to literacy that does not treat traditional reading as fundamental will make it harder to gain this skill, with the greatest impact falling on groups already at risk for reading failure. Since advantages continue to accrue to those who can read in the traditional sense, it seems essential—a moral imperative, even—to hold educators to developing children's reading and writing skills.

The Honey-Vinegar Dilemma

Early in this chapter I warned that my assessment of reading education could seem harsh. Immersion in this history makes it hard to avoid feeling like a scold, or Diana Trilling, the literary critic who was said to write letters to the *New York Times* when she felt a book review was not sufficiently negative. It is not a groundbreaking work of staggering genius to recognize deficiencies in elementary education, though solving them would be. We can all see the parallels between education and health care, massive systems that do superlatively well for many people and much less well for others, with intense disagreements about what can or should be done. And if educators' assumptions about how children learn to read are obviously flawed, why rehearse the criticisms yet again?

I cannot have done justice to every aspect of our shared educational predicament, or even the reading subpart, in one chapter. Mea culpa, sincerely. I did have a plan, however: to enable many more people to participate in the discussion without having educational ideologues pull rank. I am skeptical about holding teachers personally accountable for individual students' progress using the quantitative methods called "value-added assessment" because that equation is likely to be too imprecise. I do think that they, and more so their principals, superintendents, and college professors, can be expected to justify what happens in the classroom, respond to informed concerns, and take action to fill gaps in their training. I've said that in education as in health care, parents need to be active participants who look after their interests rather than passive consumers. I hope to have clarified why and to have provided some of the essential "background knowledge."

My other goal has been to prevent the disagreements about reading education from being dismissed as academic in the worst sense: hair-splitting arguments of no consequence that only continue because debate is oxygen in the Ivory Tower. The issues have obvious consequences, and fashioning a coherent pedagogy out of opposing views is more challenging than the injunction to "use the best of both."

For a long time I assumed that well-intentioned individuals could transcend their differences in the service of a shared goal. Barriers between education and science only exist as long as we allow them to. We can all do better jobs communicating what we do and what we have learned. Bridges are built on a foundation of mutual respect for individuals and diverse viewpoints. People are doing the best they can; neither side knows everything. I fully support creative bridge building and have engaged in it myself, but I have come to question whether good intentions and greater effort can be any more effective going forward than they have been in the past. Because I've noticed that when the facts aren't on your side, an effective defense is to kick up enough dust to create uncertainty. Say nonsensical things about dyslexia or float a conspiracy theory about attacks on education. Concerned outside parties will eventually walk away not knowing who or what to believe. As when a split decision on the Supreme Court upholds the ruling of a lower court, a draw is as good as a win if it sustains the status quo. That is how I view the history of these disputes and why I have taken them seriously.

They say you can attract more flies with honey than vinegar. In reading, that has not worked well for several decades. Maybe that is because it isn't true (Randall Munroe, xkcd.com/357).

CHAPTER 12

Reading the Future

BACK IN THE FIRST CHAPTER I asked whether the science of reading had anything to contribute to improving literacy levels in this country or elsewhere. From the outset I've tried to be a realist, recognizing the limitations of the science and the many factors beyond its scope that affect outcomes. I addressed the question by looking closely at the basic science of reading, how children are taught to read, and how well the educational system is working. Having done so, I don't think there is reason to be apologetic about the relevance of the science. Attitudes toward science helped get us into this predicament, and changing those attitudes could help get us out. It's not the only factor, but it's more significant than just getting some facts straight. The absence of a strong commitment to basic science as a source of evidence within the culture of education has had detrimental effects on reading education and, I would argue, on the entire field.

A look at the history of education as a discipline explains why a strong orientation to science did not develop. Schools of education don't get no respect. They haven't since their creation in the early twentieth-century as successors to the "normal schools," colleges dedicated to teacher training and "practical pedagogy." They were created under pressure for universities to participate in meeting the growing need for public school teachers. University administrators also envisioned the development of professional schools in education, modeled after professional schools that had been successfully established in business, medicine, and law. Schools and departments of education, however, encountered challenges to their basis for existing. The US commissioner of education noted in 1923 that they "have been built up in the face of simply unbelievable distrust and opposition in the academic world," which had lasting effects.

One obstacle was disinterest in the endeavor among other disciplines. Although some idealists held that the campus community should share responsibility for training teachers, this did not come to pass. A chemistry professor conducting original research and training the next generation of scientists will rarely have the time, interest, or skills to help educators figure out how to teach high school chemistry. Nor is there much incentive to do so. K–12 teaching is a low-status profession in this country, and in academia, teaching (what is already known) is less highly valued than research (expanding what is known). Teaching teachers is then the lowest of the low, totem pole–wise. The organization of the university contributed to the isolation of schools of education from other disciplines.

A second problem was the nagging sense that there isn't much to teach aspiring teachers. "Great teachers are born, not made," the saying goes. How much more is required than expertise in a subject area such as math or English, some knowledge of how educational institutions are organized and managed (mostly learned from having been a student for many years), a talent for teaching and dedication to the calling, and a lot of classroom experience? Asked for his thoughts on the requirements for teaching, William Bennett, then US secretary of education, said, "Teachers should demonstrate competence in their subject area, have good character, and have the interest and ability to communicate with young people."

Whether teaching teachers is either possible or necessary has been a perennial question within the academy. On his retirement in 1933, A. Lawrence Lowell, the Harvard president who oversaw creation of the graduate school of education, informed the university board of overseers that it was a "kitten that ought to be drowned." James Bryant Conant, his successor as Harvard president, wrote in a memoir, "I shared the views of the majority of my colleagues on the faculty of arts and sciences that there was no excuse for the existence of people who sought to teach others how to teach." Such concerns have periodically led universities to abandon the training of educational practitioners, as occurred at Yale and Duke (though not Harvard). Other universities that retained such programs housed them in units segregated from the higher-prestige research enterprise.

Modern schools of education have reinforced these concerns by holding that methods cannot be taught because they are contingent on the characteristics of individuals and cultures. Degree programs in teaching came to focus on the social and cultural contexts of education and the role of the teacher in promoting social justice. All the science needed could be codified from the work of the classic figures in the field; familiarity with scientific practices and modern research was not essential.

A third challenge was the perception that teaching has very little impact in comparison to larger socioeconomic forces. The Coleman Report from 1966 was a watershed event because it emphasized the greater impact of socioeconomic status, family, and peer characteristics compared to education. If "schools don't matter," as the report seemed to indicate, then neither does training the people who run them. Although that interpretation of the findings was simplistic and later research showed strong effects of school and teacher quality, the perception remains. Diane Ravitch's argument that the schools are doing well except for the impact of poverty was meant to be a defense of the profession, but it may instead corroborate the view that teaching expertise is a minor factor in the bigger scheme of things.

These conditions have undermined the status of teaching programs and the profession. If prospective teachers can't be taught proven methods, do not have to demonstrate having learned to teach, and have little impact, what is the value of a teaching degree? Students might question what their tuition dollar is buying, and school systems might resist paying a premium for people who have gained this credential. Teachers who are learning on the job might only be entitled to "learning on the job" wages. The reductio ad absurdum of this reasoning is Teach for America (TfA), where the qualifications to teach are a college degree, an interest in teaching for two years, and a short boot camp introduction to the field. I think that teaching is a hard job that requires a great deal of expertise to do well, but this is the corner that teaching has been painted into.

The plight of the teaching profession is well known. It has been discussed in academia for decades. It has been the focus of intermittent efforts to improve teacher preparation sponsored by government agencies and education-oriented foundations. Much of this history is covered in Dana Goldstein's excellent 2014 book *The Teacher Wars*. Teacher preparation is once again the focus of attention and debate.

American attitudes toward teachers are frankly incoherent. We want outstanding people to enter the field but provide little incentive to do so. We expect teachers to be able to educate every child, including ones for whom the obstacles to learning originate outside the school. They should be able to do this without adequate training, having figured it out on their own. The educational establishment—from the schools of education, to the school systems, to the state and federal government agencies, to the lawmakers—creates conditions that do not allow teachers to be as effective as they could be.

As a reading scientist, my expertise is relevant to the role of the science in reading education and identifying how it could benefit teachers and students.

However, reading also provides a detailed case study that speaks to broader concerns about K–12 education. The long-running debates about reading are mentioned in histories of the schools of education, the teaching profession, and educational reform movements, but the disconnection between education and science is not a major focus, which is itself a reflection of the problem. Reading is the case that shows just how much that disconnection matters.

Reading is an area in which there *is* a large body of modern research relevant to teaching, findings that that were not known in Chicago in 1910, Moscow in 1925, or Cambridge, Massachusetts, in 1960, ranging from theories that integrate a broad spectrum of findings, ruling out other accounts, to experiments that compare the effectiveness of methods for teaching specific skills. The research shows that there are better and worse theories of reading and learning, and methods that have better and worse effects on children's progress. Teachers complain that the schools of education emphasize highly abstract theories that are unrelated to classroom reality, leaving them underprepared. Some of the research they were looking for was over in psychology.

The distortions in how science functions in education reflect what happened when an isolated, inbred community of scholars and practitioners developed an arm's-length stance toward research. The field then turns out practitioners who lack the tools to evaluate what they are taught about issues such as how children learn to read. Absent these tools, they are forced to rely on personal experience or the assurances of authorities, whose reliability is difficult to assess. Or the treacherous Internet, where some experts teach the claims made in the Cmabrigde reading hoax as *true*. Practitioners have been misled about what is known and missed out on research relevant to achieving the goals they value. Their students bear the effects.

The failure to cultivate a scientific ethos has also been harmful to the greater field of education. There are lots of proposals for how to fix one or another problem in education. They originate in education and other disciplines. Government, business, academic entrepreneurs, parents: everyone has a plan for education. The lack of tools to ask hard empirical questions about the validity of such proposals has left the field chronically besieged and vulnerable to a good story and a hard sell.

A stronger scientific ethos could also have provided a much-needed defense against *bad* science. Some of the contemporary antipathy to education-related science in areas such as reading is a reaction to an earlier traumatic episode. In the 1960s, the radical behaviorists, the rat runners whose work mainly lives on as a *New Yorker* cartoon theme, were convinced they had solved the problem of how organisms learn. All behavior—whether rats running a maze or children learning a language—was determined by the rewards and punishments it elicited. Complex behavior could be built up from smaller parts under appropriate reinforcement contingencies. These learning principles worked well with rats and pigeons, who could be trained to perform some amazing tricks (such as learning the letters of the alphabet). Educators took up the idea that children could learn fractions, say, if the procedures were broken down into a series of simpler learning steps and behaviors were appropriately reinforced. Curricula were constructed around the idea, and primitive teaching machines that controlled the presentation of stimuli and reinforcement were manufactured. The approach was adopted before it had been shown to work. Children's behavior cannot in fact be manipulated like a rat's, and children do not adjust well to being treated like one. Although Noam Chomsky had delivered a devastating critique of behaviorism in 1959, the following decade saw its greatest impact on education. The painful excursion into the scientific psychology of the behaviorist era seems to have resulted in postdogmatic stress syndrome, the main symptom being acute distrust of psychological research other than that of the old masters.

If a strong culture of science had developed, the ideas that now dominate educational thinking might have been critically assessed from within, with far more impact than the challenges that pour in from outsiders. Frank Smith and Kenneth Goodman's influential theorizing about reading was also based on the latest and greatest science, the interdisciplinary cognitive science that emerged in the 1970s with behaviorism's demise. They were not the only cognitive scientists interested in reading, however. Their claims were contradicted early and often by researchers such as Charles Perfetti, Marilyn Adams, and Keith Stanovich, but too few people within education had the tools to weigh the evidence or the motivation to do so.

These episodes certainly contributed to resistance to modern research on reading. There is a major difference, however. The reading science I've discussed has been accruing for more than thirty-five years, conducted in many labs. The consensus theory that resulted is not an overhyped but unproved idea championed by a few academic gurus.

A stronger culture of science might have yielded better evidence about the effectiveness of specific methods, providing a basis for teaching them. Educational culture might have been less parochial and isolated. We might have had "balanced education" that made use of the best elements of scientific and sociocultural approaches.

Has the disconnection between the cultures of education and science been harmful? I rest my case. The obvious implication is that the barriers need to come down. Whether we can achieve that is not clear. The training of practitioners is the area in greatest need of reform, and for that there is at least a noteworthy precedent.

The Flexner Report: A Model for Change?

The modern approach to medical education dates from the early twentieth century. Before then, physicians were educated by small proprietary medical schools staffed by a few instructors who ran them as independent businesses. Entrance requirements were minimal, and students' education was superficial. The field was transformed by a report on the state of medical education known as the Flexner Report, after its main author, Abraham Flexner, an educator who had written a critique of American colleges. Flexner documented the ghastly practices in the proprietary medical schools and proposed reforms in medical education that became the accepted practices. The Flexner Report resulted in a radical change in physician training. Could a similar revolution take place in education?

Many changes that occurred in medical education seem directly relevant to education. The report created a professional model for training medical practitioners: students would acquire an extensive body of core knowledge through a combination of coursework and extended apprenticeships under educator-practitioners. The model asserted the scientific basis of medicine and the validity of the scientific method as a privileged source of evidence. Entry requirements were raised and included a background in science. The amount of coursework increased, and students were taught methods used in the practice of medicine.

A teacher education program along these lines would be an improvement. One could imagine a professional model that placed greater emphasis on gaining core knowledge in sciences relevant to education, including cognitive and developmental psychology, linguistics, and neuroscience. The courses would emphasize how and why the science is conducted, the questions it addresses, the methods used, the central findings and theories, and where the controversies and uncertainties lie. This basic background knowledge would allow the teacher to intelligently assess what comes down the pike about teaching practices, curricula, educational technologies, research findings, and "expert" claims.

Like a bilingual who can switch between languages, the prospective teacher would be fluent in education and science. Having courses jointly taught by scientists and educators could promote the integration of the approaches. The social, cultural, and motivational aspects of education could be approached this way equally well. Coursework could be combined with more extensive apprenticeship and mentoring than is currently provided. Teacher preparation would reflect the importance of the position rather than undermine it.

The logistical challenges to mounting such a program do not seem insuperable. On larger university campuses, most of the relevant courses are already offered; with modifications they could be integrated into teacher training. In some cases this requires merely breaking down the barriers within the school of education itself, where faculty who teach and conduct research in these areas are housed in departments such as educational psychology. Today everything can be recorded and deposited on the Internet for sharing.

The ideological and institutional impediments are formidable, however. (In a 1916 report on education, Flexner came to the same conclusion.) The proprietary medical schools were united by a business model, not a shared ideology, and they collapsed once a superior training program was formulated and implemented, with the support of universities and philanthropic foundations. This does not describe the current situation in teacher education. Modern schools of education are well-established, well-funded, trusted institutions with close connections to business (e.g., educational publishing and technology companies), philanthropy, and government. Moreover, incorporating scientific attitudes and subjects within an educational culture that has not valued them is like asking a teacher who knows whole language to teach phonics.

Other obstacles are known from previous attempts at reform. Attracting a larger number of highly qualified individuals into the teaching profession is usually high on the wish list for education reformers, and the type of program I have sketched might well serve that purpose. However, it is questionable whether if someone builds it, they will come. Traditional and alternative paths to certification might be strongly preferred over ones that are more challenging or take extra time to complete. The graduates of more rigorous programs might not realize tangible benefits when competing on the job market.

Finally, such efforts may be irrelevant because the job has changed. The demands of high-stakes testing have left less room for a teacher's creativity and expertise. The quantification of teacher effectiveness based on the quantification of students' educational progress may also prove a disincentive to entering or staying in the field. So is demonizing teachers as public employees with "part-time" jobs whose pensions and benefits are a taxpayer liability, and eliminating their collective bargaining rights, as has occurred in several states.

I have no ability to predict what will happen next. Describing some potential directions might help to stimulate research, debate, and other ideas, and I offer what follows in that spirit.

Change from Within

Schools of education could change from within, with a Flexner-like revolution in how practitioners are educated. That would entail a radical change in educational culture, which seems unlikely. The major schools are thriving, the culture is well entrenched, and the incentives favor incremental change. This type of shift might be accomplished at the less prestigious schools that face increasing competition for students and are not under pressure to protect program rankings.

The New Normal (Schools)

Responsibility for training educational professionals could be transferred from schools of education to new interdisciplinary programs that include faculty from education, the sciences, the humanities, and computer science, creating the modern equivalent of the normal schools. The professional training schools would be closely allied with area public schools (1) to allow mentored apprenticeships and provide in-service continuing education, (2) to keep research tied to issues that affect practices, and (3) to incorporate the participation of teachers and other school personnel and provide beneficial returns to them.

Call me a cockeyed optimist, but this kind of program might be realizable if only on a limited "laboratory school" basis. The problem of gaining buy-in from enough people from other disciplines with interests in educational issues may be less severe than in the past. I am encouraged by the big increase in the number of people who are working at the interface between education and the disciplines that I know—psychology, linguistics, cognitive science, cognitive neuroscience. New hybrid fields of educational cognitive science and neuroscience have coalesced. Reading is the most advanced area, but math and writing are developing along the same path, as are learning, attention, and motivation. The questions are theoretically interesting (e.g., what are the origins and varieties of mathematical knowledge?) but also closely tied to learning traditional subject matter in classroom settings. For example, what makes fractions especially difficult for people to master? Is there an optimal balance between instruction in mathematical concepts versus problem-solving procedures? How do reading and language skills affect learning math and solving word problems and the assessment of mathematical ability? Human memory is being studied in the context of how students acquire, retain, and forget new information and how courses can be structured to promote learning, retention, and "transfer"—applying what is learned to new problems and situations. Motivation is an old topic in psychology, but now there is impressive work on motivation to learn in school settings, which has led to the discovery of simple techniques that improve some kinds of performance. In contrast to the past, there is only a degree or two of separation between this science and the classroom, rather than five or six.

Experienced educators will spot the obvious reasons why this proposal it is both infeasible and flawed, so I won't list them. OK, out of many, three: several of the key elements have been tried in the past and abandoned as unsuccessful; such programs could only supply a tiny fraction of the millions of teachers who are needed; in the era of high-stakes testing, schools cannot afford to allocate teacher or class time to participate in such activities, although that restriction may be lessening somewhat in some schools.

An imaginative university administration or philanthropic organization will find a way to create an innovative program that isn't bound by school of education traditions. One successful program that attracted talented students who would not otherwise have entered the field, that offered an intellectually challenging curriculum, that provided training in useful methods, that included extended apprenticeship under close supervision, that turned out teachers who could be more successful because they were better prepared and had less to learn on the job, while minimizing student debt—perhaps that would cause other dominos to fall.

Change the Curriculum

Reforming teacher education is a moonshot project, but reforming what teachers are taught about reading shouldn't be. A chemistry teacher needs to know chemistry, and a math teacher needs to know math. What is the subject matter relevant to teaching reading in the broadest sense, from basic skills through comprehending and analyzing challenging texts? In the current zeitgeist, belief in the need to develop academic literacy is emphasized over foundational reading skills. Literacy activities for young readers are precursors to literature as it is taught in high school or college. But the subject matter for teaching reading is not children's literature or textual analysis; it is the science I've discussed in this book. This material can be taught to college students from a wide range of backgrounds; I do it every year. Add a course in child development and one in the psychology and biology of language, and a prospective teacher would have a solid foundation for gaining this expertise.

Change the Licensure Requirements

The state teacher certification or licensure exam could be used to ensure knowledge of essential facts and concepts. If it's on the test, the prospective teacher will have to know about phonemes, syllables, the alphabetic principle, the role of spoken language in learning to read, developmental reading, language and learning disorders, bilingualism, and so on. The marketplace (schools of education, for-profit schools, web entrepreneurs, and software developers) will respond to the need for relevant training.

This experiment is in progress. More rigorous certification exams in reading and language arts have been adopted in many states. The best example is the Massachusetts Tests for Educator Licensure (MTEL). Here is a sample item from the "Foundations of Reading" exam:

1. Students in an upper-elementary class examine and discuss paired words such as compete and competition, inflame and inflammation, and magic and magician. Word pairings such as these are most likely to promote students' reading development by increasing their awareness that:
   A. most phonic generalizations have at least a few exceptions.
   B. syllabification can help a reader identify the meaning of an unfamiliar word.
   C. the spelling of a word may give clues to its meaning as well as to its sound.
   D. some phonemes are represented by more than one letter combination.

Think the question is tough? I do. Here is one from the "Reading Specialist" exam:

2. A third-grade student is a proficient reader. Which of the following changes is most likely to occur as the student enters the upper-elementary grades and begins reading more complex content-area texts?
   A. The student's reading rate will become more consistent and uniform across all types of texts.
   B. The student's reading vocabulary will start to exceed the student's oral vocabulary.
   C. The student's independent reading preferences will shift from fiction texts to nonfiction texts.
   D. The student's interests will shift from oral language activities to written language activities.

The answers are 1C and 2B. The MTEL site has other sample questions. They are hard, but I bet you could do pretty well.

Test prep materials have indeed appeared; colleges and for-profit outfits offer classes. However, even a good test like the MTEL doesn't ensure that people understand *why* the concepts are important. If that hasn't been a part of their education, the exam can be seen merely as a hoop to jump through and the prep courses as "teaching to the test" rather than covering what's genuinely important.

A rigorous test to enter the profession is a valid way to avoid gaping holes in basic knowledge. However, it is not an effective mechanism for achieving fundamental changes in teacher education. First, it's coercive: learn this because it's required. Second, it's been tried before, and the history is very clear: raising the bar gives rise to the expansion of alternative paths to certification, allowing candidates to bypass the exam. The alternate routes may be introduced to circumvent the tests or simply because a state needs enough teachers to meet demand regardless of the test results. Finally, licensure requirements are subject to change in response to pressures from competing interest groups. I strongly favor intelligent certification exams for reading and math following the MTEL model, but a top-down regulatory approach isn't a viable path to systemic change.

And What Can Be Done About It—Now

Other changes in policies, practices, and allocation of resources may be possible to implement more rapidly. Here are a few, some tied to the research I've discussed, others already being undertaken for independent reasons. Some are easy to endorse; others have attractions but raise new concerns.

Reading Is Still Fundamental

In writing this book, I've assumed that being able to read traditional texts—words on the printed or digital page—is important. As Studies Have Shown©, being a skilled reader confers numerous benefits, including greater participation in health care, education, government, and business. My criticisms therefore have focused on conditions and practices that needlessly interfere with gaining this skill.

As I've noted, however, educators are more interested in a broader concept in which "literacy" takes multiple forms. Traditional reading is acknowledged as important but not necessarily privileged over other types of communication. Rather than focusing on a conjectural future in which reading is unimportant or engaging in practices that either justify and perpetuate reading gaps or turn the decline of reading into a self-fulfilling prophesy, reading educators should be held to a commitment to teach children to read.

Career Prerequisite: Linguistics 101

An introductory course in linguistics should be a permanent requirement for teaching children. Educators need to know how language works. Basic introductory courses in linguistics cover concepts I've mentioned repeatedly: morphology, phonology, syntax, and semantics; relations between spoken, signed, and written language; how language is acquired and the nature of language impairments; whether language shapes how we think; and much else. They also address issues about variability, across languages, cultures, and individuals, that are of overriding interest in schools of education. This is essential, foundational stuff for teaching.

Assistive Software

The computer technology that will revolutionize education has been coming real soon now for several decades. The current era seems different because of the growth of the educational software industry and because the culture has accepted the idea that children can learn from screens. It's an exciting time. In reading, for example, Carol Connor and colleagues have developed adaptive learning software based on the reading science I've discussed in this book. What the program does depends on the student's current level of performance. Such software seems likely to improve rapidly as it makes contact with statistical learning to optimize what gets learned when. Large quantities of venture and human capital are being invested in creating educational games that incorporate features that make *Call of Duty* or *Animal Crossing* engrossing for many people. Some theorists think games will constitute the curriculum of the future. As these games come to market, it will be important to obtain independent evidence of their effectiveness. The example of heavily hyped "brain training" software, which has yielded poor results in systematic assessments, illustrates why.

Educational software will be a boon if it allows teachers to focus on the important stuff. Say that a necessary skill (or content area) isn't being taught for some reason. Say the skill requires individualized feedback the teacher doesn't have time to provide. Say that only some children need additional instruction. Or say that a combination of direct instruction plus practice is optimal for that skill. Under the right conditions, the software could be an effective complement to teaching activities, increasing the richness of the learning environment. Teachers could concentrate on things that only they can do, like creatively convey a difficult concept or look a child in the eyes and say, "That's a great idea. Tell me more about it!" They could do the things that change children's lives, which does not include correcting their spelling tests. It will depend on the skill or material in question, what gets taken care of in the classroom, and the quality of the software, but it would be a breakthrough if teachers could be relieved of some responsibilities that could be handled by other means.

Educational software can be viewed from a different perspective, however: as a disruptive technology that diminishes the teacher's role. Large educational publishing and digital technology companies such as Houghton Mifflin Harcourt (HMH) and Pearson Education market software systems for instruction in reading, language arts, and other subjects. The availability of such software could turn educational philosophy into the study of software design and teacher training into a moot issue. Educational technology is very big business. HMH, for example, markets educational software that it acquired from Scholastic Inc. in 2015 for $575 million. Scholastic developed Read 180, which focuses on struggling readers in grades four through twelve. According to HMH, "READ 180 supports educators with a comprehensive system of curriculum, instruction, and professional development, while providing students with personalized rigorous instruction for college, career, and beyond." It has been under development since the 1980s, has a strong grounding in relevant science, is widely used, and is listed in the What Works Clearinghouse. iRead, a system for kindergartners through second graders (there's an iPad version), was released in 2015, and they offer a "suite" of related programs.

As described by HMH and in research studies, Read 180 lessens the need for teaching expertise. The teacher conducts whole-group reading activities and works with smaller groups using Read 180 instructional materials. The Read 180 software handles a lot of the heavy instructional lifting. The system is intended for use with struggling readers but can be used off-label with others. Data concerning Read 180's effectiveness are accumulating. The system may not be as effective as an outstanding teacher, but it may be much better than undertrained, inexperienced, or inept teachers. It could also reinforce the status quo in teacher preparation. If software takes care of the pedagogical challenges, teacher education could continue to focus on the social justice and cultural diversity concerns and the child's social and emotional development, for better or worse.

Open questions about the system's effectiveness aside, cost is a major concern. Read 180, like other such systems, is not a cheap proposition. The price tag includes licensing fees, hardware, support materials, technical support staff, teacher and staff training, and other expenses. Having a Read 180 license for every child in low-performing elementary schools, along with the hardware and personnel to use it, might be highly beneficial, but it would definitely be prohibitively expensive. Such software goes to the school districts that can afford it, another way to magnify the impact of SES. This fact runs counter to the emphasis on education as a vehicle for social change and the importance of equality of educational opportunity. It's a problem. The availability of open source alternatives might be a way to mitigate it.

"A comprehensive system of curriculum, instruction, and professional development" works against the goals of raising the status of the profession and attracting talented people into the field. If the product provides most of the solution, the teacher can be a paraprofessional with an alternative form of certification. The company already provides training to use the Read 180 system—for a fee.

Teachers aren't going the way of travel agents any time soon, but the idea of the teacher as a highly skilled professional could be, for some grades and subjects.

Increased Reliance on the Supplemental Educational System

A network of commercial learning centers, reading specialists, and tutors is available for those with sufficient means. Such supplemental education is an intrinsic part of children's education in high-achieving countries such as South Korea, Singapore, and China (Shanghai), though in response to a perceived need for additional schooling rather than to compensate for inadequate schools. Schools do not have to change if this parallel system picks up the slack. The objections to this solution are also obvious: (1) public schools should fulfill basic educational obligations, and (2) the practice is inherently unjust, insofar as the success of public education is predicated on the availability of resources that are unequally distributed.

The mitigating force here is the availability of free, quality online resources, such as Khan Academy, although that requires access to hardware and the Internet, and it also excludes the irreplaceable interactions between instructor and student and between students.

Outsource Teacher Education

Teachers could learn the profession by nontraditional means. Current proposals for reforming teacher education mainly emphasize ways to help teachers who are already on the job. With sufficient financial resources, school districts can take over teacher training, contracting with commercial providers. This is Teaching Recovery. It is expensive, and there would be less need for it if the individuals had been taught appropriately in the first place. However, it could work, and it could provide a path into the field for college graduates with degrees in other areas. It shifts the bill for training teachers from students to taxpayers.

A related possibility is that companies that operate charter school networks and, increasingly, public school systems will train their own teachers. For example, the KIPP company, which operates a large network of charter schools, trains teachers in its approach. Of course, such training could be done well or poorly, but it would decrease dependence on traditional schools of education. Other types of corporations provide in-house training programs tailored to their needs; so can educational corporations. The KIPP "no excuses" educational model is controversial, and teacher turnover has been high. But schools run by the private sector that provide better-educated, better-supervised, better-supported teachers who do a better job with their students? Perhaps serious competition would jar traditional programs out of their complacency.

Expand Language Enrichment and Quality, Cost-Effective Tutoring

Many children are behind in language development and reading readiness when they start school, and the impact rapidly snowballs. Children from lower-SES families, speakers of a nonmainstream dialect, and English-language learners are at higher risk. Pre-K language-enrichment programs that focus on expanding the child's knowledge of school English could be highly effective in reducing these gaps. In a preschool setting where the staff have varied language and cultural backgrounds, their own conversational speech is Big Data for the child language learner, and they would model code switching and a broader range of interactive styles and conventions.

For children who are struggling with reading in school, the pressing need is quality tutoring that a school district can afford. The excellent model for this is the Minnesota Reading Corps. As its website says, "[The program] provides what struggling readers need—individualized, data-driven instruction, one-on-one attention, well-trained tutors, instruction delivered with fidelity, and the frequency and duration necessary for student achievement." The program is conducted as part of AmeriCorps, the quasi-governmental program that engages people in intensive community-service work. It's a terrific, successful program.

The main obstacle in both areas is the availability of staff with appropriate skills and training. Here I have a specific proposal: repurpose Teach for America.

Teach for America's recruitment and training model is not adequate for the stated mission, placing teachers in schools in low-income areas. That role requires much greater expertise than can be attained with the TfA's boot camp approach. Applications to TfA have declined as the nature of the program and the experience have become clearer. Its approach could work quite well, however, if the mission were to place teachers in preschool programs and in K–1 classrooms or to place tutors for reading (and perhaps math and language) in schools. TfA recruits could make much-needed contributions to such programs, working alongside more experienced personnel. They could expand the range of linguistic, cognitive, and cultural experiences to which children are exposed, while improving child-staff ratios. They could become the American Reading and Language Corps.

This reorientation would retain the strengths of the TfA program, bringing college graduates into the educational endeavor, especially in low-income areas. Participants would still be engaged in fulfilling TfA's goal of reducing achievement and opportunity gaps in public schools. TfA recruits could be highly successful in these positions with relatively short-term, intensive training. They would be serving much-needed, rewarding, but more narrowly focused roles that do not demand the skills and experience to run a classroom. That would obviate the concern about underpreparation in the current TfA model. TfA would no longer act as an alternative route to teaching careers, an aspect of the program whose success is unclear. A positive, successful experience in these positions could also inspire people to choose an education career.

Why Are There "Wars" About Reading and Teaching?

Looking at this history left me with a question that I hope has occurred to you as well: *Can it really be that hard to decide how to teach children to read?* Like, figure it out and get on with it, people. Educators and scientists seem to have complementary kinds of expertise. They have the same goals for children. People should be able to look at the facts, weigh them in a rational manner, and make the best decisions they can. And yet groups hold diametrically opposing views on what to do. Having laid out the science, examined the conflicts, and considered some ways to move forward, I was left wondering how much plasticity is left in the system—that is, room for change.

It helps to view what has happened in education, with reading leading the way, as an example of the group polarization that has occurred over issues such as gun control, climate change, and reproductive rights—indeed, in American politics. Studies of the process show how interactions among members of an affinity group reinforce shared beliefs, solidifying group identification and coherence. Evidence that contradicts fundamental beliefs does not shift the group toward consensus with opposing views but instead strengthens existing views or shifts them to more extreme positions. Consider, for example, the belief that tax cuts produce economic growth. As journalist James Surowiecki has noted, "The empirical evidence is by now unequivocal that, with tax rates at U.S. levels, this doesn't work; cutting tax rates simply leads to lower tax revenues. . . . This message [lower taxes produce higher tax revenues] has been fact-checked and refuted over and over again, but, once something becomes an article of political faith, it's difficult to dislodge." Surowiecki then mentions research by Brendan Nyhan and Jason Reifler demonstrating that factual corrections of strongly held but erroneous beliefs can produce paradoxical "backfire effects."

That's reading.

Reading was an early instance of this phenomenon. One of the crucial ingredients is group members being able to share information and offer feedback on a large scale. Positions about reading began to polarize in the late 1980s. In that era (as today), teachers exchanged information in the old-fashioned ways: by attending workshops and conferences, by joining state education and national reading associations, by subscribing to newsletters and other special interest publications. But educators were also among the early users of the Internet. Web browsers had not been invented, but we had the precursors of modern social media: listserv (group e-mail), Usenet interest groups, and bulletin board services (BBSs). It was easy to find like-minded people to share experiences with, and it was thrilling to be able to interact so easily across long distances.

This development coincided with leaders such as Ken Goodman encouraging teachers to take control of their classrooms because they have far more contact with children and experience with what works. Theorists might be inventing the educational paradigms of the future and laboratory scientists might be collecting reaction times, but teachers were teaching. By sharing information in "communities of practice" they would develop solutions together, and with the Internet they could be updated as circumstances (e.g., passage of the No Child Left Behind Act in 2001) demanded.

This is the constructivist philosophy applied in yet another domain: how to teach reading. It is discovery learning and the social construction of knowledge par excellence, with the added fillip that information can now be widely and rapidly shared. This is a recipe for creating a tightly knit group with a strong sense of identity based on shared beliefs and experiences.

I can admire the intentions and self-reliance and yet assert that as a way to figure out how reading works and what to teach, this process is inadequate. Large numbers of people sharing observations and insights is not sufficient to converge on what is true about reading. The sharing process is subject to distortions and biases because the data are unreliable. Under these conditions, groups can converge on ideas that are logically inconsistent or demonstrably false and also supremely difficult to correct.

The science should have acted as a constraint on runaway thinking about reading. There were basic questions of fact: Is reading visual, or does it make use of phonology? Do the good or the poor readers rely on guessing? And so on. The answers left many open questions about practice that communities of practitioners could productively address. But the science wasn't allowed to serve this framing function. This experiment in the social construction of knowledge yielded consensus around ideas that were half-baked at best and often false.

I can illustrate. Every K–5 teacher has heard of something called the 3-cueing system. It is propagated through websites and slide shows, workshops and meetings. The origins of the theory are somewhat murky, but it has been circulating since the 1980s. The idea is communicated through illustrations such as Figure 12.1, typically a Venn diagram with three overlapping parts: print knowledge ("graphophonic cues"), syntax, and semantics. Marie Clay proposed a version as a way to help struggling readers who were stuck on a word: they could try different cues to figure it out (get to "meaning"). The approach is billed as what to say to a child after (or instead of) "sound it out." For example, a child having trouble reading the word HOPSCOTCH in THE KIDS PLAYED HOPSCOTCH IN THE PARK could be prompted with questions such as "What kind of word goes here?" (a noun, a "syntactic" cue) or "Can you think of games that are played in a park" (a "semantic" cue). Together with "graphophonic" cues (some letters, perhaps their sounds), the child would be able to figure out the word. The overlap in the Venn diagram is meant to convey that cues are combined. Skilled readers are thought to do the same things automatically.

The figure conveys some basic facts. Texts *are* understood using multiple types of knowledge, including print (orthography), grammar (syntax), and meaning (semantics). These are intrinsic properties of language and writing. Texts cannot be understood without these components (and others); there is no option to skip the syntax or forgo the meanings of words. The figure also incorporates the idea of solving a problem such as word recognition by combining multiple cues, which skilled readers do without conscious effort or awareness.

The problem is what's left out. The figure doesn't indicate what is in the circles (e.g., what are "syntactic cues" or "graphophonics"?) or how these types of knowledge are acquired. It doesn't address the division of labor between them or how that varies with reading skill, text difficulty, background knowledge, or

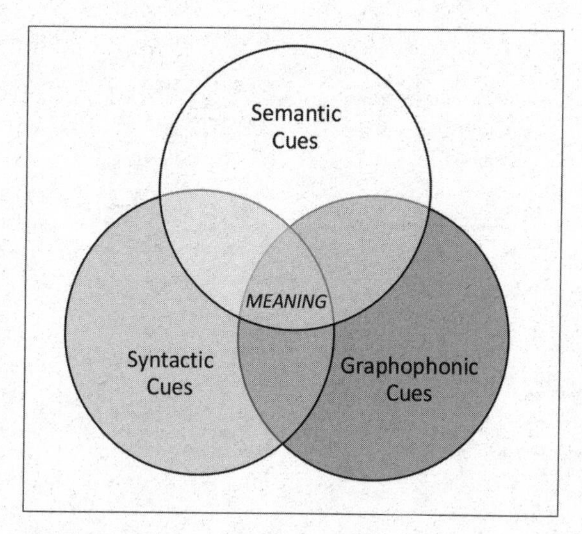

FIGURE 12.1. A generic version of the 3-cueing system.

writing system. It is therefore open to many interpretations. In fact, it is compatible with every theory of reading. One can tell a top-down, guessing story around the figure or one that emphasizes the importance of "graphophonics" as a foundation for comprehension. It is a Rorschach blot on which to project one's beliefs about reading. And that is what people did as it circulated.

Figure 12.2 shows a few other versions of the 3-cueing system found in talks and articles. Version A was discussed by Adams in 1998, who included several others from that era for comparison. Version B is clearer about the components, but it's no longer a Venn diagram, and the "intersection" is just the label "reading." Version C equates syntax with "sounding right," but it's not clear how "not sounding right" (i.e., sounding ungrammatical) helps the child identify the problem word. The graphophonic circle is now labeled "Visual (does it look right?)," which drops the phonic part and leaves open what this would yield other than a feeling about whether the child knows the word or not. Version D implies a hierarchy among four types of information, but it is unclear whether they are meant to be used simultaneously, in sequence, or some other way.

These systems are also explained using examples as proof of concept. The use of semantic cues is illustrated using an example such as "The children are playing ____ the park." The child is having difficulty reading the word in a

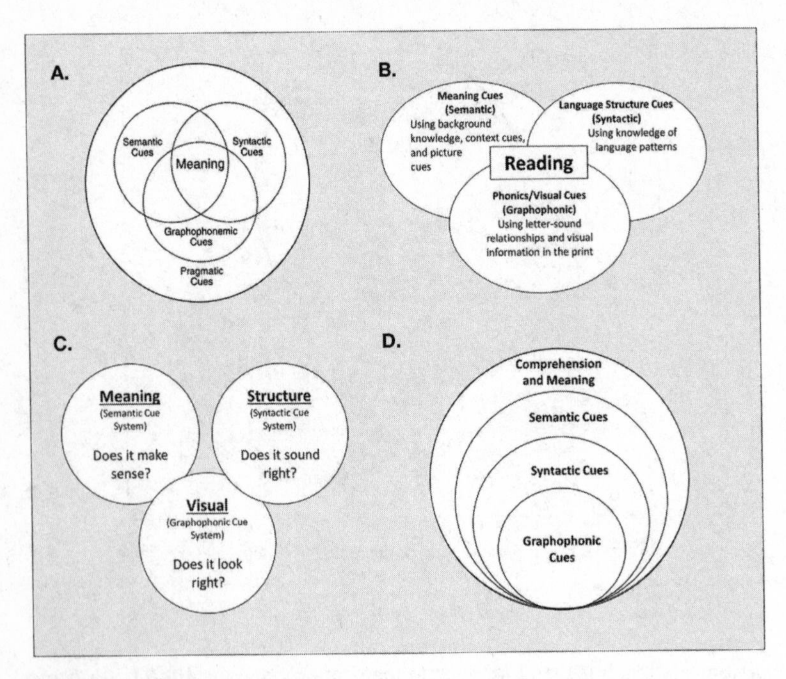

FIGURE 12.2. Other 3-cueing systems.

normal text where the blank is in the example. The child uses syntactic cues to determine that the word must be a preposition and is then encouraged to think of ones that make sense in this context. Graphophonic cues can then be used to home in on the correct word. This is the psycholinguistic guessing game, still treating reading as the act of predicting missing words, with the child given additional explicit strategies. But you can't play Twenty Questions with words and win. A child who is stuck on whether the word is IN rather than AT, AROUND, NEAR, or ON needs help with a severe decoding problem, not strategies to get around it. In reading normal texts, the most specific "cue to meaning" is the word itself. The PartiallyClips cartoon on the next page wittily captures the contrasting views about reading.

Is the 3-cueing approach valid for struggling readers, as Clay originally intended? If the child has trouble reading a word, suggest a cue; if that doesn't work, try another. The child should never be at a loss for strategies. This "more alternatives is better" idea also underlies "multisensory" approaches to learning. For example, children's comprehension of a story about barnyard animals could be facilitated by providing a physical model of the setting.

"Multisensory" is a cherished concept in early education. But is more actually better for learners who are struggling?

Neurobiological studies of developmental disorders indicate that they are associated with impairments in cognitive control, planning, and attention and in integrating multiple sources of information in areas such as the angular gyrus and visual word form area. Approaches to instruction and remediation that rely on rapidly shifting between reading strategies or emphasize multisensory integration of information thus place additional demands on systems that are already straining. For such children, narrowing the range of alternatives to one that works may be more effective than offering multiple cues.

The most worrisome aspect of 3-cueing was discovered by Adams. She realized from discussions with many teachers that they assumed children already know about syntax, semantics, and "graphophonics" but need help with the 3-cuing process. But, of course, some children are struggling because their knowledge in these areas is *insufficient*. Rather than focusing on developing this knowledge, the method jumps to teaching compensatory strategies.

The 3-cueing approach encapsulates the basic concerns I've raised about reading education. The level at which reading is understood is unacceptably low and justifies misguided practices. The theory is a gloss on some ideas picked up from the research literature that have been passed through a massive game of Internet Telephone and come out as something far removed from the original. The 3-cueing theory is the product of teachers with little knowledge of the science working with large numbers of like-minded people, under the influence of a few authorities, constructing accounts of how reading works and children gain literacy. This process yielded an amorphous theory that was compatible with existing beliefs (social construction of knowledge, guided discovery, whole language, balanced literacy) and thus within the teachers' comfort zone. It is a shallow theory but creates a strong feeling of understanding among group members. Pointing out the strengths and weaknesses of the theory, even as sympathetically as Adams did, is not well received, which produces further polarization.

The 3-cueing approach is a microcosm of the culture of education. It didn't develop because teachers lack integrity, commitment, motivation, sincerity, or intelligence. It developed because they were poorly trained and advised. They didn't know the relevant science or had been convinced it was irrelevant. Lacking this foundation, no such group could have discovered how reading works and children learn. *Because most of what goes on in reading is subconscious: we are aware of the result of having read something—that we understood it, that we found it funny, that it conveyed a fact, idea, or feeling—not the mental and neural operations that produced that outcome. That is why there is a science of reading: to understand this complex skill at levels that intuition cannot easily penetrate.*

ACKNOWLEDGMENTS

I went deep into debt writing this book.

Most of it is held by Maryellen MacDonald, my University of Wisconsin colleague, who studies how languages are spoken and comprehended and why they have the properties they do. Countless times when I was stuck on a hard question or on how to fit pieces together, when I could not see the words for the letters, she bailed me out with an original insight. I've endnoted the ones I directly incorporated, but her influence pervades the book. MacDonald is the ideal colleague in the next office, and I have benefited enormously from her brilliance and generosity.

Mark Liberman has been a helpful correspondent, but more importantly, his posts on Language Log (http://languagelog.ldc.upenn.edu/nll) are an invaluable resource. Mark is a polymath, and so it can be assumed that he has discussed with greater expertise many of the issues I've only touched upon. I've included links to several posts that reflect the most direct influences. I'm grateful to him for what he does, and what he does helped me in writing this book.

Jay McClelland introduced me to neural networks when they were still pretty simple, and working with him and in this area changed how I think about many things. I'm also grateful to him for arrangements that allowed me to spend a sabbatical year at Stanford researching and writing. As he has for as long as I've known him, Jay posed some challenging questions at timely moments that proved extremely beneficial.

I am deeply indebted to Julie Washington, my former Wisconsin colleague now at Georgia State University. The obvious point is that she introduced me to issues about dialect variation and African American English, which led to our research on African American children's reading. There is a way to decrease the achievement gap: put Julie Washington in charge of policy. Then place a Julie Washington on every university campus, after which there would be many more people conducting much-needed research. There would be Julie Washingtons at the National Institutes of Health and National Science Foundation, of course, and at the huge educational publishing and technology companies. But Julie's influence on me and on this book extends much further. She is an acute observer of language and speech, social behavior, children, schools, the educational-industrial complex, and much else. She has deep expertise and informed views. She teaches by sharing, and our interactions have been priceless. Thank you, Julie—it's a privilege and delight to know you.

Dyslexia is a hard topic. I began studying it years ago, but research has moved ahead rapidly in recent years because the methods got that much better. My own understanding of the current state of the art benefitted from discussions with and the work of several individuals. On the behavioral side, I got a great start working with my former colleagues Maggie Bruck, Gloria Waters, and Frank Manis. More recently, I've learned from Maggie Snowling, Charles Hulme, Dorothy Bishop, Jack Fletcher, and Maureen Lovett, each of whom has conducted a prodigious amount of essential research on reading acquisition and dyslexia. On the brain side, the person is Ken Pugh, the president and guiding spirit at Haskins Laboratories, where the modern study of the connections between speech and reading originated. During the period I've known Ken, the understanding of the neural and now neurochemical and genetic bases of reading impairments has leapt forward. Ken has contributed enormously to these advances, via the research that he and his many collaborators around the world have conducted and his tireless efforts to bring researchers together in the pursuit of knowledge. I'm grateful to Ken for the education, for the opportunity to work with him, and for his personal support. Don Shankweiler, one of the Haskins founders, is a personal hero, and I have treasured our conversations. Bruce Pennington generously shared his expertise about genetic influences on complex behaviors such as reading and the bases of reading impairments. Pennington's work has advanced far beyond the basic findings I discuss, but I hope to have done right by it.

The computational modeling research described in Chapters 6 and 7 was a collaborative enterprise with Jay McClelland, David Plaut, Karalyn Patterson, Michael Harm, Marc Joanisse, and Jason Zevin, in various combinations. Dream Team.

Rebecca Treiman is the world expert on spelling, and the analysis of how English spelling got that way draws heavily on her work. The discussion of eye movements in reading mainly describes research conducted by the late Keith Rayner and some of the many scientists he trained. Keith created a huge body of basic research of lasting value. He was also a straight shooter who treated people with honesty and consideration. A mensch, and greatly missed.

I have also incorporated insights that originated with several researchers who were on top of important issues early on. Keith Stanovich, Chuck Perfetti, and Phil Gough's work on learning to read continues to be influential because what they discovered proved to be true and remains relevant to the educational challenges. Louisa Moats has long studied what teachers know about language and reading. Hollis Scarborough did the groundbreaking family risk study and alerted all of us to the fact that the behaviors associated with dyslexia change over time. I've known Marilyn Adams since I was a post doc at Bolt Beranek and Newman, the Cambridge, Massachusetts, research firm where basic research on reading and other topics in cognitive science was conducted at the time. I found myself returning to several of her articles that address crucial issues with insight and verve, and their influence on several chapters is noted in the text. Marilyn later became a controversial figure among educators, and if there were a medal for bravery in the line of ideological fire, she would certainly deserve it.

Many Wisconsin colleagues were enormously helpful: David Kaplan (on the design of the PISA and NAEP assessments); Katherine Magnuson, who introduced me to studies of socioeconomic status and development; Tim Rogers, whose expertise on brain and computational models far exceeds my own; and Gary Lupyan, the very model of a modern cognitive scientist. Ed Hubbard discussed the Flexner Report and its possible relevance to reforming teacher training in talks on campus; John Rudolph broadened my view of John Dewey. I calibrated my critical sensibilities about neuroimaging working with Jeff Binder at the Medical College of Wisconsin; he has conducted particularly rigorous studies of the brain bases of language for as long as anyone. I've also learned a great deal from my own former graduate students, now established researchers conducting next-generation research, especially Jason Zevin, Marc Joanisse, Joe Devlin, Jelena Mirković, Ken McRae, Debra Jared, and Jon Willits, as well as Cammie McBride and Laura Gonnerman, with whom I also worked; and from my current students Tianlin Wang and Matt Borkenhagen. Mike Harm implemented and analyzed the Harm and Seidenberg models, and his software enabled other people to develop models as well. Mike is now at Google, naturally.

Simon Ager developed Omniglot.com, a wonderful website that is an inventory of writing systems from Linear B to Klingon, with intelligent, scholarly, agreeably brief explanatory material and links to reliable sources. The site was especially useful to consult as I was thinking through the relationships between writing systems and the languages they represent. Gene Buckley (University of Pennsylvania) generously provided the "Egypt" hieroglyphic.

Dan Willingham, Marlene Behrmann, Anne Cunningham, Elena Grigorenko, Maryanne Wolf, Fumiko Hoeft, Nicole Patton Terry, Eric Raimy, Denise Schmandt-Besserat, Richard Sproat, David Plaut, Jay Rueckl, Robin Morris, Nicole Landi, Nancy Cushen White, Victor Mair, Heidi Feldman, Anne Fernald, John Rickford, Sean Reardon, Usha Goswami, Catherine Snow, and Steve Carnevale shared expertise and perspectives. All of the people I've mentioned will appreciate my pointing out that they are not responsible for the contents of the book and that any errors are my own. Eric Cardinal, Elaine Loring, Greg Kolden, and Deborah Blum were helpful and kind, agent Eric Lupfer kept calm as deadlines passed, and TJ Kelleher shepherded the book at Basic. Crystal Hanson and Teresa Turco provided invaluable assistance in preparing the manuscript. Thanks to all of you.

I also want to express my appreciation to a crew of Wisconsin people: Steve Dykstra, John Humphries, Cheryl Ward, Dan Gustafson, Mary Newton, Pam Heyde, Marcia and Burke Henry, Jim Zellmer, Nira Scherz-Busch, Donna Hejtmanek, Julie Gocey, and everyone from the Wisconsin Reading Coalition, the Dane County Learning Difference Network, and the Wisconsin Branch of the International Dyslexia Association. Who are these people? Classroom teachers. Special education teachers. The people who deal with the most challenging kids. Parents who organize support groups and lobby to get resources for dyslexic children. Reading specialists and clinicians. Reading activists. Everyday people who have spent decades in the public schools, trying to do the right thing, often dealing with a suffocating educational bureaucracy. You were my reality check. I wrote this book with you and the many people like you in schools around the country in mind. Thank you for being there.

And then: my family. My mother Grayce, who regularly let the younger me pick out any book at Kroch's and Brentano's (I still have my copy of "The Boy [*sic*] Scientist"). Claudia and Ethan, the children who liked their parents! Your love and support made it possible to write this book. *Love your work!* Maryellen: Words do fail. Everything. So much. So lucky. Thankful. Glad.

NOTES

I mainly cite primary sources, which include technical articles that are only of interest to some. The notes include links to websites, videos, and documents that are more accessible, in both senses of the word. In the references (following the notes) readable articles of particular interest are marked with an asterisk.

I've used Google's URL shortener to condense web addresses, but they are still a pain to type. Links also have a tendency to break and require updating. All of the current links are therefore posted on the book website at www.seidenbergreading.net. That is also where demos of experiments and additional figures, illustrations, and documents can be found.

About notation: small caps are used to identify examples of words in their printed (orthographic) form. The pronunciations of words are placed in quotes. References to phonemes are set off by slashes. Thus BAT indicates the word as it is written and "bat" its pronunciation; /b/ represents the initial phoneme.

The use of "they" and "their" as gender neutral singular pronouns is not an oversight. I'm with Shakespeare, Austen, and the American Dialect Society on this one (see http://languagelog.ldc.upenn.edu/nll/?cat=27).

Chapter 1: The Problem and the Paradox

5 **Programme for International Student Assessment:** http://www.oecd.org/pisa.

5 **uncompetitive state of American education:** President Obama declared that "our generation's Sputnik moment is now" in a speech on December 6 (*New York Times*, December 6, 2010), referring to challenges in education, innovation, global competition, infrastructure, and other areas. The education activist Chester Finn, writing in the *Wall Street Journal*, called the release of the PISA results a "Sputnik moment" (December 8, 2010). The president followed up on the Sputnik theme in the State of the Union address (January 25, 2011).

5 **the second Sputnik moment:** These data are taken from transcriptions of all CNN programming during the indicated periods, gathered by Jon Willits.

5–6 **A 2003 study:** National Assessment of Adult Literacy (2003).

6 **we are turning out fewer:** For NAEP results, see http://www.nationsreportcard.gov.

7 **2012 report:** Council on Foreign Relations (2012).

7 **"The evidence is everywhere":** David Remnick, "Into the Clear," *New Yorker*, May 8, 2000.

7 **Poverty is associated:** General Accounting Office (2007).

8 **Rob Gronkowski:** "Gronkowski Talks 'Mockingbird' and Dancing on Kimmel," *Boston Globe*, http://goo.gl/hyllM9. In an earlier era, baseball player Pete Rose was said to have

read one book in his life, his autobiography (Ira Berkow, "Pain and Joy of Life as Athlete's Wife," *New York Times*, May 24, 1981).

8 **links to allow:** It may be more convenient to use the clickable links that are listed by chapter on seidenbergreading.net. Demos, pictures, and other material are also there.

8 **the ugly:** Hayes, Wolfer, & Wolfe (1996).

9 **National Endowment for the Arts:** National Endowment for the Arts (2004); on multiple literacies, see "Standards for the 21st-Century Learner," American Association of School Librarians, http://goo.gl/W57hU.

12 **late African Grey parrot:** The way that Alex learned was sadly unlike how children learn: see "Alex the Parrot," video uploaded to YouTube by Professional_Talker, June 9, 2008, http://www.youtube.com/watch?v=ldYkFdu5FJk.

12 **Minnesota children's low levels:** Garrison Keillor, "Where's St. Michael When You Really Need Him?," *A Prairie Home Companion*, January 29, 2008. http://prairiehome.publicradio.org/features/deskofgk/2008/01/29.shtml.

Chapter 2: Visible Language

17 **Not easy:** Goldin-Meadow & Mayberry (2001).

17 **online manner:** College students can accurately "shadow" (repeat back) recorded speech at lags as short as 250 milliseconds, only a syllable or so behind. They also overshadow, correcting errors intentionally placed in the experiment materials, which indicates they are comprehending what they are hearing as they go along (Marslen-Wilson 1975).

17 **For adults:** Gernsbacher, Varner, & Faust (1990).

19 **Texts for children:** Montag, Jones, & Smith (2015).

20 **Early fears:** For a linguist's perspective, see Crystal (2008). Everyone likes a comical autocorrect error, but now that such errors have become a new genre of humor, it is hard to find real ones. An early version of a popular word processing program replaced COOPERATION with CUPERTINO (as in "Could you tell us how far such policy can go under the euro zone, and specifically where the limits of this Cupertino would be?"). The headline "Homosexual Eases into 100 Final at Olympic Trials" resulted when a church group changed every instance of GAY to HOMOSEXUAL. See many posts on Cupertino effects on the invaluable Language Log (http://languagelog.ldc.upenn.edu/nll).

20 **new genres of humor:** See LOL cat attack at http://goo.gl/KJD9ce.

20 **no "spaces":** Convince yourself: see the demonstration on seidenbergreading.net.

21 **What, Norman?:** Rosanna Greenstreet, "Q&A: John Cleese," *Guardian*, October 19, 2012, http://goo.gl/onmhs1.

22 **We fill in missing information:** This "filling in" is the basis of the famous Kanizsa triangle illusion (http://goo.gl/FruOq8); "incomplete fonts" designed with missing features (e.g., http://goo.gl/MedsW3) are widely used in corporate logos.

22 **what makes a code:** These concepts were formalized in Claude Shannon's 1948 theory of communication. See Gleick (2011) for an introduction.

23 **for beginning readers:** For *Goodnight Moon*, in Hebrew, see http://goo.gl/cA7HDI. Fascinating that the cover drawing is a mirror image, so that it too "reads" right to left.

23 **It's orthographic TMI:** Adding vowels slows skilled reading of Hebrew: Bentin & Frost (1987). However, even skilled adult readers require vowels in some low-redundancy contexts, such as poems.

23 **Languages are quasiregular:** Seidenberg & McClelland (1989).

24 **"stress deafness":** Syllabic stress works differently in French and English, resulting in frequent stress-assignment errors when native speakers of one language use the other. Marking stress patterns in print might be helpful to second-language learners, but the bigger hurdle is being able to hear and pronounce the nonnative patterns. An analogous

phenomenon, "tone deafness," occurs in learning tonal languages such as Mandarin Chinese if one's first language does not employ tone.

24 **James reviewing:** Clive James, "Clive James on Brad Pitt's Chanel No5 Commercial," *Telegraph*, November 1, 2012, http://goo.gl/8gzvvo.

25 ***That* was disingenuous:** It was not that the newspaper's stylebook prohibits this use of italics.

25 **band camp:** From *American Pie*. See "American Pie (9/12) Movie CLI—One Time at Band Camp (1999) HD," video uploaded to YouTube by Movieclips, June 27, 2011, https://www.youtube.com/watch?v=MH619vxtNdo.

26 **By omitting information:** What is reported as a direct quotation is often a cleaned-up version of the utterance. This is known as "piping a quote." Alex S. Jones, "Ideas & Trends; Just How Sacrosanct Are the Words Inside Quotation Marks?," *New York Times*, January 20, 1991.

26 **"The USSR condemns":** "We Begin Bombing in Five Minutes," Wikipedia, https://en.wikipedia.org/wiki/We_begin_bombing_in_five_minutes. See it now: "1984 Soviet Union Reply to President Reagan Bombing Joke," video uploaded to YouTube by Chronos Xaris, November 17, 2012, https://www.youtube.com/watch?v=bN5wL1nw7XA.

27 **A beginning reader can:** Liberman, Shankweiler, & Liberman (1989).

28 **It encourages the fiction:** For a charming example, see "Sesame Street—That's What Reading Is All About," video uploaded to YouTube by wattamack4, July 4, 2007, https://www.youtube.com/watch?v=gzneovHIfRI.

28 **Illiterate Portuguese adults:** Morais et al. (1979).

29 **The effect is so strong:** Seidenberg & Tanenhaus (1979).

29 **Language is a virus:** Laurie Anderson, "Language Is a Virus," video uploaded to YouTube by Rebel, May 11, 2008, http://goo.gl/Hf4dhi; Frith (1998).

Chapter 3: Writing: It's All Mesopotamian Cuneiform to Me

31 **invention that MIT undergraduates:** For the MIT survey, see http://news.mit.edu/2003/lemelson.

32 **Orthography mavens are like birders:** As always on the Internet, the challenge is to differentiate the few quality sites (such as the highly recommended www.omniglot.com) from the much larger assemblage of sites that propagate misinformation.

32 **In some cultures:** The go-to site for jibberish Chinese tattoos is Hanzismatter (http://hanzismatter.blogspot.com). It is a cross-cultural phenomenon, however: http://www.engrish.com/category/tattoos.

33 **The Babylonians believed:** Schmandt-Besserat (1992).

33 **Others interpret the evidence:** On the abrupt versus gradual appearance of writing, see Daniels (1996) and Powell (2009) for differing views.

33 **domestication of cattle in Sumeria:** Walker (1990); Daniels & Bright (1996).

34 **whose name is known:** Michalowski (2003).

34 **Every element of this account:** Walker (1990).

34 **Enmebaragesi is the name:** Scholars deduced that Enmebaragesi was a king responsible for a variety of administrative innovations, but then a fragment from an Old Babylonian story that appeared in 1982 suggested that E. was actually female, a queen. Michalowski (2003) argues that E., male or female, was a fictional composite.

34 **"Passion is inversely proportional":** Benford's law of controversy, due to novelist Gregory Benford. I learned about it here: Mark Liberman, "The Long Tail of Religious Studies?," Language Log, August 5, 2010, http://languagelog.ldc.upenn.edu/nll/?p=2525.

35 **writer Janet Malcolm:** Malcolm quoted in Zoë Heller, "Cool, yet Warm," *New York Review of Books*, June 20, 2013, http://goo.gl/tzHqY2.

36 **oldest known cave paintings:** The famous French cave paintings are about 30,000 years old, but others may be older. John Noble Wilford, "Cave Paintings in Indonesia May Be Among Oldest Known," *New York Times*, October 8, 2014.

36 **modern forms of writing:** The figure is discussed by DeFrancis (1989), 84–85.

37 **Increase in abstraction:** The explanation is not very different from how the quality of one's own writing is affected by the precision of the writing instrument, the medium being written on, the amount to be written, how rapidly it has to be produced, and who has to be able to read it. Pictographs were hard to standardize and draw; writing with the wedge-shaped stylus was an improvement in those respects but worse for drawing realistic images.

37 **Writing systems that fully:** A detailed timeline is here: Geoffrey Nunberg, "Timeline of the History of Information," Berkeley School of Information, http://people.ischool.berkeley.edu/~nunberg/timeline.html.

38 **"unnatural act":** Gough & Hillinger (1980). Gough is one of the great figures in modern reading science, an iconoclast who turned out to be right about important things.

38 **"theory of mind":** Frith & Frith (2005). For a funny contemporary theory of the functions of cave paintings, see http://www.newyorker.com/cartoons/a19865.

42 **theory could be disconfirmed:** On undeciphered codes and precursors to cuneiform, see several chapters in Daniels & Bright (1996).

42 **an elegant theory:** Schmandt-Besserat (1986).

42 **Thousands of small objects:** For pictures of the tokens, with more about the theory, see "The Evolution of Writing," Denise Schmandt-Besserat, http://goo.gl/8Vv14V.

42 **world but suppositories:** Schmandt-Besserat (1978).

46 **Mark Liberman:** "What Is Writing?," University of Pennsylvania Department of Linguistics, http://www.ling.upenn.edu/courses/ling001/reading_writing.html.

47 **The pictographic information:** Walker (1990); on the "cursive" hieroglyphic, see "Ancient Egyptian Scripts," Omniglot, http://www.omniglot.com/writing/egyptian_hieratic.htm.

48 **Some characters can act:** In Figure 3.4, the HORSE radical (on the left in FOAL) is narrower than the HORSE phonetic (on the right in MOTHER). This convention reflects the greater importance of the phonetic in recognizing the character. Most words consist of two characters, representing two syllables. Monosyllabic words such as MĀ (mother) are atypical. The word almost always occurs in the disyllabic form 妈妈. The "MÀ curse" illustrates the four tones in Mandarin, which change the meanings of the MA syllables (courtesy of Tianlin Wang): mà curse: 妈妈骂马吗? Māma mà mǎ ma?
Does Mama curse the horse?

48 **In another expression:** In the modern era a number of characters containing the WOMAN radical (女) can be perceived as sexist, such as

奸 crafty, villainous, false
妄 absurd, foolish, reckless, false
妖 strange, weird, supernatural

From Joe, "Sexist Chinese Characters Discriminate Against Women," chinaSMACK, January 28, 2010, http://goo.gl/NiUYlT.

48 **A second system:** For examples of how the scripts are intermixed, see "Japanese Writing System," Wikipedia, http://goo.gl/hZFY6U. The "List of Gairaigo and Wasei-Eigo Terms," Wikipedia, http://goo.gl/vn7ruF, is entertaining.

49 **kanji's meaning depends on:** For a furigana example, see "Japanese Hiragana," Omniglot, http://www.omniglot.com/writing/japanese_hiragana.htm.

49 **A word is fully identified:** For the full sets of K-T-B words in Hebrew and Arabic, see "K-T-B," Wikipedia, https://en.wikipedia.org/wiki/K-T-B.

49 **Alphabets seem to have taken:** In fact the relationships between form and meaning are not wholly arbitrary, extending well beyond familiar examples of "sound symbolism," such as the GLISTEN-GLEAM-GLINT and SNORT-SNOT-SNIVEL clusters. Data crunching on large word lists has revealed many other nonarbitrary correspondences between sound and meaning. For example, names for males and females in English tend to have different phonological properties, and knowledge of these properties affects the interpretation of brand names; Cassidy, Kelly, & Sharoni (1999). See Dingemanse et al. (2015) for an overview.

51 **Languages with more complex:** DeFrancis (1989).

51 **major elements of Japanese phonology:** Taylor & Taylor (2014) is a masterful account of Japanese, Chinese, and Korean writing.

52 **"The depicting of objects":** Rousseau (1754), from Barton (1995).

52 **"The invention of the Greek alphabet":** From Olson (1996), an erudite analysis of Western scholars' traditional belief in the superiority of the type of writing system they happen to use, the alphabet.

52 **High praise:** Quoted by Barton (1995).

53 **vowels had surfaced earlier:** The alphabets for Semitic languages were consonantal, but a few consonants also functioned as vowels in some contexts, similar to the use of Y in English (as a consonant in YET but a vowel in LADY).

53 **world would be:** For the credo of the Quixotes at "UniSkript," see "What Is UniScript?," UniSkript, http://uniskript.org/what-is-uniskript.

54 **around 15,000 syllables:** For one such count, see Chris Barker, "How Many Syllables Does English Have?" Semanticsarchive.net, http://goo.gl/jxSLsu.

54 **spoken languages have ended up:** For some years I have been smartly issuing the epigram that "languages get the writing systems they deserve" (e.g., Seidenberg 2011). From a letter in the *New Yorker*, June 20, 2016, I learned that M. A. K. Halliday, a well-known linguist who worked on a broad range of interesting topics, made essentially the same observation. It's found in Halliday (1983, 28), where he wrote, "In the course of [writing's] long evolution, a language usually got the sort of writing system it deserved." The idea has probably occurred to others as well.

54 **Chinese speakers who keyboard:** Victor Mair, "Character Amnesia," Language Log, July 22, 2010, http://languagelog.ldc.upenn.edu/nll/?p=2473.

54 **Skeptics focus:** The standard objection is that pinyin would be hard to read because it eliminates important visual cues that characters provide. Spoken Chinese languages have a large number of homophones, which speakers often disambiguate by gesturing the character. The problem is illustrated in an exaggerated way by "The lion-eating poet in the stone den," a story consisting of the syllable "shi" spoken with different tones. This is the first line written in pinyin:

shi2 shi4 shi1 shi4 shi1 shi4 shi4 shi1. shi4 shi2 shi2 shi1. shi4 shi2 shi2 shi4 shi4

The whole poem can be seen and heard here: "The poet Shih Shih, so fond of eating lions," http://goo.gl/fxXfyy.

Richard Sproat, a scholar in Sinitic languages, writing systems, and computational linguistics, argues that pinyin or another suitable alphabet would be workable. See Sproat (2000). As in other cases like speed reading, what is tolerable in short bursts may be intolerable over longer stretches. Victor Mair (http://languagelog.ldc.upenn.edu/nll/?p=189) favors a mixed system in which characters are annotated with pinyin, much like kanji can be annotated with furigana in Japanese.

Chapter 4: The Eyes Have It

60 **prime example:** Writing allowed us to use what had been a latent capacity.

61 **The *Sorcerer's Stone* audiobook:** For a clip from the Jim Dale version, http://goo.gl/7PuOA3.

61 **The audio version:** Cranking the speed up to 1.5 times normal on your audiobook app will bring the total time down to about 5.5 hours, which is closer to the estimated reading time but hard to tolerate for that long. Getting the elapsed time down to 4.5 hours requires 1.8x playback, which is only intelligible in bursts.

A study using time-compressed instructional materials, which distorts the speech far less, found that listeners tolerated speech at 275 words per minute but comprehension was poorer than with noncompressed materials. At even higher speeds, comprehension drops precipitously. Pastore (2012).

Professional fast talker John Moschitta spoke about 360 words per minute in a famous commercial, a rare example of speech that is produced at a faster rate than most listeners can accurately comprehend. "FedEx Commercial with John Moschitta," video uploaded to YouTube by ThreeOranges, September 2, 2006, http://goo.gl/3gyjrc.

61 **Perhaps you have seen:** "Why America Isn't the Greatest Country in the World Anymore," video uploaded to YouTube by Jeff Taube, September 18, 2012. http://goo.gl/zlxI22.

62 **a baby's head:** *Mona Lisa*: "EyeTracking on Mona Lisa—Where Did the People Look at Da Vinci's Famous Painting," video uploaded to YouTube by DigitalAlchemistry, May 28, 2009, https://www.youtube.com/watch?v=e5Sa3H8QN6c; on infants, see "NYU Infant Action Lab—Infant Walking Around Our Playroom with an Eye Tracker," video uploaded to YouTube by NYU Action Lab, December 8, 2011, http://goo.gl/vcnFIc.

62 **properly designed experiment:** Most of the studies I'll describe were conducted by the late Keith Rayner, a research psychologist who was the world authority on eye movements in reading, or by his students. The *Scholarpedia* article that he coedited is an excellent resource: Keith Rayner and Monica Castelhano, "Eye Movements," *Scholarpedia* 2, no. 10 (2007): 3649, http://www.scholarpedia.org/article/Eye_movements.

63 **eye-tracking experiment:** I strongly recommend looking at an eye-movement recording. A good one is "Eye-Tracking Reading Example," video uploaded to YouTube by BUPsychTech, June 7, 2013, http://goo.gl/VnOxuT. One in which the size of the dots represents how long people looked at each fixation is "How We Read Shown Through Eyetracking," video uploaded to YouTube by digitalpolicy, March 12, 2010, http://goo.gl/96kouK. For eye movements of people looking at lots of things, see "TheDIEMProject's Videos," Vimeo, https://vimeo.com/visualcognition/videos. Fantastical claims about eye-movement observation and control, as dramatized in the 1931 movie *Svengali*, have been incorporated into new age therapies such as Neurolinguistic Programming (NLP). YouTube's indexing procedures treat the videos illustrating such claims as topically related to ones from serious eye-movement research and so they show up together in searches. This has been a consumer alert.

63 **smooth pursuit:** "Horizontal Smooth Pursuit Sitting," video uploaded to YouTube by Rehab My Patient, March 27, 2014, https://www.youtube.com/watch?v=jBY6m4YvWHk.

64 **The numbers I'll present:** These classic examples are adapted from McConkie & Rayner (1975).

65 **span is asymmetrical:** This asymmetry demonstrates an important cognitive phenomenon. The information that registers on a fixation depends on the reader's expectation about where to look next. Thus we covertly allocate attention (to the right or left) without moving our eyes. The phenomenon may be familiar from looking at a dinner companion while covertly attending to a person at the next table.

66 **space our fixations:** Short explanations and demonstrations are seen here: "What Eye Movements During Reading Reveal About Processing Speed," video uploaded to YouTube by the Children of the Code Project, April 6, 2012, https://www.youtube.com/watch?v=zQmf5TkJrJ8.

66 **Some masters:** Miellet, O'Donnell, & Sereno (2009).

71 **boost from President:** Kennedy speed reading recounted in Paston (2013).

71 **"Why have I dared":** Evelyn Wood quoted in a 1967 *Harvard Crimson* article that anticipated later controversies: Jeffrey C. Alexander, "Evelyn Wood: Most Just Waste the Money," *Harvard Crimson*, May 3, 1967, http://goo.gl/hKKzrI.

71 **In its 1970s:** "Evelyn Wood Reading Dynamics (Commercial #1, 1979)," video uploaded to YouTube by the Museum of Classic Chicago Television (www.FuzzyMemories.TV), March 6, 2008, https://www.youtube.com/watch?v=1nMP4U8JQDo.

72 **Although initially a convert:** McLuhan scene from *Annie Hall*: "Woody Allen Meets Marshall McLuhan," video uploaded to YouTube by Tralfaz666, December 1, 2011, https://www.youtube.com/watch?v=sXJ8tKRlW3E. McLuhan *aperçu* from *Understanding Media* (1964). His reading habits were unusual: he would read only the right-hand pages of books, reasoning that texts were so redundant he would not miss much (Marchand 1998).

72 **In the modern era:** "Cavuto Has World's Fastest Speed Reader Take Crack at 1500-Page Health Care Bill," video uploaded to YouTube by ReturnOfObamaSecrets, October 22, 2009, https://www.youtube.com/watch?v=620BII22Jfk.

73 **Woody Allen:** "Talk: Woody Allen," Wikiquote, https://en.wikiquote.org/wiki/Talk:Woody_Allen.

73 **1958 book:** Wood & Barrows (1958).

74 **Book of the Month Club:** Reading pacers are listed periodically on eBay. On the Book of the Month Club's reading-improvement efforts, see Wilkinson (1980).

74 **couldn't have been any evidence:** The zigzag scan is briefly seen in the opening credits for *Newsroom*'s second season, presumably to signify the fast-paced world of cable news. Unlike standard texts, many types of web pages are formatted so that specific types of content are placed in predictable locations. It works well for navigating the *Wall Street Journal*'s home page but not, say, a book like this one.

74 **for skilled deaf readers:** Bélanger et al. (2012).

75 **Serious video game players:** Green & Bavelier (2003).

75 **Using the phonological code:** The difference between subvocalization and mental phonology is easy to demonstrate. Decide if the following pairs of words rhyme. Some do (PEAR-DARE), and some don't (LATE-LAKE). Because it's such an easy task, try doing it while holding a pencil between your teeth or while saying "colacolacolacolacolacolacolacolacola" at the same time.

cloak joke, must mist, jack stack, stone blown, pint mint,
paid fade, dear wear, brain lane

You can still judge the rhymes even if you can't subvocalize (the pencil condition) or your mouth is otherwise engaged (the colacola condition), using the phonology in your head.

76 **All of these what-ifs:** For a technical but extensive review see Rayner et al. (2001). The *Scientific American* version is Rayner et al. (2002).

76 **Thairs moar two reeding:** Hat tip to Dennis, Besner, & Davelaar (1985).

78 **To remind yourself:** "Eye Tracking Reading Study," video uploaded to YouTube by Tobii Pro, May 29, 2009, https://www.youtube.com/watch?v=VFIZDZwdf-0.

78 **He reached his apotheosis:** "Berg, Howard S., in the Matter of," Federal Trade Commission, last updated June 19, 1998, http://www.ftc.gov/os/1998/06/9423278b.cmp.htm.

79 **His entry in the *Guinness Book*:** "Reading fast: 80 pages (25,000 words) per min is the supersonic 'reading' speed claimed by Howard Stephen Berg (born 1949, in Brooklyn), who

has convinced a number of TV hosts that he has comprehended and remembered what he has scanned, perhaps not the details, but the concepts, with the details left for later, slower reading. He teaches speed reading and gives lectures through North America on using the unused part of one's brain" (444–445). From which we learn that *Guinness*'s criteria for entry into their "compendium of astonishing, authenticated facts" include the judgments of TV personalities and that Mr. Berg's citation is for "reading" (their scare quotes) for gist not details (i.e., skimming). Similar controversies attend to Anne Jones, "six times winner of the World Speed Reading Championship" (http://www.speedyreader.co.uk). YouTube has videos of "world's fastest readers" from around the world. The most recent review of the science is Rayner et al. (2016).

79 **The ancient Greeks experimented:** Yakubovich (2010).

79 **"The Raven":** "The Raven (RSVP)," video uploaded to YouTube by Crutcher Dunnavant, February 19, 2008, http://www.youtube.com/watch?v=4mlS3UCV5SY.

79 **free websites:** Spreeder (http://www.spreeder.com); WordFlashReader (http://goo.gl/BepCzm) for Windows and Linux.

80 **In laboratory studies:** Potter (1984).

81 **a 2014 RSVP app:** Spritz (http://spritzinc.com).

Chapter 5: F u cn rd ths, u cn gt a gd jb n rdng rsch

85 **a massive hoard:** 250 billion photos, 350 million uploaded per day, according to a Facebook announcement in 2013.

85 **"deep learning":** For an overview, see "The Code That Runs Our Lives," video uploaded to YouTube by the Agenda with Steve Paikin, March 3, 2016, https://www.youtube.com/watch?v=XG-dwZMc7Ng; for the graduate seminar, see "Geoffrey Hinton: 'Introduction to Deep Learning & Deep Belief Nets,'" video uploaded to YouTube by Institute for Pure and Applied Mathematics (IPAM), August 24, 2015, https://www.youtube.com/watch?v=GJdWESd543Y; for the spooky part, see Raffi Khatchadourian, "The Doomsday Invention," *New Yorker*, November 23, 2015.

85 **Stylometry:** "Stylometry Methods and Practices: Home," Temple University, http://guides.temple.edu/stylometryfordh.

85 **profiles of texts:** A statistical profile of the book was compiled using a few easily countable linguistic properties (e.g., the one hundred most frequently used words, the most frequent two-word combinations). These statistics were then compared to the ones for books written by some other female British mystery writers and a *Harry Potter* book. The match between the Galbraith and Potter books was much closer than for any other pair, suggesting they were written by the same person. Rowling outed herself as the author soon thereafter. For the tale as told by Patrick Juola, one of the sleuths, see Ben Zimmer, "The Science That Uncovered J. K. Rowling's Literary Hocus-Pocus," *Wall Street Journal*, July 16, 2013, http://goo.gl/AEIfi.

85 **Similar methods showed:** Markowitz & Hancock (2014).

86 **The most poignant application:** About Murdoch, see Dwight Garner, "Review: 'Living on Paper,' Seven Decades of Letters from Iris Murdoch," *New York Times*, January 5, 2016. Byatt quote from Garrard et al. (2005).

86 **Truman Capote famously:** *The David Susskind Show*, January 18, 1959. From Battaglio (2011).

86 **People have been tabulating:** For a newspaper article about an early Bible concordance compiled by computer, a Univac, see E. C. Keissling, "Faith and Univac," *Milwaukee Journal*, July 14, 1957, Google News, http://goo.gl/Gh2rSK.

87 **They begin in utero:** Jusczyk (2000).

87 Later, reading becomes: Romberg & Saffran (2010).

88 As the text circulated: The people of the Internet created many versions of this text. Mine is from Snopes.com ("Can You Raed Tihs?," Snopes, http://www.snopes.com/language/apocryph/cambridge.asp). See the history here: "Aoccdrnig to Rscheearch . . . ," Know Your Meme, http://goo.gl/QntGJs. For an account written by an actual Cambridge reading researcher, see http://goo.gl/q4PUcj.

88 Some people think: For one such example, see Lidor Wyssocky, "The Magic Button," *Creativity Post*, June 29, 2016, http://goo.gl/aDiXr4.

89 The range of possibilities: Adams (1980) is a primary source for this section.

90 property is known as redundancy: See Gleick (2011) or "Intro to Information Theory: Claude Shannon, Entropy, Redundancy, Data Compression, and Bits," *Cracking the Nutshell*, http://goo.gl/bKeqXV. "Redundancy" is one of a set of related concepts, including information, mutual information, uncertainty, entropy, and others.

90 Captcha security systems: Captchas keep getting harder in part because bots are getting better. More importantly, they can be defeated by cheap human labor paid to decode them: Motoyama et al. (2010).

91 English is redundant: At one time the Apple iTunes store censored the Gilbert and Sullivan song "T*t Willow," exactly like that. I clipped an image of it on March 9, 2010, available on seidenbergreading.net. It was not censored on British iTunes.

91 Peter Norvig: "English Letter Frequency Counts: Mayzner Revisited or ETAOIN SRH LDCU," Norvig.com, http://norvig.com/mayzner.html.

92 all that spelling data: Just as organisms are merely vehicles for passing along their selfish genes, books are merely the vehicles for passing along orthographic statistics!

92 pigeon reading: Blough (1982).

92 A study in *Science*: Grainger et al. (2012).

92 "baboons can read!": Goldberg (2012).

93 tournament Scrabble players: Fatsis (2001).

93 Scrabble skills do not carry: Tuffiash, Roring, & Ericsson (2007); Hargreaves et al. (2012).

94 "The Knowledge": For a famous study of those taxi drivers, see Maguire et al. (2000).

94 Unlike Scrabblists: Maguire (2006).

95 A utility for generating Cmabrigde: "Scramble a Word," 4umi, http://goo.gl/K8N6bI.

95 The only broad generalization: The amount of information needed to identify a given word isn't fixed; more is needed in noisy contexts (ones in which the letters are obscured or hard to see) than in clear ones.

96 how well a book will sell: Ashok, Feng, & Choi (2013) took a stab at it.

96 president uses personal pronouns: Obama critics repeatedly cast aspersions about his unusually frequent use of first-person pronouns (e.g., http://goo.gl/Qufkqf). They hadn't actually counted, but Mark Liberman at Language Log did: http://languagelog.ldc.upenn.edu/nll/?p=14625.

97 Sudoku: Seidenberg & MacDonald (1999). The mechanisms that I am describing informally have been rendered in more rigorous computational and quantitative terms. See, e.g., Piantadosi, Tily, & Gibson (2012), Chater, Tenenbaum, & Yuille (2006), and Flusberg & McClelland (2014).

97 Working back and forth: David Rumelhart described reading as an interactive process in a classic 1977 article. McClelland & Rumelhart (1981) then took the next step, implementing a computational model of interactive processes in word and letter recognition. I have informally redescribed a bit of this hugely influential work.

98 The answer could be wrong: The representation of words by combinations of semantic and phonological cues in writing systems is based on this principle.

99 *Wheel of Fortune:* "Wheel of Fortune: Amazing One-Letter Solve!," video uploaded to YouTube by Wheel of Fortune, March 26, 2012, http://goo.gl/bjU1Hz.

Chapter 6: Becoming a Reader

103 assessment for eighth graders: NAEP (2011), 58.

103 [Individual demonstrates ability]: From "English Language Arts Standards » Reading: Literature » Grade 11–12," Common Core State Standards Initiative, http://www.corestandards.org/ELA-Literacy/RL/11-12.

105 Two esteemed reading researchers: Beck & Juel (1995).

106 Louis Goldstein and Cathy Browman: Browman & Goldstein (1990).

106 The translation from one code: Werker & Tees (1999). Phonology is also a branch of linguistics that addresses a much more extensive range of phenomena than the pronunciations and sounds of words.

106 acoustic blobs: Infants rapidly acquire strong language-specific sensitivity to differences between initial sounds in syllables such as "bin" and "din." They nonetheless treat the syllables alike in word-learning experiments (Stager & Werker 1997). The sounds can be distinguished phonetically (based on smaller features) but aren't yet treated as discrete phonemic segments.

107 "lexical restructuring": Metsala & Walley (1998).

107 Phonemes are abstractions: Phonemes (and the letters that represent them) involve a further abstraction because they ignore systematic variation in how they are pronounced. Both "pin" and "spin" contain the phoneme /p/, written with the same letter, but the sound is aspirated in "pin" and unaspirated in "spin." These variants of the phoneme (called allophones) are not represented in this writing system. In other languages, the allophones are distinct phonemes with their own letters.

107 speech consists of phonemes: Fowler (1991).

108 Studies in English: McCardle, Scarborough, & Catts (2001).

108 Letter songs may be common: Horace Mann (1844, 91, 92) also found letter names perplexing: "I am satisfied that our greatest error, in teaching children to read, lies in beginning with the alphabet;—in giving them what are called the 'Names of the Letters,' *a,b, c,* &c." Later: "Although in former reports and publications I have dwelt at length upon what seems to me the absurdity of teaching to read by beginning with the alphabet, yet I feel constrained to recur to the subject again,—being persuaded that no thorough reform will ever be effected in our schools until this practice is abolished."

108 The names teach the child: Some educators now favor associating letters with sounds rather than names on the view that the sound "b" is more relevant to reading than the name "bee." The initial sound in "bat" is not pronounced "bee" to be sure, but it is not the "buh" we conventionally produce as the letter's pronunciation either. Consonants cannot be pronounced in isolation; they require a vowel, which is truncated when we say "b." A recording of the sound "b" produced in isolation, spliced onto a recording of "at," yields a weird syllable, not "bat."

108 beginning reader has to learn: The font on the left in Figure 6.1 was created by David Rumelhart, a brilliant and inspirational figure in modern cognitive science (Rumelhart & Siple 1974).

109 Recognizing letters: Lake, Salakhutdinov, & Tenenbaum (2015). See a model learn to recognize handwritten digits here: http://www.cs.toronto.edu/~hinton/adi/index.htm. Explanation here: "Explanation of the Digit Movies," University of Toronto Computer Science, http://goo.gl/AbNxE2. For a fascinating discussion of how a deep learning network handled the letter categorization problem for fifty fonts, see Erik Bernhardsson, "Analyzing 50k Fonts Using Deep Neural Networks," erikbern.com, January 20, 2016, http://goo.gl/e8S2w5.

109 role of names in forming categories: Lupyan, Rakison, & McClelland (2007).

110 narrow sense of knowing: The Peabody Picture Vocabulary Test is a widely used assessment of this type; Dunn & Dunn (2015).

110 Charles Perfetti: Perfetti (2007).

111 Much of our knowledge: MacDonald, Pearlmutter, & Seidenberg (1994).

111 The cup could hold coffee: Seidenberg (1997). **The fur cup:** http:/mo.ma./2j8vvqk/works/80997

111 Five- to six-year-old English learners: Beck & McKeown (1991).

111 This question assumes great importance: Hoff (2003).

111 effective teaching procedures: Beck, McKeown, & Kucan (2013).

112 "You shall know a word": Firth (1957).

112 of little use: The limited relevance of language statistics was the point of Chomsky's famous sentence "Colorless green ideas sleep furiously." The statistical properties of the word sequences were said to be so low as to be indistinguishable from "Furiously sleep ideas green colorless," which is word salad. Chomsky was wrong, however. See this post for an overview of the research by computational linguist Fernando Perreira: "Colorless Green Probability Estimates," Language Log, October 4, 2003, http://goo.gl/45Hzyt.

We also implemented a simple computational model that distinguished between the two by tracking expected semantic properties rather than specific words: Allen & Seidenberg (1999).

112 a landmark study: Landauer & Dumais (1997). Trigrams are used here to illustrate their general idea.

113 According to the statistical learning theory: The "bootstrapping" concept was introduced by Lila Gleitman (1990) in a highly influential account of how children learn about the syntactic structures associated with words, but it applies to learning many things.

113 This account of word learning also explains: Carey (1978).

113 children are born knowing: The innate concepts theory comes from Fodor (1975). The statistical learning account doesn't exclude the possibility that human biology biases infants' learning toward certain types of generalizations.

113 Computational analyses of the performance: Zhu (2005). It should be possible to use statistical methods to determine which words should be selected for explicit instruction to achieve the greatest impact, given the current state of a child's vocabulary. Words can be represented in a multidimensional semantic space, with the distance between them indicating degree of similarity. This space can be mapped for samples of words (e.g., the ones that a third grader should know). Call this the goal. Learning a word also affects other words with which it overlaps. Given an index of the current state of a third grader's vocabulary, it could be determined which word would have the biggest impact on moving the state of their knowledge toward the goal. This would be a good thing because vocabulary instruction is time-consuming, and teaching time is limited.

114 The late Martin Gardner: Douglas Martin, "Martin Gardner, Puzzler and Polymath, Dies at 95," *New York Times*, May 23, 2010, http://www.nytimes.com/2010/05/24/us/24gardner.html.

115 Kanzi, a bonobo: Savage-Rumbaugh et al. (1986).

115 The scientists who conducted: Seidenberg & Petitto (1987).

115 Research has clearly established: Justice & Ezell (2002).

115 Some children do learn: Given the overwhelming evidence that children benefit from reading instruction and are hindered by poor or absent instruction, it astonishes me when individuals with no discernible expertise but a popular platform declare, "Children teach themselves to read." Gray (2010).

115 Reading to children: This keen observation is due to Maryellen MacDonald, who shared it with me some years ago, as well as the duckling and hungry caterpillar examples. Montag & MacDonald (2015), Montag et al. (2015), and Cameron-Faulkner & Noble (2013) document that children's books contain a higher proportion of complex sentence types (passives, relative clauses, conjoined clauses) and more diverse vocabulary than does parental speech to children. Montag & MacDonald (2015) found that children's reading experience affected how often they used certain complex sentence structures when speaking.

116 At thirty words: On children's speech and parental speech to children, see Snow (1977); Huttenlocher et al. (1991).

116 You talk better too: See *Calvin and Hobbes*: http://www.gocomics.com/calvinand hobbes/1995/06/10.

116 A low-income family: Neuman & Celano (2001).

116 Reach Out and Read program: http://www.reachoutandread.org.

117 Kindergarten is sometimes: Bassok & Rorem (2014) report data indicating increases in teachers' emphasis on reading in kindergarten. Whether children benefit from this earlier emphasis hasn't been established.

117 Within the pre-K to grade two: Snow et al. (1998); Cunningham & Zibulsky (2013). These findings run counter to the intuition that starting as early as possible is advantageous. From 2008 to 2012, credulous parents could purchase "Your Baby Can Read," devised by kinesiologist Robert Titzer. This was "Mega-Reading" for children, sold on a late-night infomercial. Eventually the FTC caught up with it, decided it had no educational value, and enjoined Titzer and his company from selling products that claimed to teach babies to read. Titzer was fined $185 million, equal to the gross sales for "Your Baby Can Read" (!), with the stipulation that the fine would be suspended after $300,000 was paid (!!). For the FTC judgment, see "Your Baby Can LLC, et al.," FTC, http://goo .gl/WliG4L. As in the Mega-Reading case, the plaintiff continues to ply his trade, selling products including "Your Baby Can Learn!," "Your Child Can Read!," and "Your Baby/ Child Can Discover!" Caveat emptor.

117 Fred Morrison: Morrison, Alberts, & Griffith (1997).

118 Asked to pick out: Ehri (1995).

118 "a simple view of reading": Hoover & Gough (1990); Gough, Hoover, & Peterson (1996). The main weakness in Gough's theory is that it did not make sufficient room for the ways that the components influence each other. Vocabulary, for example, is jointly determined by spoken language and reading. Vocabulary can also be considered a component of both basic skills and comprehension.

120 Such difficulties are commonly reported: Hulme et al. (2007).

121 At the neural level: Munakata & McClelland (2003).

121 It is also important to appreciate: Escoffier & Di Giacomo "Take Away the A" (2014) is a phonemic awareness alphabet book. "SNOW without the s falls now," it says, but "snow" without the "s" says "no."

Chapter 7: Reading: The Eternal Triangle

123 Writing systems vary: Frost, Katz, & Bentin (1987).

123 myriad inconsistencies: The English Spelling Society wants to modify spelling (http:// spellingsociety.org). For PG-17-rated spelling humor, watch "'The Impotence of Proofreading,' by Taylor Mali," video uploaded to YouTube by Taylor Mali, August 14, 2008, http://goo.gl/PdZlYZ.

127 His first, landmark article: Van Orden (1987).

127 The subjects in these studies: The figure illustrates conditions in the series of experiments reported in Van Orden (1987) and Van Orden, Johnston, & Hale (1988). They weren't all included in a single experiment, and some control conditions are omitted.

127 I've created a toy version: On seidenbergreading.net.

128 reliably make more erroneous: Van Orden, Johnston, & Hale (1988).

129 unreplicable "false positives": On replications in psychology, see the several articles in *Perspectives on Psychological Science*(7[6], 2012).

129 Phonological effects: Pollatsek et al. (1992) and Lee et al. (1999) are representative studies.

129 THE TAXIS DELIVERED THE TOURISTS: Filik & Barber (2011); McCutchen & Perfetti (1982).

129 And so on: Ellis (1979); Jared, Levy, & Rayner (1999).

129 Van Orden effect: Goswami et al. (2001).

129 third- and fifth-grade children: Perfetti & Hogaboam (1975).

130 Is skill in reading aloud: Spurious correlation fun! "Spurious Correlations," tylervigen .com, http://www.tylervigen.com/spurious-correlations.

130 The second Perfetti experiment: Perfetti & Roth (1981).

130 Good readers were faster: Perfetti and his colleagues conducted many such studies in this period, including ones in which children either heard the context over headphones or read it themselves; the target words appeared at intermittent intervals in a story; and the words were preceded by unrelated words or read in isolation. They also looked at factors such as word length and frequency. The conclusions rest on this set of studies and others they inspired.

131 a good reader rapidly identifies: Keith Stanovich put these pieces together in 1980 (Stanovich 1980).

131 "the eye of the tiger": For "Roar," see "Katy Perry—Roar (Official)," video uploaded to YouTube by KatyPerryVEVO, September 5, 2013, https://www.youtube.com/watch ?v=CevxZvSJLk8. Pointed out to me by Maryellen MacDonald.

131 Korean alphabet: On Korean spelling reform, see "Korean," Omniglot, http://www.omniglot.com/writing/korean.htm.

131 The problem is: Joshi et al. (2008) is a fine, accessible discussion of English spelling.

133 SIGN is an oddity: Chomsky & Halle (1968).

133 Skilled readers are aware: The classic work on this topic is J. Winter, "How I Met My Wife," *New Yorker*, July 25, 1994. It begins, "It had been a rough day, so when I walked into the party I was feeling very chalant, despite my efforts to appear gruntled and consolate." For cartoonist Mark Stivers's witty illustrations of morphological inconsistencies, see http://goo.gl/qiIdeV.

135 Some of the most interesting work: Spencer and Hanley (2003); Hanley et al. (2004).

135 Researchers have concluded: Hoxhallari, van Daal, & Ellis (2004).

135 As comprehension is clearly the goal: Hanley et al. (2004).

136 Spelling-sound inconsistencies: Seidenberg (2011).

138 An alternative approach: Thomas & McClelland (2008); Rogers (2009).

138 My colleagues and I: For overviews, see Seidenberg (2005) and Plaut (2005).

139 The models I'll describe: Deep learning models are solving hard problems such as speech recognition and scene analysis, and what matters is how well they work, not whether they work the same way as people. Our goal was to try to understand some basic properties of human behavior and their brain bases, and for that purpose the simpler models have proved helpful.

139 Basic properties of these neural networks: Flusberg & McClelland (2014).

139 Think of the card section: A card section creates a distributed representation of a word: http://goo.gl/Oe9cT2. Each card is a binary "unit" that is either on or off (light or dark). Each letter is represented by a unique pattern of on and off units. Each card contributes to the representations of many letters. Words can be represented by using cards to represent successive letters.

144 Rule-governed but inconsistent words: The breakthrough study was Glushko (1979), which launched many others.

144 the system isn't rule governed: Seidenberg (2012).

144 Language is said to be rule governed: Our first model could correctly pronounce simple nonwords like NUST but mispronounced hard ones like JINJE. The errors arose from the fact that it didn't adequately represent phonology. Once that was improved, so did generalization.

145 The idea that generalization: The application of these models to language was controversial: Seidenberg & Plaut (2014).

146 This developmental progression: Pugh et al. (2010).

146 It has long been thought: Baron & Strawson (1976) called them "Chinese" and "Phoenician" readers. At the time it was thought that Chinese characters could only be read visually, but we now know that Chinese readers make use of the phonological cues in the radical + phonetic characters.

147 At a given level of experience: The empirical jury is still out on how much skilled readers vary. Behavioral research has tended to focus on a narrow demographic (the "convenience sample" of mainly white, middle-class good readers) that does not represent the full range of backgrounds and experience. The models also focused on only a few of the factors that affect outcomes. Neuroanatomical variation relevant to skilled reading is only starting to be studied, but here is one provocative finding. In a recent study (Graves et al. 2014), we found individual differences among highly skilled readers in their use of the orthography→semantics→phonology pathway in reading aloud. Reliance on this pathway was found to be correlated with neuroanatomical variation in the degree of connectivity (density of white matter tracts) between brain regions for semantics and phonology. Further studies of this sort should settle questions about how much skilled readers vary in division of labor. Then we will have to figure out why.

147 The most interesting evidence: Yang et al. (2009).

147 Such sequential models exist: Thomas & McClelland (2008).

Chapter 8: Dyslexia and Its Discontents

149 Shifts in the understanding: A good source for basic information is "Dyslexia at a Glance," International Dyslexia Association, http://eida.org/dyslexia-at-a-glance.

150 combination of a high-interest topic: Researcher Dorothy Bishop on how to distinguish credible interventions from quackery: "Pioneering Treatment or Quackery? How to Decide," BishopBlog, December 4, 2011, http://goo.gl/UfC1JJ. Her blog is an exceptionally good, highly recommended source of information on developmental disorders and many other topics.

151 For example, impaired phonological processing: Pennington & Bishop (2009) is an excellent review.

153 Definitions vary: See Fletcher (2009) for a history of the concept of dyslexia.

153 American Academy of Pediatrics provides: The American Academy of Pediatrics (2009) statement is also very clear about why vision problems are rarely the cause of dyslexia. "Learning Disabilities, Dyslexia, and Vision," *Pediatrics* 124, no. 2 (August 2009), http://pediatrics.aappublications.org/content/124/2/837.full.

154 definitions used in such legislation: On IDEA language, see "Topic: Identification of Specific Learning Disabilities," ED.gov, http://goo.gl/L9bTjs.

154 It defines a general category: For an explanation of the DSM-5 categorization scheme, see Rosemary Tannock, "DSM-5 Changes in Diagnostic Criteria for Specific Learning

Disabilities (SLD)1: What Are the Implications?," International Dyslexia Association, https://eida.org/dsm5-update.

154 Researchers have their own accounts: Thomas Insel, former director of the National Institute for Mental Health (NIMH), launched an initiative to replace the DSM with a new classification system based on behavioral, neural, and genetic data. As yet there isn't sufficient evidence to sort conditions on these bases, however. See Thomas Insel, "Director's Blog: Transforming Diagnosis," NIMH, April 29, 2013, http://goo.gl/3oNVfY.

154 children have spoken-language impairments: Dyslexia is far more common than comprehension disorder. Very few children with good basic skills fail to comprehend. When this occurs, the first question is whether the child can understand the text if it is read to them. Instructional and motivational issues aside, comprehension deficits are not reading-specific. The extreme example is hyperlexia, a rare condition seen in some children on the autism spectrum for whom reading words aloud is their area of specialized expertise, but their comprehension of the same words is poor. Good decoding with poor comprehension also occurs in some children with developmental language impairments. See Snowling & Frith (1986) and Snowling (2013).

154 World Federation of Neurology defined dyslexia: Critchley (1970).

155 US Department of Education definitions: The history is in Hammill et al. (1988).

155 Reading ability and intelligence fall: On the history of the IQ discrepancy criterion, see Fletcher (2009).

155 inherent arbitrariness of the boundary: Keith Stanovich conducted the critical research on the flaws in the IQ discrepancy criterion. See Stanovich (2005).

155 cutoffs were initially set: The logic here is similar to the focus on the relatively rare cases of brain-injured patients with highly selective impairments in neuropsychology.

156 heritability of a trait: The discussion of height in Lai (2006) is helpful.

156 use of discrepancies: Stanovich (2005); Fletcher (2009).

156 correlations are modest: Peterson & Pennington (2012).

156 Patterns of brain activity: Tanaka et al. (2011).

156 primary question: The IQ discrepancy criterion was flawed because it excluded lower-IQ children whose reading was indistinguishable from that of higher-IQ dyslexics. Dyslexia is now more often defined by the discrepancy between the child's reading level and what is expected for their age or grade. Bruce Pennington, a distinguished figure in this field, has pointed out that this procedure will overlook high-IQ children whose reading is within the normal range but below what might be expected given their intelligence. Children with this type of discrepancy may well merit additional attention at school, which could easily be overlooked. However, given their level of reading skill, categorizing these children as "dyslexic" would probably create further confusion because the category would no longer be limited to children who read poorly.

157 individuals can be studied: Pennington & Lefly (2001). A bump appears at the low end of the distribution if such children are included.

158 Dyslexia is thus the *purposeful*: As trenchantly expressed to me by Judith Levine.

158 boundary between having: Also paralleling dyslexia, changes in the criteria for identifying hypertension have created controversy; see James et al. (2014); Candy Sagon, "New Blood Pressure Guidelines Draw Fire," *AARP Bulletin*, March 2014, http://goo.gl/RGDZGx.

159 "There is no such thing": Arin Gencer, "National Reading Expert Comes to Baltimore County," *Baltimore Sun*, February 13, 2009, http://goo.gl/L8Dp2t. Also: "I don't happen to believe that dyslexia or learning disabilities exist, dyspedagogia yes, but not the mythological dyslexia." Anthony Rebora, "Responding to RTI," *Education Week*, April 9, 2010, http://goo.gl/mIBQ8M.

159 Some of the biggest excuses: Lindsay Fiori, "Getting Children to Read Books Must Supersede Disinterest, Limited Funding, Learning Disorders," *Journal Times*, February 7, 2012, http://goo.gl/L8n7x1; Allington (2012). For the British version, see "Dyslexia Is NOT a Disease. It Is an Excuse for Bad Teachers," *Mail Online*, March 2, 2014, http://goo.gl/BeyLtt.

159 Whereas a 2009 cover article: Gabrieli (2009); Elliott & Gibbs (2008); see also Elliott & Grigorenko (2014).

159 Keith Stanovich: Stanovich (1994).

159 Extreme skepticism about dyslexia: For somewhat different views, see Dorothy Bishop, "My Thoughts on the Dyslexia Debate," BishopBlog, March 20, 2014, http://goo.gl/bAs79p, and Elliott & Grigorenko (2014).

161 dyslexia and ADHD were invented: A colleague from education has suggested that Allington's rhetoric was in the service of a constructive goal, dissuading teachers from abandoning responsibility for their poorest-performing children because they have a disability. The ends are not justified by dissembling about the existence of the condition, however.

161 Given this behavioral overlap: As Willcutt et al. (2010a) stated in a review of the etiology of reading disability (dyslexia), math disability, and ADHD, "There is currently no valid genetic test for [these conditions], and it is unlikely that a definitive diagnostic test will be developed in the near future. Because most developmental disorders have polygenic, multifactorial etiologies in which each risk factor confers only a small increase in susceptibility, it is unlikely that any specific risk factor will have sufficient predictive power to be useful as a diagnostic measure."

162 elegant solution: Fletcher & Vaughn (2009).

163 For dyslexia they are questions: See Peterson & Pennington (2012).

164 Repeatedly it has been found: Goswami (2015).

165 low-vision problem: Frith & Happé (1998).

166 impairments do not just co-occur: See Peterson & Pennington (2012) for discussion.

166 Co-occurring conditions can also result: Bishop (2009).

167 researchers have powerful quantitative methods: Willcutt et al. (2010a), for example.

168 The pioneering experiment was conducted: Scarborough (1990).

169 One study followed at-risk children: Muter & Snowling (2009).

169 Maggie Bruck's landmark studies: Bruck (1990, 1992).

169 later study from this group: Hulme et al. (2015).

169 Picture babies: Linda Geddes, "The Big Baby Experiment," November 4, 2015, http://www.nature.com/news/the-big-baby-experiment-1.18701.

170 Speech sounds elicit: Richardson et al. (2003).

170 exposed to maternal smoking: Key et al. (2007).

170 anomalous evoked potentials: Espy et al. (2004).

172 hidden unit activations: The figures show the contributions of each hidden unit to the activation of a phonetic feature critical for the "ee" vowel as in "eat."

172 The dyslexic model was trained: We also analyzed the similarity of the hidden unit representations for the -EAT words using multidimensional scaling. In the normal model, the representations of the rhyming -EAT words clustered together, and the nonword GEAT was in there too. In the dyslexic model, they were more spread out, indicating that the model hadn't picked up the pattern for this rime. The figures are on seidenberg.reading.net.

172 phonological impairment: The phonological impairment was implemented in several ways that had similar effects on the model's behavior. One of the methods instantiated the idea that dyslexics benefit less from each learning experience. The weights in the network would change in the normal manner on each training trial, but the changes were not wholly retained. Neurobiological mechanisms that may have this type of effect have recently begun to be identified. Pugh et al. (2014).

173–174 Consider this behavioral study: Perrachione et al. (2011).

174 Similar effects have been observed: See, for example, Pugh et al. (2014).

175 In the real world: In a series of studies, Anne Sperling, Frank Manis, Zhong-Lin Lu, and I found evidence for noisy visual processing in dyslexia. For example, in one study (Sperling et al. 2006) we created visual stimuli for a motion-detection task that were presented with high or low levels of visual noise. (An example can be found on seidenbergreading.net.) Dyslexics performed as well as nondyslexics with low noise levels but much worse at higher noise levels, meaning they were less able to ignore or inhibit the irrelevant visual noise. Our other studies suggested that noisy processing may be internal to the visual system (Sperling et al. 2005). The research did not address how often such deficits occur in dyslexics and nondyslexics, how the impairments develop, and whether they co-occur with other impairments, however.

175 brain solves the problem: If letter names are critical to the formation of letter categories, multiple representations for syllables such as "bee" and "eff" would also lead to a less efficient solution, though this observation is entirely conjectural.

176 documented in multiple languages: Ziegler & Goswami (2005).

177 Some years ago a team: Paulesu et al. (2001).

178 dyslexia spectrum: Snowling (2006).

180 A small number of neuroimaging studies: Diehl et al. (2014).

180 Knowing that a celebrity: Gladwell (2013).

181 He describes a study: Diemand-Yauman et al. (2011).

182 For evidence, he relies: Logan (2009).

184 My guess: For a far more revealing and informative case study, I recommend the renowned artist Chuck Close, starting with this long interview: "Oral History Interview with Chuck Close, 1987 May 14–September 30," Archives of American Art, http://goo.gl/1mXoit. Chuck Close identifies as dyslexic and prosopagnosic and is forceful in asserting that his disabilities shaped his life and art. The Close interview provides a wealth of detail. For example, his description of his dyslexia matches that of many other dyslexics. He describes discovering coping strategies such as getting more time for assignments, avoiding timed tests, and having someone check his spelling. He became an outstanding student. Tellingly, he mentions that his daughter is also dyslexic. His prosopagnosia seems closely related to his art; however, his life history is as complex as every other. He describes intense interest in drawing from a young age, before a reading impairment would have manifested. Art came easily to him; reading did not. Art was rewarded; reading was a source of embarrassment and humiliation. Close is deeply insightful about his conditions and their impact on his work and life history, and probably a clearer example of how an impairment can enable an unusually talented person to achieve extraordinary success.

Chapter 9: Brain Bases of Reading

188 Complexity is intimidating: For some beautiful brain tractography, visit http://goo.gl/3jzKxd.

188 The truly exciting prospect: Much as our view of another familiar phenomenon, peptic ulcers, radically changed with the discovery that most are caused by bacterial infection. See PubMed Health, "Helicobacter Pylori," http://goo.gl/2Ou1wn.

188 It's worth a look: David Van Essen's circuit diagram of the visual system, UCSD Neurosciences, http://goo.gl/X9ms3p.

188 this conflict poses a challenge: For another approach to this challenge, see http://www.gocomics.com/calvinandhobbes/2013/06/18.

188 It is very hard to write well: A figure reproduced on many quality websites (it can be seen here on Medline Plus: http://goo.gl/jwIsNF) illustrates the concern. It mixes basic anatomical facts, such as the locations of the major lobes, with simplistic assignments of functions to specific areas (e.g., Wernicke's) and one pure fiction: there is no "reading comprehension area."

189 A good theory: Putnam (1972).

189 forty-three major anatomically distinguishable: Quick introduction, McGill University, The Brain, http://goo.gl/udSioK.

189 modern Brodmann map: Glasser et al. (2016). For a video about the study, see "The Ultimate Brain Map," uploaded to YouTube by Nature Video, July 20, 2016, http://www.youtube.com/watch?v=UHDfvfYCY0U.

190 tuberculosis in the literature: As discussed in Susan Sontag's classic *Illness as Metaphor* (Sontag 1978).

191 The late Oliver Sacks: "I am a clinical ontologist, one for whom the diagnostic question is: How are you? How do you be?" Sacks as quoted by Lawrence Weschler, "A Rare, Personal Look at Oliver Sacks's Early Career," *Vanity Fair*, April 28, 2015, http://goo.gl/WbXIbh. In the same article, Weschler writes, "[Sacks] respects facts, he tells me, and he has a scientist's passion for precision. But facts, he insists, must be embedded in stories. Stories—people's stories—are what really have him hooked."

192 found that she regularized: Marlene Behrmann (personal communication).

194 We then developed models: The first-generation model: Seidenberg & McClelland (1989); the second-generation model with division of labor: Plaut et al. (1996).

194 The impairment is due: Hodges et al. (1992). The cause is usually Pick's disease, a type of frontotemporal dementia.

194 Also as predicted: Patterson et al. (2006).

195 Responding with a word: Funnell (1983).

196 One patient was reported: Saffran & Marin (1977).

197 David Plaut, Tim Shallice, and Geoff Hinton: Hinton, Plaut, & Shallice (1993) is an accessible and entertaining overview.

199 As a British neurologist noted: Leff (2004).

201 "pure alexia" is not pure: Behrmann, Nelson, & Sekuler (1998). This type of lesion would also affect an illiterate's performance on such tasks.

203 Connections from areas in the temporal lobe: Presenting the components in this way omits crucial facts that are hard to diagram: that there are feedback in addition to feedforward connections, which create much more complex neural dynamics, and that the functions of the parts depend on and are changed by each other because they are interconnected.

203 "Language" was located: My colleague Tim Rogers has pointed out that language may have been situated beneath the eyes because of a patient with a language impairment who also exhibited exophthalmos, the eye-bulging symptom of Graves' disease.

203 visual word form area is one: Behrmann & Plaut (2013).

204 The evolutionary story: Caramazza & Mahon (2003).

204 Swiss army knife: Pinker (2003).

204 increasingly sophisticated methods: Behrmann & Plaut (2013).

205 Spelling gets routed: On Stanislaus Dehaene's neuronal recycling hypothesis, see Dehaene (2010).

205 study of eighty patients: Hillis et al. (2005).

205 processing of visual word forms: For a case study of a patient who could read following surgical excision of the VWFA, see Seghier et al. (2012).

205 it is activated by objects: Price & Devlin (2003).

206 study of nonimpaired readers: Mano et al. (2012).

206 More sensitive methods: Cox, Seidenberg, & Rogers (2015).

206 other brain areas also respond: McCandliss, Cohen, & Dehaene (2003).

206 More recently Behrmann and Plaut: Behrmann & Plaut (2013).

206 The term "visual word form area": Sandak et al. (2004) referred to it as a "skill zone," but the term is perhaps too general.

208 Holmes nearly said to Watson: "You could not possibly have come at a better time, my dear Watson," he said cordially. Arthur Conan Doyle, "The Red-Headed League," East of the Web, http://www.eastoftheweb.com/short-stories/UBooks/RedHead.shtml.

208 subjects are quicker: Ziegler & Ferrand (1998).

209 The experiment was then repeated: Stoeckel et al. (2009).

209 The technique may seem odious: A British journalist experiences TMS: "Michael Mosley Has Areas of His Brain Turned Off. The Brain: A Secret History. BBC Four," video uploaded to YouTube by BBC, December 24, 2010, https://www.youtube.com/watch?v=FMR_T0mM7Pc. Some methods of applying the stimulation improve rather than interfere with performance.

209 Signatures of developmental reading impairments: For recent reviews, see Peterson & Pennington (2012); Pennington & Bishop (2009); Goswami (2015).

210 The effect is to interfere: Turkeltaub et al. (2003).

210 poorer performance on visual tasks: The visual deficit hypothesis is discussed in the reviews I've cited here.

210 Suggestive evidence is emerging: Yeatman et al. (2012).

210 Visual impairments may be: This evidence in no way validates "visual training" programs sold by optometrists and others as treatments for reading difficulties. Anomalies in these white matter tracts may be the source of the noisy visual processing in the Sperling et al. (2005, 2006) experiments.

211 The initial left hemisphere: Hoeft et al. (2011).

211 The global impact of these conditions: Rueckl et al. (2015).

211 The product of the newer: Ashkenazi et al. (2013) summarize structural brain anomalies associated with reading and math impairments.

212 Two studies have reported: Several studies have reported changes in white matter in dyslexics following short-term interventions. The findings are suggestive, but see Bishop (2013) for a clear-eyed assessment.

212 increases in white matter volume: Myers et al. (2014).

212 brain research is converging: Pugh et al. (2014).

213 Migration errors produce ectopias: See Peterson & Pennington (2012). For neural migration explained, with animations, see "Research Interests," Rakic Lab, http://rakiclab.med.yale.edu/research. The occurrence of migration errors in dyslexia was discovered in the 1980s by neurologist Albert Galaburda, who conducted postmortem studies of the brains of dyslexics who had recently died. Now, decades later, such anomalies can be observed, using structural neuroimaging, in dyslexics who are very much alive. See Galaburda et al. (2006) for the story of this research.

214 behavioral measures could provide: Robin Morris (Georgia State University) described this use of behavioral and imaging data in a meeting of the scientists involved in a large project on dyslexia and comorbid conditions that he leads (June 24, 2016).

Chapter 10: How Well Does America Read?

217 Mann was deeply impressed: Mann (1844). The original documents are available through "The Online Books Page," http://goo.gl/DOsoxW.

217 two large-scale programs: "About PISA," OECD, http://www.oecd.org/pisa/aboutpisa; NAEP: "National Assessment of Education Progress," National Center for Education Statistics, https://nces.ed.gov/nationsreportcard.

218 "high-stakes testing": NCLB mandated that each state measure every child's progress in reading and math in grades three through eight and at least once during grades ten through twelve. See "Introduction: No Child Left Behind," US Department of Education, http://www2.ed.gov/nclb/overview/intro/index.html.

219 congressionally mandated program: The first NAEP data were collected on a trial basis in 1969. The assessments went through several major revisions before reaching the modern format in 1992. The data from the early assessments are rarely cited because they are not directly comparable to the post-1992 results. See "Grading the Nation's Report Card: Evaluating NAEP and Transforming the Assessment of Educational Progress," National Academies Press, http://www.nap.edu/catalog/6296.html.

219 Participating schools and children: Sampling methods: "About the 2015 Reading Assessment," Nation's Report Card, http://www.nationsreportcard.gov/reading_math_2015/#reading/about?grade=4.

220 The NAEP website also provides: All data cited here are from two Department of Education websites. A simple overview of the basic results is found on the Nation's Report Card website (http://www.nationsreportcard.gov). The National Center for Education Statistics website has tools for exploring the data, as well as summaries and technical reports: http://nces.ed.gov/nationsreportcard/naepdata.

220 *Really?* reflex: Seth Meyers and Amy Poehler, "Weekend Update: Really!?! Congress' Birth Control Hearing," NBC.com, February 18, 2012, http://goo.gl/usdXaP.

220 Claims that depend: Trust, but verify, as they said in the Reagan era: https://en.wikipedia.org/wiki/Trust,_but_verify.

223 I mainly focus: For the blog, see www.dianeravitch.net. The blog is also a forum for like-minded critics. The book is *Reign of Error* (Ravitch 2013). For the talk, see "Conversation: Diane Ravitch," uploaded to YouTube by The Nation, July 19, 2010, http://goo.gl/UXAaq0.

223 "The educational foundations": National Commission on Excellence in Education (1983).

223 two reports: National Endowment for the Arts (NEA) (2004, 2007).

223 resulted from a toxic combination: Ravitch: "Venture capitalists and for-profit firms are salivating over the exploding $788.7 billion market in K–12 education" (Lee Fang, "Venture Capitalists Are Poised to 'Disrupt' Everything About the Education Market," *Nation*, September 25, 2014, http://goo.gl/xCg5yP).

223 reports on the state of reading: The NEA analyses of trends in reading were not competently conducted. The authors did not understand basic data handling and analysis (http://futureofthebook.org/blog/2007/11). The data were from surveys about reading habits; the declines were in "leisure" and "literary" reading, ignoring reading on digital platforms, amount of reading for school or employment, and other reading. Whereas the first two reports reported alarming decreases in reading, the third one, "Reading on the Rise" (National Endowment for the Arts 2009), documented a sudden turnaround, "a significant turning point in recent American cultural history," some slight upticks in the amount of literary reading in the survey data. These conclusions cannot be taken at face value.

224 independent operations: NAEP is federally mandated and funded but operates independently. NCLB mandated that schools receiving Title I funding participate in the

NAEP's assessments if selected under its sampling procedure. The purpose was to be able to use the NAEP to benchmark the states' own tests. See Hombo (2003).

224 Hence the many attempts to explain: Examples: Martin Carnoy and Richard Rothstein, "What Do International Tests Really Show About U.S. Student Performance?," Economic Policy Institute, January 28, 2013, http://www.epi.org/publication/us-student-performance-testing; Nick Pearce, "PISA Panic: Being Honest About What PISA Really Shows," Left Foot Forward, December 3, 2013, http://leftfootforward.org/2013/12/being-honest-about-what-pisa-really-shows; Diane Ravitch, "David Berliner on PISA and Poverty," Diane Ravitch's Blog, April 12, 2014, http://dianeravitch.net/2014/04/12/david-berliner-on-pisa-and-poverty.

225 "Don't believe anyone": Ravitch (2013), 50.

225 Flynn effect: Flynn (1987).

225 GED holders: Heckman & LaFontaine (2008) note that GEDs account for 15 to 20 percent of high school diplomas each year and that those with GEDs perform at the level of dropouts in the US labor market. By inflating graduation rates, "The GED program conceals major problems in American society." See also Heckman & LaFontaine (2010).

225 The US graduation rate: See Murnane (2013). The author also documents a true (non-GED-dependent) but unexplained uptick in high school graduation rates between 2000 and 2010.

226 "In my recent book": Diane Ravitch, "Four Lessons on New PISA Scores—Ravitch," *Washington Post*, December 3, 2013. See also Ravitch (2013), 70–72.

226 Hanushek and Woessmann: Hanushek & Woessmann (2012).

226 FISS, FIRS, SIMS, SISS, SIRS, TIMSS, and PIRLS: Mmmmm, you could Google them (try "_________ cross national assessment"). They are identified in the Hanushek & Woessmann article, p. 304, and on seidenbergreading.net.

226 analyses indicated "a close relationship": Hanushek & Woessmann (2012), 267.

226 Hanushek and others have pursued: See, e.g., Murnane et al. (2000) and Hanushek & Woessmann (2008); for a different view, see Ramirez et al. (2006).

226 A five- or eight-point change: Sean Cavanagh & Kathleen Kennedy Manzo, "NAEP Gains: Experts Mull Significance," *Education Week*, October 3, 2007.

227 They matter enough: Buckley (2011).

228 A smaller percentage of US students: OECD (2013b). For the lowest level, 1, scores are further divided into three sublevels. The < 2 refers to all level 1 data.

228 A comparison of the results: Amazingly, a large representative cohort of students took both the NAEP (in 2007 when they were thirteen) and the PISA (in 2009 when they were fifteen). This allowed Peterson et al. (2011) to make closer comparisons between the tests. They concluded that the NAEP's proficiency standard was higher for reading, but the PISA standard was higher for math.

228 "It is fair to say": Diane Ravitch, "Every State Left Behind," *New York Times*, November 7, 2005.

229 "That 27 percent": Garrison Keillor, "Where's St. Michael When You Really Need Him?" *Prairie Home Companion*, January 29, 2008, http://goo.gl/s2dSC8.

229 only a rough index: The National School Lunch Program, established by President Harry S. Truman in 1946, is a federally assisted meal program operated in public and private nonprofit schools and residential child-care centers. Students from households with an income at or below 130 percent of the federal poverty guideline are eligible for free lunch; they qualify for subsidized lunch if the household income is between 130 and 185 percent of the guideline. Peterson et al. (2011). For 2013–2014, children from families of four with incomes less than $30,515 qualified for free lunch; the cutoff for subsidized lunch was $43,568. Children above this cutoff are in the highest SES group, which includes many of modest means.

230 Low achievement in reading: See "Closing the Opportunity Gap: What America Must Do to Give Every Student a Chance," Stanford Center for Opportunity Policy in Education, https://edpolicy.stanford.edu/projects/373. The press release provides a good summary.

230 Gaps in educational opportunity: For a grim reminder of poverty's impact on education, see Jaclyn Zubrzycki, "Detroit Studies Illuminate Problem of Lead Exposure," *Education Week*, September 26, 2012.

231 Every PISA assessment includes: To get a fuller picture of how factors such as parental education, educational spending, and percentage of low-income participants influence country scores, I recommend looking at pp. 34–36 in a 2010 PISA in-depth report (http://goo.gl/UDOaTW, also posted on seidenbergreading.net).

231 example from close to home: Correction for SES and US versus Canada data: OECD (2011, 34–35). According to this document, 17 percent of the variation in student performance is explained by students' socioeconomic background in the US, compared to 9 percent in Canada (also Japan). The report also notes that in the US, "The relationship between socio-economic background and learning outcomes is far from deterministic. For example, some of the most socio-economically disadvantaged schools match the performance of schools in Finland. Furthermore . . . a quarter of American 15-year-olds enrolled in socio-economically disadvantaged schools reach the average performance standards of Finland, one of the best-performing education systems."

231 "U.S. students lag": For Duncan's remarks about the 2009 results, see "Secretary Arne Duncan's Remarks at OECD's Release of the Program for International Student Assessment (PISA) 2009 Results," US Department of Education, December 7, 2010, http://goo.gl/PAKnO. Many educators took issue with them and with his comments on the 2012 results. See Cameron Brenchley, "Duncan Calls for Higher Standards and Expectations Following PISA Results," Homeroom, December 3, 2013, http://goo.gl/JUS5n8; Christopher H. Tienken, "Problems with PISA," Chris Tienken, February 11, 2014, http://christienken.com/2014/02/11/pisa-problems; and Diane Ravich, "David Berliner on PISA and Poverty," Diane Ravitch's blog, April 12, 2014, https://dianeravitch.net/2014/04/12/david-berliner-on-pisa-and-poverty.

232 2013 report from the Economic Policy Institute: Carnoy & Rothstein (2013).

232 "damaging world-wide effects": On the letter from educators, see "OECD and PISA tests Are Damaging Education Worldwide—Academics," *Guardian*, May 6, 2014, http://goo.gl/nl4epU. For a response from OECD, see Peter DeWitt, "What is PISA?," *Education Week*, September 25, 2015, http://goo.gl/gGXCA0.

232 much easier way to improve: Kay McSpadden, "Public Schools Aren't Failing," *Charlotte Observer*, January 30, 2015.

233 "Less than 25 percent": Kena et al. (2015).

234 "resilient" children: OECD (2011), 37.

234 "One of the basic principles": "Educational Policy in Finland," Finnish Ministry of Education and Culture, http://goo.gl/EUw3I.

235 particularly disturbing: On eligibility for free/reduced-price lunch, see "Percent Low Income Students: 2013–2014," ED.gov, http://goo.gl/oP5FC0.

235 educational redlining: "Redlining is the practice of arbitrarily denying or limiting financial services to specific neighborhoods, generally because its residents are people of color or are poor": D. Bradford Hunt, "Redlining," Encyclopedia of Chicago, http://goo.gl/J7sshD. I am familiar with the concept from growing up on the South Side of Chicago at a time when the practice was rampant. Educator Linda Darling-Hammond has also described ESSA's retention of high-stakes testing for the lowest-performing schools, which are in low-income communities, as a form of educational redlining: "Why Is Washington Redlining Our Schools?," *Nation*, January 30, 2012.

236 Many factors contribute: Several perspectives on the achievement gap: Hoff (2013); Richardson (2003); Magnuson & Duncan (2006); Barton & Coley (2009); Washington, Terry, & Seidenberg (2013); Seidenberg (2013).

236 "East Versus West": Sharon Begley, "East Versus West: One Sees Big Picture, Other Is Focused," *Wall Street Journal*, March 28, 2003.

237 A good place to look is the reading data: Motoko Rich, Amanda Cox, and Matthew Bloch, "Money, Race and Success: How Your School District Compares," *New York Times*, April 29, 2016, http://goo.gl/yVrTRF.

239 Studies of African American youth: Gosa & Alexander (2007).

239 These kinds of data: In analyses using a large longitudinal survey, Fryer & Levitt (2006) observed that the achievement gap present in kindergarten can be accounted for by a small number of factors related to SES. However, the gap also increased through the first several years of school for reasons they could not identify. It may be that standard SES measures are less valid for African Americans than whites (Rothstein & Wozny 2013) or that effects of SES multiply as school demands increase. It may also be that other factors such as language background start to assume more importance in school.

240 Children from lower-income families: Hart & Risley (1995). They used the labels "professional," "working-class," and "welfare," which I've renamed as SES levels. The follow-up study on reading achievement: Walker et al. (1994).

240 Variability in spoken-language acquisition: Fernald, Marchman, & Weisleder (2013).

240 Recognition of the importance: Duncan & Sojourner (2013).

240 a study of fifty-three low-income mothers: Weizman & Snow (2001).

240 Similar variation: Kena et al. (2015); Horton-Ikard & Miller (2004); Gosa & Alexander (2007).

240 "Parents' verbal engagement": Fernald & Weisleder (2015). The study of the impact of teachers' speech to children is Dickinson & Sprague (2003).

241 Most African American children: Wolfram (2004).

241 The question is not: AAE is linguistically unremarkable, an example of dialect variation as it occurs in language. The breakthrough research establishing this fact was conducted by sociolinguist William Labov (1972) and extended by many others.

241 Part of the "achievement gap": These conditions were first pointed out to me by Julie Washington.

241 This occurs for several reasons: In fact the conditions governing the deletions of final phonemes are complex, and deletions are always optional. Basic facts about how often phonemes are omitted for a given word or by individual speakers are not known because the questions have not been adequately studied.

242 For an AAE speaker: Some of these phenomena also occur in the Southern White English dialect, and the impact on reading would be expected to extend to those speakers. See Wolfram (1974); Oetting & Kent (2004).

242 The impact of such pronunciation: Brown et al. (2015).

242 Acquiring this additional linguistic expertise: Bialystok et al. (2009).

242 A speaker of a minority dialect: It is also more difficult to identify AAE speakers with developmental spoken-language impairments because features that are indicative of an impairment in MAE can be grammatical in AAE: Oetting & McDonald (2001).

242 For verbally proficient children: Terry et al. (2010).

243 The impact of dialect differences: The issue was examined in the 1970s and 1980s, but the findings were inconclusive and the methods used do not hold up by modern standards. In recent years researchers have started to investigate the issue again, building on advances in research methods and findings about reading and language.

243 The evidence that speaking: Some representative studies: Terry et al. (2010); Charity, Scarborough, & Griffin (2004); Terry et al. (2012).

243 The findings are particularly sensitive: History of racist characterizations of black speech: Baratz & Baratz (1970).

243 I have used computational models: See Brown et al. (2015).

244 Both Australia and Canada: Siegel (2010). For similar issues in Arabic, see Levin et al. (2008), and in Finland, see Latomaa & Nuolijärvi (2002).

244 Preservice teacher education rarely includes: Moats (1999, 2009).

244 negative perceptions of AAE: Blake & Cutler (2003).

244 Having to learn two dialects: From a biography of Condoleezza Rice (Mabry 2008, 13): "Then as now, many African American parents told their children, 'You have to be twice as good.' Meaning, they had to be twice as good as white people to receive the same level of respect, opportunity or status. 'You were taught that you were good enough, but you might have to be twice as good given you're black,' Condoleezza Rice often recalled." New data from the National Bureau of Economic Research (http://www.nber.org/papers/w21612) shows this to be essentially true with respect to getting and keeping jobs: Gillian B. White, "Black Workers Really Do Need to Be Twice as Good," *Atlantic*, October 7, 2015, http://goo.gl/oDZjzz.

244 Compelling children to learn MAE: Dudley-Marling & Lucas (2009).

244 "Often powerful": Steiner & Rozen (2004).

245 with particular attention: For an example of casual bias, see Jennifer Holladay, "The Character of Our Content: A Parent Confronts Bias in Early Elementary Literature," *Rethinking Schools* 27, no. 2 (winter 2012–2013), http://goo.gl/aFuCH. On tenacity, perseverance, and grit, see Duckworth et al. (2007). On the impact of poverty and racial bias, see Wilson (2011).

245 agents of social justice: Sleeter (1996, 152), for example, sees multicultural education as "a collaborative process involving dialog and bonding across racial and ethnic boundaries for the purpose of forging greater equality and social justice." See also Stinson et al. (2012), who discuss pedagogical practices that link the teaching of mathematics to social justice issues.

245 "Using a theoretical approach": Hassett (2006).

246 For lower-SES individuals: Heritability of reading depends on SES: Hart et al. (2013). Similar effects have been found for parental education (Friend, DeFries, & Olson 2008) and educational quality (Taylor et al. 2010). Note that although heritability is a measure of genetic influence on behavior, these heritability differences do not arise from the genome but rather from the impact of the environment on gene expression, brain, and behavior.

Chapter 11: The Two Cultures of Science and Education

247 "skeleton-shaped, bloodless": The Mann quote is from an 1844 report in which he advocated what was later called the "look and say" method, which involves memorizing words as patterns, without regard to the functions of the component letters. The Boston educators favored a "phonic" teaching method. Their takedown of Mann's "new method" was thorough and incisive but settled nothing. See Mann (1844), which is easily accessible online.

247 reading wars: Nicholas Lemann's 1997 *Atlantic* article is still a good introduction to this conflict: "The Reading Wars," *Atlantic*, November, 1997, http://goo.gl/KLL5Yp.

248 In practice, that usually means: Moats (2000); another critic of balanced literacy: Ravitch (2011).

248 The unresolved issues: Wilson (2009).

249 I also know: Mehta & Doctor (2013); American Federation of Teachers (2012); Darling-Hammond (2014).

250 Teachers need to know: Walsh (2013); Steiner & Rozen (2004); Walsh, Glaser, & Wilcox (2006).

250 "Teaching requires": Mehta (2013).

251 "Across the studies": Cochran-Smith & Zeichner (2005).

251 what the Internet is for: On constructivism, see http://goo.gl/i8c1SW; on social constructivism, see http://goo.gl/MQz8ng; on postmodernist theory in education, see www.tpress.free-online.co.uk/post.pdf. These approaches differ in ways that are not important in the present context.

251 my nonexpert eye: Crews (2006); Sokal (1996).

252 "If we accept constructivist theory": For a thoughtful discussion by a museum educator, see George E. Hein, "Constructivist Learning Theory," Exploratorium, http://goo.gl/f3QRyh. Similarly, "Critical literacy views readers as active participants in the reading process and invites them to move beyond passively accepting the text's message to question, examine, or dispute the power relations that exist between readers and authors" (McLaughlin & DeVoogd 2004, 14).

252 Some classic research by Jerome Bruner: Bruner & Goodman (1947).

252 Since learning is the process: See Schmeck (1988) for an introduction to learning styles and strategies; see Stanovich & Stanovich (2003) for a critical review.

253 "Sociocultural theory describes": From Risko et al. (2008), 254. I have omitted the references cited in the original text.

253 "Teacher educators now view": Walsh (2013).

253 I have no direct experience: Lifton (1961).

254 The goal is: Examples include "Writing a Philosophy of Teaching Statement," Ohio State University, http://goo.gl/89hzmf; "Teaching Statements," Vanderbilt University, http://goo.gl/GRwQXV.

254 "Review current theories": The text has been lightly modified from the original to keep the focus on the content, which reflects widely held views, rather than identifying it with a particular individual.

255 Children learn via: Cochran-Smith & Lytle (2006, 688) characterize traditional teaching as a "technical transmission activity," referring to the transmission of knowledge from the head of the expert into the head of the learner.

> During the 1980s, the technical view of teaching and the training view of teacher development were to a large extent rejected, at least partly because of their overly simplified view of teaching and learning that not only ignored teacher cognition, but also ignored the dynamic, social, moral, and political aspects of teaching. . . . [T]eacher development generally shifted from teacher training to teacher learning, which meant examining the kinds of knowledge, attitudes, and beliefs teachers brought with them to teacher-preparation and professional-development opportunities (as well as how these changed over time); how teachers learned (and generated) the knowledge, skills, and dispositions needed to teach; how they made professional decisions inside and outside the classroom; how they learned about their students and their cultures; and how they interpreted and connected their experiences in courses, workshops, and learning communities to their work in schools and classrooms.

See also Woolfolk (2007), a standard textbook in the field.

255 Educational practice has evolved: The source of this widely known witticism is unknown.

255 "Education through experience": On life at the Dewey school, see Mayhew & Edwards (1936); for early history, see Harms & DePencier (1996).

255 Some of the concepts: According to an American Federation of Teachers report (AFT 2012, 7), "Fewer than half of new teachers describe their training as very good, and more say that on-the-job learning or assistance from other teachers was more helpful than their formal training; 1 in 3 new teachers reports feeling unprepared on his or her first day. The top problem experienced by teachers in their own training was a failure to prepare them for the challenges of teaching in the 'real world.'"

255 Such findings have low status: The Common Core State Standards may themselves be problematic, but the same concern arises for other modern math and reading curricula.

256 The Philosophical Lexicon: "The Lexicon," Philosophical Lexicon, http://www.philosophicallexicon.com/#LEXICON. As my former professor the late Sidney Morganbesser said of Dewey's pragmatic philosophy, "It's all very well in theory, but it doesn't work in practice." Morganbesser, whose wit is legend, was the John Dewey Professor of Philosophy at Columbia for many years. His serious account of Dewey's philosophy is Morganbesser (1977).

256 approach probably worked: See "Elementary Geography Class. Laboratory School," University of Chicago Centennial Catalogues, http://goo.gl/lpSlk4.

256 It is a great school: Dewey's ideas about education were revolutionary in their time, as suggested by the contrast between a room in his school (http://goo.gl/lpSlk4) and a public school classroom twenty miles away (http://goo.gl/7vw86u). Those classrooms differed in more than educational philosophy, however. One was a classroom in a public school with many students and few resources. The other was a private school with fewer students and the resources to mount a variety of activities, created with a large endowment from private philanthropy (Mayhew & Edwards 1936, 12). It was never a sustainable model for public education. According to Menand (2001, 320), Dewey viewed it not as a model school or educational experiment but as a "philosophy laboratory," a place to "work out in the concrete, instead of merely in the head or on paper, a theory of the unity of knowledge."

256 new teachers: Although the Dewey school model was a failure, several of Dewey's foundational principles have been retained and embellished, one being that classrooms are places in which to experiment with practices to determine what works (see Mayhew & Edwards 1966, vi). Another is that "the child, not the lesson, is the center of the teacher's attention; each student has individual strengths which should be cultivated and grown" (Harms & DePencier 1996, chap. 1). A third is that learning that occurs as the by-product of the child's own authentic, meaningful activity is deeper and more valuable than what can be learned from instruction, the basis of "discovery learning" in the modern era.

Dewey's own attitudes about learning were more nuanced than the deweyite caricature. For example, he was highly critical of teaching science by having students engage in mock scientific practices, judging science labs in chemistry and physics to be shallow exercises. He advocated teaching students about the scientific method and its particular claims on truth rather than the body of facts and standard methods in a science (Rudolph 2005).

256 one noteworthy: What's with the Finns? For their own perceptions of the strengths of their educational system, see Kupiainen, Hautamäki, & Karjalainen (2009) and Sahlberg (2011). Finland is extensively discussed in the OECD's 2011 report "Lessons from PISA for the United States" and in the accompanying video, "Finland—Strong Performers and Successful Reformers in Education," video uploaded to YouTube by EduSkills OECD, January 24, 2012, https://www.youtube.com/watch?v=ZwD1v73O4VI. There are useful facts to be gleaned from these sources and especially from "Educational Policy Outlook Finland," http://www.oecd.org/edu/highlightsfinland.htm, along with some broad generalizations that are hard to verify. Whether any factor that accounts for success in one country is transferrable to another country or would have the same impact is an open question. For unknown reasons Finnish PISA reading scores have been trending downward: 2000, 546; 2003, 543; 2006, 547; 2009, 536; 2012, 519. Finnish students also report relatively low happiness levels on the PISA survey.

256 It may also be related: Gray & Taie (2015). In a study of teacher training and student achievement, Harris & Sass (2011) found that informal, on-the-job training increased teachers' effectiveness, with most of the gains being realized during the first five years. Preservice professional development was not related to student achievement, however.

256 "Educational courses have always": Clifford & Guthrie (1988), 25. A 2007 survey of 2,237 teachers about to take their first jobs found that they rated themselves as underprepared in the five components of beginning reading identified by the National Reading Panel (Salinger et al. 2010), and they demonstrated weak knowledge of the concepts on a short test.

257 "Poverty, writes Ravitch": From John Buntin, "A Battle over School Reform: Michelle Rhee vs. Diane Ravitch," Governing, January 2014, http://goo.gl/8Ru8qw.

258 Vygotsky thought: Au (1998), 300; Coles (2000).

259 Nor is there acknowledgment: As summarized in an incisive history of the reading wars (Kim 2008), Keith Stanovich, a leading researcher and theorist with high credibility in both science and education, observed that the whole-language theorists who controlled educational practices "had failed to respond to evidence and enact norms of practice rooted in scientific research. In short, whole-language theorists and advocates left the teaching profession vulnerable to intrusive legislative mandates by failing to police itself" (99–100).

260 educational computer games: Granic, Lobel, & Engels (2014).

260 John Bransford: Bransford, Brown, & Cocking (2000).

261 "zone of proximal development": The "zone of proximal development" is the sweet spot for learning: something within the zone exceeds one's current knowledge but not by so much that it can't be learned.

261 Inferences based on observation: Kahneman (2011).

262 The clearest illustration: National Institute of Child Health and Human Development (2000). The NRP also published a readable summary of the main findings for parents and teachers, "Putting Reading First," downloadable here: http://goo.gl/cdlFyQ.

262 That justified ignoring: The National Education Association, the labor union, convened a task force to respond to the NRP, "Report of the NEA Task Force on Reading 2000," NEA, http://www.nea.org/home/18301.htm. Its report echoed the complaint that the NRP had erred in emphasizing experimental rather than observational studies, thereby omitting important findings. "For example, Piaget and Vygotsky, two influential contributors to the understanding of learning in young children, did no experimental studies" (5). The statement is grossly misinformed because both figures are known for their innovative and highly influential experiments with young children. Modern versions of Piaget's famous conservation experiments are seen here: http://goo.gl/1dN3P0. A Vygotsky experiment is seen here: http://goo.gl/VgBl62.

262 These lobbyists target: See, e.g., Allington & Woodside-Jiron (1999). The authors asserted that many of the research findings that contradicted their own views were the product of research funded by Reid Lyon, an official at the National Institute of Child Health and Human Development (NICHD), as part of an anti-education political agenda. The founding document for this political movement, they claim, is Grossen (1997), an obscure, minor twenty-two-page catalog of thirty years of reading research funded by NICHD. Allington and Woodside-Jiron focus their paranoia so keenly on NICHD that they ignore the mass of similar findings from research conducted in other countries. It would be easier to dismiss Allington's campaign against reading science (see also Allington 2002) were he not a leading figure in reading education, former president of the International Reading (now Literacy) Association, former president of the National Reading Conference, and a member of the "Reading Hall of Fame" (http://www.readinghalloffame.org).

264 phonics can be taught: The definitive response to objections to direct instruction and practice, from Anderson, Reder, & Simon (1999), three giants in the study of learning and cognition (Simon a Nobel prize winner):

> It is sometimes argued that direct instruction leads to "routinization" of knowledge and drives out understanding. . . . An extension of this argument is that excessive practice will also drive out understanding. This criticism of practice (called "drill and kill," as if this phrase constituted empirical evaluation) is prominent in constructivist writings. Nothing flies more in the face of the last 20 years of research than the assertion that practice is bad. All evidence, from the laboratory and from extensive case studies of professionals, indicates that real competence only comes with extensive practice. . . . In denying the critical role of practice one is denying children the very thing they need to achieve real competence. The instructional task is not to "kill" motivation by demanding drill, but to find tasks that provide practice while at the same time sustaining interest.

266 Balanced literacy allowed: The overwhelming evidence demonstrating the importance of phonology in reading does not in any way entail, nor should it be taken to imply, that only phonics instruction is involved. Every rational account of learning to read includes it as one of several components, necessary but not sufficient.

267 A teacher needs: A story told to me by a parent who is a highly educated professional (as is the spouse). Their bright first grader was having difficulty with reading, beginning to avoid it and acting out in school. The parent brought the child to a reading specialist who worked on her basic skills, including phonics. The child soon caught on and began reading with enthusiasm and interest. At the end of the school year, the parent asked the child's teacher why phonics had not been taught in the classroom. The teacher's reply: "Your child was not in school that day." (Anecdotes aren't evidence, of course; they are "small batch artisanal data." Sara Pikelet, http://goo.gl/DUkF2W.)

267 Smith's influential books: See Adams (2004) for background on Smith's work.

267 as Smith assures readers: The sixth edition of Smith's book is downloadable here: http://goo.gl/H1b0yJ.

267 "The first alternative": Smith (2006) is the fourth edition of his 1978 book.

268 "psycholinguistic guessing game": Goodman (1976).

269 Cloze procedure: You can try the Cloze procedure on seidenbergreading.net. Every fifth word in a news story has been deleted. The task is simply to guess each missing word as you read along. What happens is pretty surprising.

270 Philip Gough (the "simple view"): Gough (1983).

270 "It is often incorrectly assumed": Stanovich & Stanovich (2003). Prediction is one of the central concepts in modern theories in cognitive science and neuroscience, but it is a more general statistical mechanism than Goodman had envisioned. Today many researchers think of the brain as continually engaged in assessing bottom-up sensory information against top-down expectations (Clark 2013). Expectations take the form of probability distributions—for example, the probabilities associated with many possible upcoming words, grammatical categories (noun, verb, and so forth), or semantic properties (e.g., animate, inanimate). These predictions are important for language comprehension, but they do not make individual words highly predictable or permit the text sampling that Goodman advocated (as the eye-movement data showed). Children learn these distributions via statistical learning over large samples of text and speech, not by being taught to guess words.

270 Guessing provides no advantage: Given this understanding of Goodman's idea, it should be clear how to turn a good reader into a poor reader: make it harder to read individual words by reducing the visual quality of the stimulus (present it only briefly, in Captcha style or in low contrast). Degrading the target forces the good reader to rely more on context. For poor readers, it is as though all words are "degraded" because of their limited basic skills.

270 Children might manage: Children who are used to predicting and sampling texts may also be at a disadvantage on standardized assessments that require close reading of passages and comprehension questions and math story problems, though I know of no relevant studies.

270 It is still presented: Wray (2004).

271 It's right there: Commission on Reading of the National Council of Teachers of English (NCTE), "On Reading, Learning to Read, and Effective Reading Instruction: An Overview of What We Know and How We Know It," NCTE, http://www.ncte.org/positions/statements/onreading.

271 The original guessing game article: Goodman's (1967) article can be downloaded: http://goo.gl/c5C8wa. It includes figures said to represent Chomsky's ideas about reading comprehension. There must have been a misunderstanding somewhere. The theory presented is not Chomsky's, and I am not aware of any publication in which he endorsed anything like it. No record of Chomsky's long-ago talk exists, unfortunately. Goodman seems to be referring to a theory of perception from that period called "sophisticated guessing."

271 By 1980 numerous studies: Leu (1982).

271 The theory was one: Educator David Pearson quoted in Kim (2008, 97): "Never have I witnessed anything like the rapid spread of the whole-language movement. Pick your metaphor—an epidemic, wildfire, manna from heaven—whole language has spread so rapidly throughout North America that it is a fact of life in literacy curriculum and research."

272 Several reading researchers: Many are described in Britton & Graesser (2014).

274 For children who have acquired: For academic vocabulary, a useful resource is "The Academic Word List," Victoria University of Wellington School of Linguistics and Applied Language Studies, http://www.victoria.ac.nz/lals/resources/academicwordlist. For insightful discussion of background knowledge, see Willingham (2006).

274 Reading skills depend: Adams (2011).

275 Phonics programs of marginal: For the FTC settlement with the Hooked on Phonics company, see http://goo.gl/3Lqx3G.

275 Phonics is on its way: Children who successfully learn phonics out of school may inadvertently perpetuate the illusion that classroom practices are effective. Teachers may not realize that a child's reading greatly improved because they had extensive outside instruction.

276 A drug is not marketed: Whether biomedical clinical trials are held to sufficiently high standards is a separate, though worrisome, issue.

276 it has not worked well: McArthur (2008).

276 Effects of instructional practices: They are also subject to several artifacts that do not arise in medical RCTs: Dorothy Bishop, "Three Ways to Improve Cognitive Test Scores Without Intervention," BishopBlog, August 14, 2010, http://goo.gl/BndBlZ. Methods that are not as restrictive as RCTs can yield comparable results in some cases (Cook & Steiner 2009).

276 WWC has also run afoul: Eric Westervelt, "There Is No FDA for Education. Maybe There Should Be," NPR.org, March 7, 2016, http://goo.gl/Grw7yU.

277 Reading Recovery provides: Reading Recovery Council of North America: http://readingrecovery.org.

277 The Reading Recovery organization: William Tunmer is an excellent reading researcher who happens to live in New Zealand, where Reading Recovery originated and was adopted as part of a national literacy effort. See Tunmer et al. (2013). For support for Reading Recovery, see Shanahan & Barr (1995). For Reading Recovery on the What Works Clearinghouse, see http://goo.gl/SuwdTz. On short- but not long-term gains, see Center et al. (1995).

The program is narrowly focused on an age group in which performance is highly variable and many children do catch up. RR is offered in addition to regular classroom activities, and so children who receive both should do better compared to children who have the usual classroom experience. The What Works Clearinghouse found 202 studies that investigated the effectiveness of RR. Three met their criteria for valid experimental designs. Those studies were conducted by researchers with close ties to the Reading Recovery organization.

277 That is what teachers thought: Baumann et al. (2000).

278 "Literacy is the ability": "Why Literacy," International Literacy Association, http://literacy worldwide.org/why-literacy. Reading, writing, and communicating are mentioned next.

278 lists of literacies: American Association of School Librarians (2009).

278 The new screen-based technology: "Television literacy" is an entry from the *Oxford English Dictionary.*

278 Compare the experience: *New York Times* 1914: "New York Times Front Page, July 29, 1914. Wikipedia, http://goo.gl/c2OOi7.

279 "In addition to reading": Kellner (2006).

279 *Moby Dick*: Sort of: Fred Beneson, "Emoji Dick," Kickstarter, https://www.kickstarter.com/projects/fred/emoji-dick/posts/1002713.

279 Multimedia software may: Karemaker, Pitchford, & O'Malley (2010).

280 Diana Trilling: On Trilling's letters, see Lehman (1998), 192.

281 Maybe that is because: Randall Munroe explains: http://imgs.xkcd.com/comics/flies.png.

Chapter 12: Reading the Future

283 A look at the history: The history in this section is mainly from Clifford & Guthrie (1988), including the quotations.

284 unbelievable distrust and opposition: Clifford & Guthrie (1988), 136.

284 The organization of the university: See Goldstein (2014). Teaching was a lower-status profession than the other fields, undertaken by lower-status individuals (women) who were paid lower-status wages. The status of a college or university or a specific program is mainly determined by its "inputs," characteristics of the individuals who enroll, rather than "outputs," how the graduates subsequently fare (thus the race to qualify as "highly selective" on the measures that determine the specious *U.S. News & World Report* rankings). Ed schools have lower-quality "inputs" in this country, though not in Finland.

284 "Teachers should demonstrate": Bennett quoted in Clifford & Guthrie (1988), 16.

284 "kitten that ought to be drowned": As quoted by Clifford & Guthrie (1988, 137). This ancient history does not reflect the current standing of the Harvard Graduate School of Education.

284 Degree programs in teaching: Adams, Bell, & Griffin (2007); Tyson & Park (2006).

285 Although that interpretation: Coleman (1966). The report documented large differences in educational opportunity (quality of teachers, school resources, and so forth) associated with race and socioeconomic status. It noted, "It is known that socioeconomic factors bear a strong relation to academic achievement." The first finding reported under "relation of achievement to school characteristics" was that "when socioeconomic factors are statistically controlled . . . it appears that differences between schools account for only a small fraction of differences in pupil achievement" (21). This result does not indicate that "schools don't matter," but many interpreted it that way. The opposite analysis would be equally informative: examine how much SES accounts for student performance after differences in school quality and other educational resources have been "statistically controlled." See Clifford & Guthrie (1988) for discussion.

286 Teachers complain: The complaint arises from the fact that the educational theory is taught by professors whose research focuses on alternative conceptions of learning and learning environments that might revolutionize education. The approaches are intended to be radically different from current practices and thus are of secondary interest to the prospective teacher.

286 Their students bear: A charismatic, seemingly authoritative figure whose ideas are bunk can do serious damage, but the effects are greatly magnified when combined with assurances that personal experience is as valid a source of evidence as systematic investigation. Ken Goodman conveyed that message in the guise of teacher self-empowerment. It encourages a populist epistemology that greatly undermines the utility of science.

In *Denialism*, his book about the growth of antiscience belief systems, Michael Specter analyzed the impact of Dr. Andrew Weil, the famous guru of health and holistic medicine, in similar terms. Weil's controversial ideas are also coupled with an emphasis on the primacy of personal experience. Spector notes, "It is much easier to dismiss a complete kook—there are thousands to choose from—than a respected physician who, interspersed with disquisitions about life forces and energy fields, occasionally has something useful to say." As quoted by Janet Maslin, "Firing Bullets of Data at Cozy Anti-science," *New York Times*, November 6, 2009, http://goo.gl/HaPkSh.

286 They originate: My favorite "everybody's an expert" example is from Richard Feynman (2014, 116), one of the most celebrated and revered scientists of modern times (www.feynman.com):

> All the time you hear the question, "Why can't Johnny read?" And the answer is, because of the spelling. The Phoenicians, 2,000; more, 3,000, 4,000 years ago, somewhere around there, were able to figure out from their language a scheme of describing the sounds with symbols. It was very simple. Each sound had a corresponding symbol, and each symbol, a corresponding sound. So that when you could see what the symbols' sounds were, you could see what the words were supposed to sound like. It's a marvelous invention. And in the period of time things have happened, and things have gotten out of whack in the English language. Why can't we change the spelling? Who should do it if not the professors of English? If the professors of English will complain to me that the students who come to the universities, after all those years of study, still cannot spell "friend," I say to them that something's the matter with the way you spell "friend."

The comments convey strong beliefs but little understanding about spelling, reading, or English professors. If English were written in the Phoenician system, it would be unreadable because the Phoenician alphabet had no vowels. FRIEND is not notably difficult, because it is a lexical isolate, a high-frequency word with only one, lower-frequency neighbor, FIEND. Such words are easy to discriminate from other words and thus easy to remember. And English professors are the culprits? Surely Mr. Feynman was joking.

286 lack of tools: In-service teachers eventually hear about methods from speakers who are brought in for lectures and workshops. "Experts" travel state and national Chautauqua circuits pitching their systems and sharing special insights. No quality controls to speak of. Some examples: novel theories: http://www.the2sisters.com, http://goo.gl/xCQcc4;

vision therapists: http://vtresource.weebly.com/covd-2016.html; Wisconsin high school teacher Doug Buehl: http://goo.gl/ytavh0.

288 Flexner Report: Flexner, Pritchet, & Henry (1910). The plan for medical education synthesized and codified practices that had begun to be implemented at major medical schools. The report is discussed by Clifford & Guthrie (1988).

288 courses would emphasize: The readings could include Stanovich & Stanovich (2003), a highly readable guide to "using research and reason" in education.

289 a 1916 report: Flexner & Bachman (1916). Education was said to consist of "simple practical problems, which would quickly yield to experience, reading, common sense, and a good general education" (Flexner, 1930, 118).

289 So is demonizing: "A certain casual demonization of teachers has become sufficiently culturally prevalent that it passes for uncontroversial." Rebecca Mead, "Chicago's Teacher Problem, and Ours," *New Yorker*, September 11, 2012, http://goo.gl/xsJ7bC.

290 New hybrid fields: For the British Royal Society's view of neuroscience and education, see http://goo.gl/B3qVmj.

290 Motivation is an old topic: See Carol Dweck, "The Power of Believing You Can Improve," TED talk, November 2014, http://goo.gl/DXNali; Hulleman & Harackiewicz (2009).

291 One successful program: A new type of teacher education program: Goldie Blumenstyk, "After Years Lambasting Teacher-Ed Programs, Art Levine Creates One," *Chronicle of Higher Education*, June 16, 2015, http://goo.gl/q5SNbb.

291 Add a course: There are plenty of excellent resources for teachers to learn what they missed in ed school, if their circumstances allow. *American Educator*, a publication of the American Federation of Teachers, is top-notch. It has readable articles on current issues of importance to teachers written by educators and researchers. For several years it has included a column, *Ask the Cognitive Scientist*, by Daniel Willingham, whose explanations of basic science are models of clarity and accuracy, and his assessments of what is known are reliably sensible. He also writes books, such as Willingham (2015). Cunningham & Zibulsky (2014) is equally good, research based but very oriented to helping children read. Stanovich & Stanovich (2003) is an accessible introduction to how to think about scientific findings and make use of them in planning curricula. The National Reading Panel issued a pamphlet that explained the basic findings and recommended practices in accessible terms; the pdf is downloadable here: http://goo.gl/1mkrYP. My book was preceded by Wolf (2007) and Dehaene (2010), both excellent.

292 best example: For the MTEL "Foundations of Reading" questions, see "Field 90: Foundations of Reading Sample Multiple-Choice Questions," MTEL Test Information Guide, http://goo.gl/S9VURR; for the "Reading Specialist" questions, see "Field 08: Reading Specialist Sample Multiple-Choice Questions," MTEL Test Information Guide, http://goo.gl/OaW2K2.

293 alternate routes: See Clifford & Guthrie (1988).

294 Linguistics 101: A course like this one, for example: http://goo.gl/wQ7i9k.

294 In reading, for example: Connor et al. (2007).

294 Some theorists think: Gee (2003).

294 heavily hyped "brain training": Owen et al. (2010); Melby-Lervåg & Hulme (2013).

295 HMH, for example, markets: For HMH's Read 180, see "About Read 180 Universal," HMH, http://goo.gl/ay26lh.

296 Current proposals for reforming: Darling-Hammond & Richardson (2009); the New Teacher Project, http://tntp.org.

297 KIPP "no excuses": Rachel Monahan, "Charter Schools Try to Retain Teachers with Mom-Friendly Policies," *Atlantic*, November 11, 2014, http://goo.gl/6HN6ET.

297 The excellent model: See "About Minnesota Reading Corps," Minnesota Reading Corps, http://goo.gl/ZpWJGb.

297 **"[The program] provides":** For information about AmeriCorps, see "AmeriCorps," Corporation for National and Community Service, http://www.nationalservice.gov/programs/americorps.

298 **Applications to TfA:** On challenges facing TfA see Stephen Sawchuk, "At 25, Teach for America Enters Period of Change," *Education Week*, January 15, 2016, http://goo.gl/Xx5aDf; Olivia Blanchard, "I Quit Teach for America," *Atlantic*, September 23, 2013, http://goo.gl/JnzGuh; Michael Zuckerman, "Is Teach for America Good for America?," *Harvard Magazine*, December 18, 2013, http://goo.gl/y7U1bw; Kerry Kretchmar and Beth Sondel, "Organizing Resistance to Teach for America," *Rethinking Schools* 28 (spring 2014), http://goo.gl/g5SNGt. Teach for America is a major source of teachers for KIPP schools.

298 **Evidence that contradicts:** On polarization, see Sunstein (1999). For a short overview, see Hastie & Sunstein (2015).

299 **"The empirical evidence":** James Surowiecki, "The Campaign of Magical Thinking," *New Yorker*, March 21, 2016, http://goo.gl/5eJw34.

299 **Brendan Nyhan and Jason Reifler:** Nyhan & Reifler (2010).

299 **we had the precursors:** Teachers Advocating Whole Language was an active BBS, remnants of which are here: http://www.ncte.org/wlu/tawl#groups.

300 **origins of the theory:** Adams (1998).

301 **Figure 12.2:** Many versions of the figure can be found by typing this URL into your browser: http://goo.gl/3M69nQ.

302 **If the child has trouble:** The approach is explained and demonstrated by a top practitioner, Catherine Compton-Lilly, here: https://www.youtube.com/watch?v=4YYpFl2PRvY.

302 **children's comprehension of a story:** Glenberg et al. (2004). The story in this study is so unusually written that it can't be easily comprehended without obtaining additional information from somewhere.

303 **Neurobiological studies:** Hahn, Foxe, & Molholm (2014).

303 **most worrisome aspect:** The 3-cueing method is reminiscent of the fruit-yellow-round combination of probabilistic constraints described back in Chapter 5. The difference is that a skilled reader combines a very large number of very small, unlabeled statistical constraints, automatically and unconsciously. I used games to illustrate the combination of probabilistic constraints precisely because the process occurs with explicit awareness and the constraints can be described linguistically.

There is a connection between these explicit (games) and implicit (lexical statistics) processes nonetheless: David Rumelhart, a pivotal figure in psychology, artificial intelligence, education, and neuroscience. The 3-cueing approach can be traced to Rumelhart (1977), a highly influential article that described reading comprehension as "the process of applying simultaneous constraints at all levels." Having described the general idea, Rumelhart then led the revival of the neural network approach, developing (with McClelland and Hinton) a general framework for applying such networks to human cognition. That framework was the basis for the reading models I've described. The 3-cueing method is like doing constraint satisfaction without a net—that is, without a theory that specifies what "cues" are, how they are learned, how they are combined, and how they function in reading words and comprehending texts.

REFERENCES

Readable articles of particular interest are marked with *

Adams, M. J. (1980). What good is orthographic redundancy? Center for the Study of Reading Technical Report 192. ERIC. http://eric.ed.gov/?id=ED199663.

Adams, M. J. (1998). The three-cueing system. In F. Lehr & J. Osborn (Eds.), *Literacy for all: Issues in teaching and learning*. New York: Guilford Press.

Adams, M. J. (2004). Why not phonics and whole language? In W. Wray (Ed.), *Literacy: Major themes in education*, Vol. 2. London: RoutledgeFalmer.

*Adams, M. J. (2011). Advancing our students' language and literacy: The challenge of complex texts. *American Educator, 34*, 3–11.

Adams, M., Bell, L. A., & Griffin, P. (Eds.). (2007). *Teaching for diversity and social justice*. New York: Routledge.

Allen, J., & Seidenberg, M. S. (1999). Grammaticality judgment and aphasia: A connectionist account. In B. MacWhinney (Ed.), *The emergence of language*. Hillsdale, NJ: Erlbaum.

Allington, R. L. (2002). *Big Brother and the National Reading Curriculum: How ideology trumped evidence*. Portsmouth, NH: Heinemann.

Allington, R. L. (2012). *What really matters for struggling readers: Designing research-based programs*. New York: Pearson.

Allington, R. L., & Woodside-Jiron, H. (1999). The politics of literacy teaching: How "research" shaped educational policy. *Educational Researcher, 28*, 4–13.

American Academy of Pediatrics. (2009). Learning disabilities, dyslexia, and vision. *Pediatrics, 124*, 837–844.

*American Association of School Librarians. (2009). *Standards for the 21st-century learner in action*. American Library Association. http://www.ala.org/aasl/standards/in-action.

American Federation of Teachers (AFT). (2012). *Raising the bar: Aligning and elevating teacher preparation and the teaching profession*. AFT. http://www.aft.org/sites/default/files/news/raisingthebar2013.pdf.

*Anderson, J. A., Reder, L., & Simon, H. (1999). Applications and misapplications of cognitive psychology to mathematics education. Memory Lab, Carnegie Mellon University. http://memory.psy.cmu.edu/publications/Applic.MisApp.pdf.

Ashkenazi, S., Black, J. M., Abrams, D. A., Hoeft, F., and Menon, V. (2013). Neurobiological underpinnings of math and reading learning disabilities. *Journal of Learning Disabilities, 46*, 549–569.

Ashok, V. G., Feng, S., & Choi, Y. (2013). Success with style: Using writing style to predict the success of novels. *Proceedings of the 2013 Conference on Empirical Methods in Natural Language Processing*, 1753–1764. Seattle, WA: Association for Computational Linguistics.

Au, K. H. (1998). Social constructivism and the school literacy learning of students of diverse backgrounds. *Journal of Literacy Research, 30*, 297–319.

Baratz, S., & Baratz, J. (1970). Early childhood intervention: The social science base of institutional racism. *Harvard Educational Review*, *40*, 29–50.

Baron, J., & Strawson, C. (1976). Use of orthographic and word-specific knowledge in reading words aloud. *Journal of Experimental Psychology: Human Perception and Performance*, *4*, 207–214.

Barton, D. (1995). Some problems with an evolutionary view of written language. In S. Puppel (Ed.), *The biology of language*. Philadelphia: John Benjamins Publishing.

Barton, P. E., & Coley, R. J. (2009). *Parsing the achievement gap II: Policy information report*. ERIC. http://goo.gl/cRGmDl.

Bassok, D., & Rorem, A. (2014). Is kindergarten the new first grade? The changing nature of kindergarten in the age of accountability. Amherst Full Day Kindergarten. http://goo.gl/6Fcfxd.

Battaglio, S. (2011). *David Susskind: A televised life*. New York: St. Martin's Griffin.

Baumann, J. F., Hoffman, J. V., Duffy-Hester, A. M., & Ro, J. M. (2000). The first r yesterday and today: US elementary reading instruction practices reported by teachers and administrators. *Reading Research Quarterly*, *35*(3), 338–377.

*Beck, I. L., & Juel, C. (1995). The role of decoding in learning to read. *American Educator*, *19*, 3.

*Beck, I. L., & McKeown, M. G. (1991). Social studies texts are hard to understand: Mediating some of the difficulties. *Language Arts*, *68*, 482–490.

Beck, I. L., McKeown, M. G., & Kucan, L. (2013). *Bringing words to life: Robust vocabulary instruction*. New York: Guilford Press.

Behrmann, M., Nelson, J., & Sekuler, E. B. (1998). Visual complexity in letter-by-letter reading: Pure alexia is not pure. *Neuropsychologia*, *36*, 1115–1132.

Behrmann, M., & Plaut, D. C. (2013). Distributed circuits, not circumscribed centers, mediate visual recognition. *Trends in Cognitive Sciences*, *17*, 210–219.

Bélanger, N. N., & Rayner, K. (2015). What eye movements reveal about deaf readers. *Current Directions in Psychological Science*, *24*, 220–226.

Bélanger, N. N., Slattery, T. J., Mayberry, R. I., & Rayner, K. (2012). Skilled deaf readers have an enhanced perceptual span in reading. *Psychological Science*, *23*, 816–823.

Benford, G. (1980). *Timescape*. New York: Bantam Books.

Bentin, S., & Frost, R. (1987). Processing lexical ambiguity and visual word recognition in a deep orthography. *Memory and Cognition*, *15*, 13–23.

*Bialystok, E., Craik, F. I. M., Green, D. W., & Gollan, T. H. (2009). Bilingual minds. *Psychological Science in the Public Interest*, *10*, 89–129.

Bishop, D. V. M. (2009). Genes, cognition, and communication. *Annals of the New York Academy of Sciences*, *1156*, 1–18.

Bishop, D. V. M. (2013). Neuroscientific studies of intervention for language impairment in children: Interpretive and methodological problems. *Journal of Child Psychology and Psychiatry*, *54*, 247–259.

Blake, R., & Cutler, C. (2003). AAE and variation in teachers' attitudes: A question of school philosophy? *Linguistics and Education*, *14*, 163–194.

Blough, D. S. (1982). Pigeon perception of letters of the alphabet. *Science*, *218*, 397–398.

Bransford, J. D., Brown, A. L., & Cocking, R. R. (Eds.). (2000). *How people learn: brain, mind, experience, and school*. Washington, DC: National Academy Press.

Britton, B. K., & Graesser, A. C. (2014). *Models of understanding text*. New York: Psychology Press.

Browman, C. P., & Goldstein, L. (1990). Gestural specification using dynamically-defined articulatory structures. *Journal of Phonetics*, *18*, 299–320.

Brown, M. C., Sibley, D. E., Washington, J. A., Rogers, T. T., Edwards, J. R., MacDonald, M. C., & Seidenberg, M. S. (2015). Impact of dialect use on a basic component of learning to read. *Frontiers in Psychology*, 6, article 00196. http://goo.gl/7bdhup.

Bruck, M. (1990). Word-recognition skills of adults with childhood diagnoses of dyslexia. *Developmental Psychology*, *26*, 439–454.

Bruck, M. (1992). Persistence of dyslexics' phonological awareness deficits. *Developmental Psychology, 28,* 874–886.

Bruner, J. S., & Goodman, C. C. (1947). Value and need as organizing factors in perception. *Journal of Abnormal and Social Psychology, 42,* 33–44.

Buckley, J. (2011). A closer look at verifying the integrity of NAEP. National Assessment Governing Board Meeting, August 5, 2011. DocSlide. http://docslide.us/download/link/a-closer-look-at-verifying-the-integrity-of-naep-commissioner-jack-buckley.

Cameron-Faulkner, T., & Noble, C. (2013). A comparison of book text and child directed speech. *First Language, 33,* 268–279.

Caramazza, A., & Mahon, B. Z. (2003). The organization of conceptual knowledge: The evidence from category-specific semantic deficits. *Trends in Cognitive Sciences, 7,* 354–361.

Caravolas, M., & Samara, A. (2015). Learning to read and spell words in different writing systems. In A. Pollatsek & R. Treiman (Eds.), *The Oxford handbook of reading.* Oxford: Oxford University Press.

Carey, S. (1978). The child as word learner. In M. Halle, J. Bresnan & G. A. Miller (Eds.), *Linguistic theory and psychological reality.* Cambridge, MA: MIT Press.

Carnoy, M., & Rothstein, R. (2013). *What do international tests really show about US student performance?* Economic Policy Institute. http://www.epi.org/publication/us-student-performance-testing.

Cassidy, K. W., Kelly, M. H., & Sharoni, L. (1999). Inferring gender from name phonology. *Journal of Experimental Psychology: General, 128,* 362–381.

Center, Y., Wheldall, K., Freeman, L., Outhred, L., & McNaught, M. (1995). An experimental evaluation of reading recovery. *Reading Research Quarterly,* 30, 240–263.

*Charity, A. H., Scarborough, H. S., & Griffin, D. M. (2004). Familiarity with school English in African American children and its relation to early reading achievement. *Child Development, 75,* 1340–1356.

Chater, N., Tenenbaum, J. B., & Yuille, A. (2006). Probabilistic models of cognition: Conceptual foundations. *Trends in Cognitive Sciences, 10,* 287–291.

Chomsky, N. & Halle, M. (1968). *The sound pattern of English.* New York: Harper & Row.

Clark, A. (2013). Whatever next? Predictive brains, situated agents, and the future of cognitive science. *Behavioral and Brain Sciences, 36,* 181–204.

Clay, M. M. (1991). *Becoming literate: The construction of inner control.* Taylor & Francis.

Clay, M. M. (1993). *An observation survey of early literacy achievement.* Portsmouth, NH: Heinemann.

Clay, M. M. (1998). *An observation of early literacy achievement.* Auckland, NZ: Heineman.

Clifford, G. J., & Guthrie, J. W. (1988). *Ed school: A brief for professional education.* Chicago: University of Chicago Press, p. 136.

Cochran-Smith, M., & Lytle, S. (2006). Troubling images of teaching in No Child Left Behind. *Harvard Educational Review, 76,* 668–697.

Cochran-Smith, M., & Zeichner, K. M. (Eds.). (2005). *Studying teacher education: The report of the AERA panel on research and teacher education.* Mahway, NJ: Erlbaum.

Coleman, J. S., US Office of Education, & National Center for Education Statistics. (1966). *Equality of Educational Opportunity.* Center for Educational Statistics Report OE-38001. Washington, DC: US Department of Health, Education, and Welfare, Office of Education.

Coles, G. (2000). *Misreading reading: The bad science that hurts children.* Portsmouth, NH: Heinemann.

*Connor, C. M., Morrison, F. J., Fishman, B. J., Schatschneider, C., & Underwood, P. (2007). Algorithm-guided individualized reading instruction. *Science, 315,* 464–465.

Cook, T. D., & Steiner, P. M. (2009). Some empirically viable alternatives to random assignment. *Journal of Policy Analysis and Management, 28,* 165–166.

Council on Foreign Relations (CFR). (2012). *U.S. education reform and national security*. Independent Task Force Report No. 68. CFR. http://www.cfr.org/united-states/us-education-reform-national-security/p27618.
Cox, C. R., Seidenberg, M. S., & Rogers, T. T. (2015). Connecting functional brain imaging and parallel distributed processing. *Language, Cognition and Neuroscience*, *30*, 380–394.
Craig, H. K., Zhang, L., Hensel, S. L., & Quinn, E. J. (2009). African American English–speaking students: An examination of the relationship between dialect shifting and reading outcomes. *Journal of Speech, Language, and Hearing Research*, *52*, 839–855.
*Crews, F. (2006). *Postmodern Pooh*. Evanston, IL: Northwestern University Press.
Critchley, M. (1970). *The dyslexic child*. Springfield, IL: Charles C. Thomas, 1970.
*Crystal, D. (2008). *Txtng: The gr8 db8*. Oxford: Oxford University Press.
*Cunningham, A. E., & Zibulsky, J. (2014). *Book smart: How to develop and support successful, motivated readers*. Oxford: Oxford University Press.
Daniels, P. T. (1996). The first civilizations. In P. T. Daniels & W. Bright, *The world's writing systems*. Oxford: Oxford University Press.
Daniels, P. T., & Bright, W. (1996). *The world's writing systems*. Oxford: Oxford University Press.
*Darling-Hammond, L. (2014). One piece of the whole: Teacher evaluation as part of a comprehensive system for teaching and learning. *American Educator*, *38*, 4.
Darling-Hammond, L., & Richardson, N. (2009). Research review/teacher learning: What matters. *Educational Leadership*, *66*, 46–53.
*DeFrancis, J. (1989). *Visible speech: The diverse oneness of writing systems*. Honolulu: University of Hawaii Press.
*Dehaene, S. (2010). *Reading in the brain: The new science of how we read*. New York: Penguin.
Dennis, I., Besner, D., & Davelaar, E. (1985). Phonology in visual word recognition: Their is more two this than meats the I. In D. Besner, T. Waller, & G. MacKinnon (Eds.), *Reading Research: Advances in Theory and Practice*, 5:170–195. New York: Academic Press.
Dickinson, D. K., & Sprague, K. E. (2003). The nature and impact of early child care environment on the language and early literacy development of children from low-income families. In S. B. Neuman & D. K. Dickinson (Eds.), *Handbook of early literacy research* (New York: Guilford Press), 1:263–280.
Diehl, J. J., Frost, S. J., Sherman, G., Mencl, W. E., Kurian, A., Molfese, P., Landi, N., Preston, J., Soldan, A., Fulbright, R. K., Rueckl, J., Seidenberg, M. S., & Pugh, K. R. (2014). Neural correlates of language and non-language visuospatial processing in adolescents with reading disability. *NeuroImage*, *101*, 653–666.
Diemand-Yauman, C., Oppenheimer, D. M., & Vaughan, E. B. (2011). Fortune favors the bold (and italicized): Effects of disfluency on educational outcomes. *Cognition*, *118*, 111–115.
Dingemanse, M., Blasi, D. E., Lupyan, G., Christiansen, M. H., & Monaghan, P. (2015). Arbitrariness, iconicity, and systematicity in language. *Trends in Cognitive Sciences*, *19*, 603–615.
Duckworth, A. L., Peterson, C., Matthews, M. D., & Kelly, D. R. (2007). Grit: Perseverance and passion for long-term goals. *Journal of Personality and Social Psychology*, *92*, 1087–1101.
Dudley-Marling, C., & Lucas, K. (2009). Pathologizing the language and culture of poor children. *Language Arts*, *86*(5), 362–370.
Duncan, G. J., & Sojourner, A. J. (2013). Can intensive early childhood intervention programs eliminate income-based cognitive and achievement gaps? *Journal of Human Resources*, *48*, 945–968.
Dunn, L. M., & Dunn, D. M. (2015). *Peabody Picture Vocabulary Test: PPVT 4*. New York: Pearson.
Ehri, L. C. (1995). Phases of development in learning to read words by sight. *Journal of Research in Reading*, *18*, 116–125.

Elliott, J. G., & Gibbs, S. (2008). Does dyslexia exist? *Journal of Philosophy of Education, 42*, 475–491.

Elliott, J. G., & Grigorenko, E. L. (2014). *The dyslexia debate*. Cambridge: Cambridge University Press.

Ellis, A. W. (1979). Slips of the pen. *Visible Language, 13*, 265–282.

Escoffier, M., & Di Giacomo, K. (2014). *Take Away the A*. Enchanted Lion Books.

Espy, K. A., Molfese, D. L., Molfese, V. J., & Modglin, A. (2004). Development of auditory event-related potentials in young children and relations to word-level reading abilities at age 8 years. *Annals of Dyslexia, 54*, 9–38.

*Fatsis, S. (2001). *Word freak: Heartbreak, triumph, genius, and obsession in the world of competitive Scrabble players*. Boston: Houghton Mifflin Harcourt.

Fernald, A., Marchman, V. A., & Weisleder, A. (2013). SES differences in language processing skill and vocabulary are evident at 18 months. *Developmental Science, 16*, 234–248.

*Fernald, A., & Weisleder, A. (2015). Twenty years after "meaningful differences," it's time to reframe the "deficit" debate about the importance of children's early language experience. *Human Development, 58*, 1–4.

Feynman, R. P. (2014). *The meaning of it all: Thoughts of a citizen-scientist*. New York: Basic Books.

Filik, R., & Barber, E. (2011). Inner speech during silent reading reflects the reader's regional accent. *PLOS ONE, 610*, e25782.

Firth, J. R. (1957). A synopsis of linguistic theory, 1930–1955. In Philological Society (Ed.), *Studies in linguistic analysis*. Oxford: Blackwell.

Fletcher, J. M. (2009). Dyslexia: The evolution of a scientific concept. *Journal of the International Neuropsychological Society, 15*, 501–508.

Fletcher, J. M., & Vaughn, S. (2009). Response to intervention: Preventing and remediating academic difficulties. *Child Development Perspectives, 3*, 30–37.

Flexner, A. (1930). *Universities: American, English, German*. New York: Oxford University Press.

Flexner, A., & Bachman, F. P. (1916). *Public education in Maryland: A report to the Maryland Educational Survey Commission*. New York: General Education Board.

Flexner, A., Pritchet, H., & Henry, S. (1910). *Medical education in the United States and Canada*. New York: Carnegie Foundation for the Advancement of Teaching.

Flusberg, S., & McClelland, J. L. (2014). Connectionism and the emergence of mind. Oxford Handbooks Online. http://goo.gl/02BU8G.

Flynn, J. R. (1987). Massive IQ gains in 14 nations: What IQ tests really measure. *Psychological Bulletin, 101*, 171–191.

Fodor, J. A. (1975). *The language of thought*. Cambridge, MA: Harvard University Press.

Fowler, A. E. (1991). How early phonological development might set the stage for phoneme awareness. In S. Brady & D. P. Shankweiler (Eds.), *Phonological processes in literacy: A tribute to Isabelle Y. Liberman*. Hillsdale, NJ: Erlbaum.

Friend, A., DeFries, J. C., & Olson, R. K. (2008). Parental education moderates genetic influences on reading disability. *Psychological Science, 19*, 1124–1130.

*Frith, C., & Frith, U. (2005). Quick guide: Theory of mind. *Current Biology, 15*, R644–R645.

*Frith, U. (1998). Literally changing the brain. *Brain, 121*, 1011–1012.

Frith, U., & Happé, F. (1998). Why specific developmental disorders are not specific: On-line and developmental effects in autism and dyslexia. *Developmental Science, 1*, 267–272.

Frost, R., Katz, L., & Bentin, S. (1987). Strategies for visual word recognition and orthographical depth: A multilingual comparison. *Journal of Experimental Psychology: Human Perception and Performance, 13*, 104–115.

Fryer, R. G., & Levitt, S. D. (2006). The black-white test score gap through third grade. *American Law and Economics Review, 8*(2), 249–281.

Funnell, E. (1983). Phonological processes in reading: New evidence from acquired dyslexia. *British Journal of Psychology, 74*, 159–180.

Gabrieli, J. D. E. (2009). Dyslexia: A new synergy between education and cognitive neuroscience. *Science, 325*, 280–283.

*Galaburda, A. M., LoTurco, J., Ramus, F., Holly Fitch, R., & Rosen, G. D. (2006). From genes to behavior in developmental dyslexia. *Nature Neuroscience, 9*, 1213–1217.

Garan, E. (2001). Beyond the smoke and mirrors: A critique of the National Reading Panel Report on phonics. *Phi Delta Kappan, 82*, 500–506.

Garrard, P., Maloney, L. M., Hodges, J. R., & Patterson, K. (2005). The effects of very early Alzheimer's disease on the characteristics of writing by a renowned author. *Brain, 128*, 250–260.

*Gee, J. (2003). *What video games have to teach us about learning and literacy*. New York: Palgrave Macmillan.

General Accounting Office (GAO). (2007). *Poverty in America*. GAO. http://www.gao.gov/products/GAO-07-344.

Gernsbacher, M. A, Varner, K. R., & Faust, M. (1990). Investigating differences in general comprehension skill. *Journal of Experimental Psychology: Learning, Memory, and Cognition, 16*, 430–445.

Gladwell, M. (2013). *David and Goliath: Underdogs, misfits and the art of battling giants*. Boston: Little, Brown & Co.

Glasser, M. F., Coalson, T. S., Robinson, E. C., Hacker, C. D., Harwell, J., et al. (2016). A multimodal parcellation of human cerebral cortex. *Nature*, advance online publication, July 20, http://dx.doi.org/10.1038/nature18933.

*Gleick, J. (2011). *The information: A history, a theory, a flood*. New York: Pantheon.

Gleitman, L. R. (1990). The structural sources of verb meanings. *Language Acquisition, 1*, 3–55.

Glenberg, A. M., Gutierrez, T., Levin, J. R., Japuntich, S., & Kaschak, M. P. (2004). Activity and imagined activity can enhance young children's reading comprehension. *Journal of Educational Psychology, 96*, 424.

Glushko, R. J. (1979). The organization and activation of orthographic knowledge in reading aloud. *Journal of Experimental Psychology: Human Perception and Performance, 5*, 674–691.

Goldberg, Y. (2012). Do baboons really care about letter-pairs? Ben-Gurion University of the Negev, http://www.cs.bgu.ac.il/~yoavg/uni/bloglike/baboons.html.

Goldin-Meadow, S., & Mayberry, R. I. (2001). How do profoundly deaf children learn to read? *Learning Disabilities Research & Practice, 16*, 222–229.

*Goldstein, D. (2014). *The Teacher Wars*. New York: Doubleday.

Goodman, K. S. (1974). Effective teachers of reading know language and children. *Elementary English, 51*, 823–828.

Goodman, K. S. (1976). Reading: A psycholinguistic guessing game. In H. Singer & R. B. Ruddell (Eds.), *Theoretical models and processes of reading*, 497–508. Newark, DE: International Reading Association.

Goodman, K. S. (1986). *What's whole in whole language*. Richmond Hill, Ontario: Scholastic.

Goodman, K. S. (1994). Reading, writing, and written texts: A transactional sociopsycholinguistic view. In R. B. Ruddell, M. R. Ruddell, & H. Singer (Eds.), *Theoretical models and processes of reading*. Newark, DE: International Reading Association.

Goodman, K. S. (1996). *Ken Goodman on reading: A common-sense look at the nature of language and the science of reading*. Portsmouth: Heinemann.

Goodman, K., & Goodman, Y. (1981). Twenty questions about teaching language. *Educational Leadership*, 38, 437–442.

Gosa, T., & Alexander, K. (2007). Family (dis)advantage and the educational prospects of better off African American youth: How race still matters. *Teachers College Record, 109*(2), 285–321.

Goswami, U. (2015). Sensory theories of developmental dyslexia: Three challenges for research. *Nature Reviews Neuroscience, 16*, 43–54.

Goswami, U., Ziegler, J. C., Dalton, L., & Schneider, W. (2001). Pseudohomophone effects and phonological recoding procedures in reading development in English and German. *Journal of Memory and Language, 45*, 648–664.

Gough, P. B. (1983). Context, form, and interaction. In K. Rayner (Ed.), *Eye movements in reading: Perception and language processes*. New York: Academic Press.

*Gough, P. B., & Hillinger, M. L. (1980). Learning to read: An unnatural act. *Annals of Dyslexia, 30*, 179–196.

Gough, P. B., Hoover, W. A., & Peterson, C. L. (1996). Some observations on a simple view of reading. In C. Cornoldi & J. Oakhill (Eds.), *Reading comprehension difficulties: Processes and intervention*, 1–13. Erlbaum.

Grainger, J., Dufau, S., Montant, M., Ziegler, J. C., & Fagot, J. (2012). Orthographic processing in baboons (*Papio papio*). *Science, 336*, 245–248.

Granic, I., Lobel, A., & Engels, R. C. (2014). The benefits of playing video games. *American Psychologist, 69*, 66–78.

Graves, W. W., Binder, J. R., Desai, R. H., Humphries, C., Stengel, B. C., & Seidenberg, M. S. (2014). Anatomy is strategy: Skilled reading differences associated with structural connectivity differences in the reading network. *Brain & Language, 133*, 1–13.

Gray, L., & Taie, S. (2015). Public school teacher attrition and mobility in the first five years: Results from the first through fifth waves of the 2007–08 Beginning Teacher Longitudinal Study. National Center for Education Statistics. https://nces.ed.gov/pubsearch/pubsinfo.asp?pubid=2015337.

Gray, P. (2010). Children teach themselves to read. *Psychology Today*. February 24, 2010. http://goo.gl/H4JYpa.

Green, C. S., & Bavelier, D. (2003). Action video game modifies visual selective attention. *Nature, 423*, 534–537.

Grossen, B. (1997). 30 years of research: What we now know about how children learn to read. ERIC. http://www.eric.ed.gov/PDFS/ED415492.pdf.

Guinness. (1990). *Guinness Book of World Records*. New York: Bantam Books.

Hahn, N., Foxe, J. J., & Molholm, S. (2014). Impairments of multisensory integration and cross-sensory learning as pathways to dyslexia. *Neuroscience & Biobehavioral Reviews, 47*, 384–392.

Halliday, M. A. K. (1983). Ideas about language. In F. B. Agard, G. Kelley, A. Makkai, & V. Makkai (Eds.), *Essays in honor of Charles F. Hockett*. Leiden: Brill.

Hammill, D. D., Leigh, J. E., McNutt, G., & Larsen, S. C. (1988). A new definition of learning disabilities. *Learning Disability Quarterly, 11*, 217–223.

Hanley, J. R., Masterson, J., Spencer, L., & Evans, D. (2004). How long do the advantages of learning to read a transparent orthography last? *Quarterly Journal of Experimental Psychology, 57*, 1393–1410.

Hanushek, E. A., & Woessmann, L. (2008). The role of cognitive skills in economic development. *Journal of Economic Literature, 46*, 607–668.

Hanushek, E. A., & Woessmann, L. (2012). Do better schools lead to more growth? Cognitive skills, economic outcomes, and causation. *Journal of Economic Growth, 17*, 267–321.

Hargreaves, I. S., Pexman, P. M., Zdrazilova, L., & Sargious, P. (2012). How a hobby can shape cognition: Visual word recognition in competitive Scrabble players. *Memory and Cognition, 40*, 1–7.

Harm, M., & Seidenberg, M. S. (1999). Reading acquisition, phonology, and dyslexia: Insights from a connectionist model. *Psychological Review, 106*, 491–528.

Harm, M., McCandliss, B., & Seidenberg, M. S. (2003). Modeling the successes and failures of interventions for disabled readers. *Scientific Studies of Reading, 7*, 155–182.

Harm, M., & Seidenberg, M. S. (2004). Computing the meanings of words in reading: Division of labor between visual and phonological processes. *Psychological Review, 111*, 662–720.

Harms, W., & DePencier, I. (1996). 100 years of learning at the University of Chicago Laboratory Schools. University of Chicago Laboratory Schools. http://goo.gl/ddfGl1.

Harris, D. N., & Sass, T. R. (2011). Teacher training, teacher quality, and student achievement. *Journal of Public Economics, 95*, 798–812.

*Hart, B., & Risley, T. R. (1995). *Meaningful differences in the everyday experience of young American children*. Baltimore: Paul H. Brookes Publishing.

Hart, S. A., Soden, B., Johnson, W., Schatschneider, C., & Taylor, J. (2013). Expanding the environment: gene × school-level SES interaction on reading comprehension. *Journal of Child Psychology and Psychiatry, 54*, 1047–1055.

Hassett, D. D. (2006). Signs of the times: The governance of alphabetic print over "appropriate" and "natural" reading development. *Journal of Early Childhood Literacy, 6*, 77–103.

*Hastie, R., & Sunstein, C. (2015). Polarization: One reason groups fail. *Chicago Booth Review*. July 21, 2015. http://goo.gl/FnuHdo.

Hayes, D., Wolfer, L. T., & Wolfe, M. F. (1996). Schoolbook simplification and its relation to the decline in SAT-verbal scores. *American Educational Research Journal, 33*, 489–508.

Heckman, J. J., & LaFontaine, P. A. (2008). The declining American high school graduation rate: Evidence, sources, and consequences. National Bureau of Economic Research. http://www.nber.org/reporter/2008number1/heckman.html.

Heckman, J. J., & LaFontaine, P. A. (2010). The American high school graduation rate: Trends and levels. *Review of Economics and Statistics, 92*, 244–262.

Hillis, A. E., Newhart, M., Heidler, J., Barker, P., Herskovits, E., & Degaonkar, M. (2005). The roles of the "visual word form area" in reading. *Neuroimage, 24*, 548–559.

*Hinton, G. E., Plaut, D. C., & Shallice, T. (1993). Simulating brain damage. *Scientific American, 269*, 76–82.

Hodges, J. R., Patterson, K., Oxbury, S., & Funnell, E. (1992). Semantic dementia: Progressive fluent aphasia with temporal lobe atrophy. *Brain, 115*, 1783–1806.

Hoeft, F., McCandliss, B. D., Black, J. M., Gantman, A., Zakerani, N., Hulme, C., et al. (2011). Neural systems predicting long-term outcome in dyslexia. *Proceedings of the National Academy of Sciences, 108*, 361–366.

Hoff, E. (2003). The specificity of environmental influence: Socioeconomic status affects early vocabulary development via maternal speech. *Child Development, 74*, 1368–1378.

Hoff, E. (2013). Interpreting the early language trajectories of children from low-SES and language minority homes: Implications for closing achievement gaps. *Developmental Psychology, 49*, 4–14.

Hombo, C. M. (2003). NAEP and No Child Left Behind: Technical challenges and practical solutions. *Theory into Practice, 42*, 59–65.

Hoover, W. A., & Gough, P. B. (1990). The simple view of reading. *Reading and Writing, 2*, 127–160.

Hornickel, J., & Kraus, N. (2013). Unstable representation of sound: A biological marker of dyslexia. *Journal of Neuroscience, 33*, 3500–3504.

Horton-Ikard, R., & Miller, J. F. (2004). It is not just the poor kids: The use of AAE forms by African-American school-aged children from middle SES communities. *Journal of Communication Disorders, 37*, 467–487.

Hoxhallari, L., van Daal, V. H. P., & Ellis, N. C. (2004). Learning to read words in Albanian: A skill easily acquired. *Scientific Studies of Reading, 8*, 153–166.

*Hulleman, C. S., & Harackiewicz, J. M. (2009). Promoting interest and performance in high school science classes. *Science, 326,* 1410–1412.

Hulme, C., Goetz, K., Gooch, D., Adams, J., & Snowling, M. J. (2007). Paired-associate learning, phoneme awareness, and learning to read. *Journal of Experimental Child Psychology, 96,* 150–166.

Hulme, C., Nash, H. M., Gooch, D., Lervåg, A., & Snowling, M. J. (2015). The foundations of literacy development in children at familial risk of dyslexia. *Psychological Science, 26,* 1877–1886.

Huttenlocher, J., Haight, W., Bryk, A., Seltzer, M., & Lyons, T. (1991). Early vocabulary growth: Relation to language input and gender. *Developmental Psychology, 27,* 236.

James, P. A., Oparil, S., Carter, B. L., Cushman, W. C., Dennison-Himmelfarb, C., et al. (2014). 2014 evidence-based guideline for the management of high blood pressure in adults: Report from the panel members appointed to the Eighth Joint National Committee (JNC 8). *Journal of the American Medical Association, 311,* 507–520.

Jared, D., Levy, B. A., & Rayner, K. (1999). The role of phonology in the activation of word meanings during reading: Evidence from proofreading and eye movements. *Journal of Experimental Psychology: General, 128,* 219–264.

*Jones, N. (2014). The learning machines. *Nature, 505,* 146–148.

*Joshi, R. M., Treiman, R., Carreker, S., & Moats, L. C. (2008). How words cast their spell. *American Educator, 32,* 6–16.

Jusczyk, P. W. (2000). *The discovery of spoken language.* Cambridge, MA: MIT Press.

Justice, L. M., & Ezell, H. K. (2002). Use of storybook reading to increase print awareness in at-risk children. *American Journal of Speech-Language Pathology, 11,* 17–29.

*Kahneman, D. (2011). *Thinking, Fast and Slow.* New York: Macmillan.

Karemaker, A., Pitchford, N. J., & O'Malley, C. (2010). Enhanced recognition of written words and enjoyment of reading in struggling beginner readers through whole-word multimedia software. *Computers & Education, 54*(1), 199–208.

Kellner, D. (2006). Technological transformation, multiple literacies, and the re-visioning of education. In J. Weiss, J. Hunsinger, J. Nolan, & P. Trifonas (Eds.), *The international handbook of virtual learning environments* 14, 241–268. Netherlands: Springer.

Kena, G., Musu-Gillette, L., Robinson, J., Wang, X., Rathbun, A., et al. (2015). The condition of education 2015. NCES 2015-144. National Center for Education Statistics. https://nces.ed.gov/pubsearch/pubsinfo.asp?pubid=2015144.

Key, A. P., Ferguson, M., Molfese, D. L., Peach, K., Lehman, C., & Molfese, V. J. (2007). Smoking during pregnancy affects speech-processing ability in newborn infants. *Environmental Health Perspectives, 115,* 623–629.

*Kim, J. (2008). Research and the reading wars. In F. M. Hess (Ed.), *When research matters: How scholarship influences education policy.* Cambridge, MA: Harvard Education Press.

Kramer, S. N. (1963). *The Sumerians: Their history, culture, and character.* Chicago: University of Chicago Press.

Kupiainen, S., Hautamäki, J., & Karjalainen, T. (2009). *The Finnish education system and PISA,* vol. 46. Norberto Bottani Website. http://www.oxydiane.net/IMG/pdf_opm46.pdf.

Labov, W. (1972). *Language in the inner city: Studies in the black English vernacular.* Philadelphia: University of Pennsylvania Press.

*Lai, C. Q. (2006). How much of human height is genetic and how much is due to nutrition. *Scientific American,* December 11, 2006.

Lake, B. M., Salakhutdinov, R., & Tenenbaum, J. B. (2015). Human-level concept learning through probabilistic program induction. *Science, 350,* 1332–1338.

Landauer, T. K., & Dumais, S. T. (1997). A solution to Plato's problem: The latent semantic analysis theory of acquisition, induction, and representation of knowledge. *Psychological Review, 104,* 211–240.

Landerl, K., Wimmer, H., & Frith, U. (1997). The impact of orthographic consistency on dyslexia: A German-English comparison. *Cognition, 63,* 315–334.

Latomaa, S., & Nuolijärvi, P. (2002). The language situation in Finland. *Current Issues in Language Planning, 3,* 95–202.

Lee, Y. A., Binder, K. S., Kim, J. O., Pollatsek, A., & Rayner, K. (1999). Activation of phonological codes during eye fixations in reading. *Journal of Experimental Psychology: Human Perception and Performance, 25,* 948–964.

*Leff, A. (2004). Cognitive primer: Alexia. *Advances in Clinical Neuroscience and Rehabilitation, 4,* 18–22.

Lehman, D. (1998). *The last avant-garde: The making of the New York school of poets.* New York: Doubleday.

*Lesgold, A., & Welch-Ross, M. (2012). *Improving adult literacy instruction: Options for practice and research.* Washington, DC: National Research Council.

Leu, D. J., Jr. (1982). Oral reading error analysis: A critical review of research and application. *Reading Research Quarterly, 17*(3), 420–437.

Levin, I., Saiegh-Haddad, E, Hende, N., & Ziv, M. (2008) Early literacy in Arabic: An intervention study among Israeli Palestinian kindergartners. *Applied Psycholinguistics, 29,* 413–436.

*Liberman, I. Y., Shankweiler, D., & Liberman, A. (1989). The alphabetic principle and learning to read. In D. Shankweiler & I. Y. Liberman (Eds.), *Phonology and Reading Disability: Solving the Reading Puzzle.* International Academy for Research in Learning Disabilities Monograph Series, 6: 1–33. Ann Arbor: University of Michigan Press. http://files.eric.ed.gov/fulltext/ED427291.pdf.

Lifton, Robert J. (1961). Thought reform and the psychology of totalism: A study of "brainwashing" in China. New York: W. W. Norton. Reprint: http://goo.gl/FxWKvm.

Logan, J. (2009). Dyslexic entrepreneurs: The incidence; their coping strategies and their business skills. *Dyslexia, 1,* 328–346.

Lupyan, G., Rakison, D. H., & McClelland, J. L. (2007). Language is not just for talking: Redundant labels facilitate learning of novel categories. *Psychological Science, 18,* 1077–1083.

Mabry, M. (2008). *Twice as Good: Condoleezza Rice and Her Path to Power.* New York: Modern Times.

MacDonald, M. C., Pearlmutter, N., & Seidenberg, M. S. (1994). The lexical nature of syntactic ambiguity resolution. *Psychological Review, 101,* 767–703.

Magnuson, K. A., & Duncan, G. J. (2006). The role of family socioeconomic resources in the black-white test score gap among young children. *Developmental Review, 26*(4), 365–399.

Maguire, E. A., Gadian, D. G., Johnsrude, I. S., Good, C. D., Ashburner, J., Frackowiak, R. S., & Frith, C. D. (2000). Navigation-related structural change in the hippocampi of taxi drivers. *Proceedings of the National Academy of Sciences, 97,* 4398–4403.

*Maguire, J. (2006). *American bee: The national spelling bee and the culture of word nerds.* Emmaus, PA: Rodale Books.

Mann, H. (1844). Mr. Mann's seventh annual report: Education in Europe. *Common School Journal, 6,* 72.

Mano, Q. R., Humphries, C., Desai, R., Seidenberg, M. S., Osmon, D. C. et al. (2013). The role of the occipitotemporal cortex in reading: Reconciling stimulus, task, and lexicality effects. *Cerebral Cortex, 23*(4), 988-1001.

Marchand, P. (1998). *Marshall McLuhan: The medium and the messenger.* Cambridge, MA: MIT Press.

Markowitz, D. M., & Hancock, J. T. (2014). Linguistic traces of a scientific fraud: The case of Diederik Stapel. *PLOS ONE, 9,* e105937.

Marslen-Wilson, W. D. (1975). Linguistic structure and speech shadowing at very short latencies. *Nature, 244,* 522–523.

*Mayer, R. E. (2004). Should there be a three-strikes rule against pure discovery learning? *American Psychologist, 59,* 14–19.

Mayhew, K. C., & Edwards, A. C. (1936). *The Dewey School: The laboratory school of the University of Chicago 1896–1903.* New York: D. Appleton-Century Company.

McArthur, G. (2008). Does What Works Clearinghouse work? A brief review of Fast ForWord®. *Australian Journal of Public Education* 32, 101–107. http://goo.gl/VxKirz.

McCandliss, B. D., Cohen, L., & Dehaene, S. (2003). The visual word form area: Expertise for reading in the fusiform gyrus. *Trends in Cognitive Sciences, 7,* 293–299.

*McCardle, P., Scarborough, H. S., & Catts, H. W. (2001). Predicting, explaining, and preventing children's reading difficulties. *Learning Disabilities Research & Practice, 16,* 230–239.

McClelland, J. L., Mirman, D., Bolger, D. J., & Khaitan, P. (2014). Interactive activation and mutual constraint satisfaction in perception and cognition. *Cognitive Science, 38,* 1139–1189.

McClelland, J. L., & Rumelhart, D. E. (1981). An interactive activation model of context effects in letter perception: I. An account of basic findings. *Psychological Review, 88,* 375–407.

McConkie, G. W., & Rayner, K. (1975). The span of the effective stimulus during a fixation in reading. *Perception & Psychophysics, 17,* 578–586.

McCutchen, D., & Perfetti, C. A. (1982). The visual tongue-twister effect: Phonological activation in silent reading. *Journal of Verbal Learning and Verbal Behavior, 21,* 672–687.

McLaughlin, M., & DeVoogd, G. L. (2004). *Critical literacy: Enhancing students' comprehension of text.* New York: Scholastic Inc.

McLuhan, M. (1964). *Understanding media.* New York; McGraw-Hill.

*Mehta, J. (2013). Teachers: Will we ever learn? *New York Times.* http://www.nytimes.com/2013/04/13/opinion/teachers-will-we-ever-learn.html.

*Mehta, J., & Doctor, J. (2013). Raising the bar for teaching. *Phi Delta Kappan, 94,* 8–13.

Melby-Lervåg, M., & Hulme, C. (2013). Is working memory training effective? A meta-analytic review. *Developmental Psychology, 49*(2), 270–291.

Menand, L. (2001). *The metaphysical club: A story of ideas in America.* New York: Farrar, Straus & Giroux.

Metsala, J. L., & Walley, A. C. (1998). Spoken vocabulary growth and the segmental restructuring of lexical representations: Precursors to phonemic awareness and early reading ability. In J. Metsala & L. Ehri (Eds.), *Word recognition in beginning literacy.* Mahwah, NJ: Erlbaum.

Michalowski, P. (2003). A man called Enmebaragesi. In W. Sallaberger, K. Volk, and A. Zgoll (Eds.), *Literatur, Politik und Recht in Mesopotamien: Festschrift für Claus Wilcke.* Heidelberg: Harrassowitz.

Michelli, N., & Earley, P. (2011). Teacher education policy context. In P. M. Earley, D. G. Imig & N. M. Michelli (Eds.), *Teacher education policy in the United States: Issues and Tensions in an era of evolving expectations.* New York: Taylor & Francis, 1–13.

Miellet, S., O'Donnell, P. J., & Sereno, S. C. (2009). Parafoveal magnification visual acuity does not modulate the perceptual span in reading. *Psychological Science, 20,* 721–728.

*Moats, L. C. (1999). *Teaching reading is rocket science: What expert teachers of reading should know and be able to do.* Washington, DC: American Federation of Teachers. http://www.aft.org/search/site/moats.

*Moats, L. C. (2000). *Whole language lives on: The illusion of "balanced" reading instruction.* Washington, DC: Thomas B. Fordham Institute.

Montag, J. L., & MacDonald, M. C. (2015). Text exposure predicts spoken production of complex sentences in 8- and 12-year-old children and adults. *Journal of Experimental Psychology: General, 144,* 447.

Montag, J. L., Jones, M. N., & Smith, L. B. (2015). The words children hear: Picture books and the statistics for language learning. *Psychological Science, 26,* 1489–1496.

Morais, J., Cary, L., Alegria, J., & Bertelson, P. (1979). Does awareness of speech as a sequence of phones arise spontaneously? *Cognition, 7,* 323–331.

Morganbesser, S. (1977). Introduction. In S. Morgenbesser & J. Dewey, *Dewey and his critics: Essays from the Journal of Philosophy.* Indianapolis: Hackett Publishing.

Morrison, F. J., Alberts, D. M., & Griffith, E. M. (1997). Nature-nurture in the classroom: Entrance age, school readiness, and learning in children. *Developmental Psychology, 33*, 254–262.

Motoyama, M., Levchenko, K., Kanich, C., McCoy, D., Voelker, G. M., & Savage, S. (2010). Re: CAPTCHAs—Understanding CAPTCHA-Solving Services in an Economic Context. *USENIX Security Symposium, 10*, 3–8.

*Munakata, Y., & McClelland, J. L. (2003). Connectionist models of development. *Developmental Science, 6*, 413–429.

Murnane, R. J. (2013). *US high school graduation rates: Patterns and explanations*. NBER Working Paper No. 18701. National Bureau of Economic Research. http://www.nber.org/papers/w18701.

Murnane, R. J., Willett, J. B., Duhaldeborde, Y., & Tyler, J. H. (2000). How important are the cognitive skills of teenagers in predicting subsequent earnings? *Journal of Policy Analysis and Management, 19*, 547–568.

Muter, V., & Snowling, M. J. (2009). Children at familial risk of dyslexia: Practical implications from an at-risk study. *Child and Adolescent Mental Health, 14*, 37–41.

Myers, C. A., Vandermosten, M., Farris, E. A., Hancock, R., Gimenez, P., et al. (2014). White matter morphometric changes uniquely predict children's reading acquisition. *Psychological Science, 25*, 1870–1883.

NAEP. (2011). *Reading 2011: National assessment of educational progress at grades 4 and 8*. NCES 2012–457. Washington, DC: US Department of Education.

National Assessment of Adult Literacy (NAAL). (2003). What is NAAL? http://nces.ed.gov/naal.

National Commission on Excellence in Education. (1983). *A nation at risk: The imperative for educational reform: A report to the nation and the secretary of education*. Washington, DC: US Department of Education. http://www2.ed.gov/pubs/NatAtRisk/risk.html.

National Endowment for the Arts (NEA). (2004). *Reading at risk: A survey of literary reading in America*. NEA. https://www.arts.gov/publications/reading-risk-survey-literary-reading-america-0.

National Endowment for the Arts. (2007). *To read or not to read: A question of national consequence*. NEA. https://www.arts.gov/publications/read-or-not-read-question-national-consequence-0.

National Endowment for the Arts. (2009). *Reading on the rise: A new chapter in American literacy*. NEA. https://www.arts.gov/publications/reading-rise-new-chapter-american-literacy.

National Institute of Child Health and Human Development (NICHD). (2000). *Report of the National Reading Panel: An evidence-based assessment of the scientific research literature on reading and its implications for reading instruction: Reports of the subgroups*. NIH Publication No. 00–4754. Washington, DC: US Government Printing Office.

Neuman, S. B., & Celano, D. (2001). Access to print in low-income and middle-income communities: An ecological study of four neighborhoods. *Reading Research Quarterly, 36*(1), 8–26.

Nyhan, B., & Reifler, J. (2010). When corrections fail: The persistence of political misperceptions. *Political Behavior, 32*(2), 303–330. http://www.dartmouth.edu/~nyhan/nyhan-reifler.pdf.

*Oetting, J. B., & Kent, R. (2004). Dialect speakers. In R. D. Kent (Ed.), *MIT Encyclopedia of Communication Disorders*, 294–296.

Oetting, J. B., & McDonald, J. L. (2001). Nonmainstream dialect use and specific language impairment. *Journal of Speech, Language, and Hearing Research, 44*, 207–223.

*Olson, D. R. (1996). *The world on paper: The conceptual and cognitive implications of writing and reading*. Cambridge: Cambridge University Press.

Organisation for Economic Co-operation and Development (OECD). (2010). *PISA 2009 results: Overcoming social background: Equity in learning opportunities and outcomes (volume II)*. OECD. http://goo.gl/AXnBq9.

Organisation for Economic Co-operation and Development (OECD). (2011). *Lessons from PISA for the United States: Strong performers and successful reformers in education*. OECD. https://www.oecd.org/pisa/46623978.pdf.

Organisation for Economic Co-operation and Development. (2012). *PISA 2012 results in focus*. OECD. https://www.oecd.org/pisa/keyfindings/pisa-2012-results-overview.pdf.

Organisation for Economic Co-operation and Development. (2013a). *PISA 2012 results*, Vol. 1: *What students know and can do*. OECD. https://www.oecd.org/pisa/keyfindings/pisa-2012-results-volume-I.pdf.

Organisation for Economic Co-operation and Development. (2013b). *PISA 2012 results*, Vol. 2: *Excellence through equity: Giving every student the chance to succeed*. https://www.oecd.org/pisa/keyfindings/pisa-2012-results-volume-ii.htm.

Orton, S. T. (1925). "Word-blindness" in school children. *Archives of Neurology and Psychiatry, 14*, 581.

*Owen, A. M., Hampshire, A., Grahn, J. A., Stenton, R., Dajani, S., et al. (2010). Putting brain training to the test. *Nature, 465*, 775–778.

Paston, A. (Ed.). (2013). *The Smithsonian book of presidential trivia*. Washington, DC: Smithsonian Books.

Pastore, R. (2012). The effects of time-compressed instruction and redundancy on learning and learners' perceptions of cognitive load. *Computers & Education, 58*, 641–651.

Patterson, K., Ralph, M. A. L., Jefferies, E., Woollams, A., Jones, R., Hodges, J. R., & Rogers, T. T. (2006). "Presemantic" cognition in semantic dementia: Six deficits in search of an explanation. *Journal of Cognitive Neuroscience, 18*, 169–183.

Paulesu, E., Démonet, J. F., Fazio, F., McCrory, E., Chanoine, V., et al. (2001). Dyslexia: Cultural diversity and biological unity. *Science, 291*, 2165–2167.

Pearson, P. D. (1991). Developing expertise in reading comprehension: What should be taught? How should it be taught? Center for the Study of Reading Technical Report 512. IDEALS, University of Illinois Urbana-Champaign. http://goo.gl/tEHIuY.

Pennington, B. F., & Bishop, D. V. (2009). Relations among speech, language, and reading disorders. *Annual Review of Psychology, 60*, 283–306.

Pennington, B. F., & Lefly, D. L. (2001). Early reading development in children at family risk for dyslexia. *Child Development, 72*, 816–833.

Perfetti, C. (2007). Reading ability: Lexical quality to comprehension. *Scientific Studies of Reading, 11*, 357–383.

Perfetti, C. A., & Hogaboam, T. (1975). Relationship between single word decoding and reading comprehension skill. *Journal of Educational Psychology, 67*, 461–469.

Perfetti, C. A., & Roth, S. (1981). Some of the interactive processes in reading and their role in reading skill. In A. M. Lesgold & C. A. Perfetti (Eds.), *Interactive processes in reading*. Hillsdale, NJ: Erlbaum.

Perrachione, T. K. (2012). Impaired learning of phonetic consistency and generalized neural adaptation deficits in dyslexia. PhD thesis, MIT Department of Brain and Cognitive Sciences.

Perrachione, T. K., Del Tufo, S. N., & Gabrieli, J. D. (2011). Human voice recognition depends on language ability. *Science, 333*, 595.

Peterson, P. E., Woessmann, L., Hanushek, E. A., & Lastra-Anadón, C. X. (2011). Globally challenged: Are US students ready to compete? The latest on each state's international standing in math and reading. PEPG 11-03. Program on Education Policy and Governance, Harvard University, http://goo.gl/ji63Ro.

Peterson, R. L., & Pennington, B. F. (2012). Developmental dyslexia. *Lancet, 379*, 1997–2007.

Piantadosi, S. T., Tily, H., & Gibson, E. (2012). The communicative function of ambiguity in language. *Cognition, 122,* 280–291.

Pinker, S. (2003). *The Blank Slate: The Modern Denial of Human Nature.* New York: Penguin.

*Plaut, D. C. (2005). Connectionist approaches to reading. In M. J. Snowling & C. Hulme (Eds.), *The science of reading: A handbook.* Oxford: Blackwell.

Plaut, D. C., McClelland, J. L., Seidenberg, M. S., & Patterson, K. E. (1996). Understanding normal and impaired word reading: Computational principles in quasiregular domains. *Psychological Review, 103,* 56–115.

Poldrack, R. A. (2011). Inferring mental states from neuroimaging data: From reverse inference to large-scale decoding. *Neuron, 72,* 692–697.

Pollatsek, A., Lesch, M., Morris, R., & Rayner, K. (1992). Phonological codes are used in integrating information across saccades in word identification and reading. *Journal of Experimental Psychology: Human Perception and Performance, 18,* 148–162.

Potter, M. C. (1984). Rapid serial visual presentation (RSVP): A method for studying language processing. *New Methods in Reading Comprehension Research, 118,* 91–118.

Powell, B. B. (2009). *Writing: Theory and history of the technology of civilization.* Hoboken, NJ: John Wiley & Sons.

Price, C. J., & Devlin, J. T. (2003). The myth of the visual word form area. *Neuroimage, 19,* 473–481.

Pugh, K. R., Frost, S. J., Rothman, D. L., Hoeft, F., Del Tufo, S. N., et al. (2014). Glutamate and choline levels predict individual differences in reading ability in emergent readers. *Journal of Neuroscience, 34,* 4082–4089.

Pugh, K. R., Frost, S. J., Sandak, R., Landi, N., Moore, D., et al. (2010). The neural basis of reading. In P. L. Cornelissen, P. Hansen, M. Kringelbach, & K. R. Pugh (Eds.), *Mapping the word reading circuitry in skilled and disabled readers.* New York: Oxford University Press. 281–305.

Putnam, H. (1972). Psychology and reduction. *Cognition, 2,* 131–146.

Ramirez, F. O., Luo, X., Schofer, E., & Meyer, J. W. (2006). Student achievement and national economic growth. *American Journal of Education, 113,* 1–29.

Ravitch, D. (2011). *The death and life of the great American school system: How testing and choice are undermining education.* New York: Basic Books.

Ravitch, D. (2013). *Reign of error.* New York: Knopf.

Rayner, K., Foorman, B. R., Perfetti, C. A., Pesetsky, D., & Seidenberg, M. S. (2001). How psychological science informs the teaching of reading. *Psychological Science in the Public Interest, 2,* 31–74.

*Rayner, K., Foorman, B. R., Perfetti, C. A., Pesetsky, D., & Seidenberg, M. S. (2002). How should reading be taught? *Scientific American, 286,* 84–91.

Rayner, K., Schotter, E. R., Masson, M. E., Potter, M. C., & Treiman, R. (2016). So much to read, so little time: How do we read, and can speed reading help? *Psychological Science in the Public Interest, 17,* 4–34.

Richardson, E. B. (2003). *African American literacies.* New York: Psychology Press.

Richardson, U., Leppänen, P. H., Leiwo, M., & Lyytinen, H. (2003). Speech perception of infants with high familial risk for dyslexia differ at the age of 6 months. *Developmental Neuropsychology, 23,* 385–397.

Risko, V. J., Roller, C. M., Cummins, C., Bean, R. M., Block, C. C., et al. (2008). A critical analysis of research on reading teacher education. *Reading Research Quarterly, 43,* 252–288.

*Rogers, T. T. (2009). Connectionist models. In L. R. Squire (Ed.), *Encyclopedia of neuroscience,* vol. 3. Oxford: Academic Press.

*Romberg, A. R., & Saffran, J. R. (2010). Statistical learning and language acquisition. *Wiley Interdisciplinary Reviews: Cognitive Science, 1,* 906–914.

Rothstein, J., & Wozny, N. (2013). Permanent income and the black-white test score gap. *Journal of Human Resources, 48*, 510–544.

Rudolph, J. L. (2005). Epistemology for the masses: The origins of "the scientific method" in American schools. *History of Education Quarterly, 45*(3), 341–376.

Rueckl, J. G., Paz-Alonso, P. M., Molfese, P. J., Kuo, W. J., Bick, A., et al. (2015). Universal brain signature of proficient reading: Evidence from four contrasting languages. *Proceedings of the National Academy of Sciences, 112*, 15510–15515.

Rumelhart, D. E. (1977). Toward an interactive model of reading. In S. Domic (Ed.), *Attention and Performance VI*. Hillsdale, NJ: Erlbaum.

Rumelhart, D. E., & Siple, P. (1974). Process of recognizing tachistoscopically presented words. *Psychological Review, 81*, 99.

Saffran, E. M., & Marin, O. S. (1977). Reading without phonology: Evidence from aphasia. *Quarterly Journal of Experimental Psychology, 29*, 515–525.

*Sahlberg, P. (2011). The professional educator: Lessons from Finland. *American Educator, 35*, 34–38.

Salinger, T., Mueller, L., Song, M., Jin, Y., Zmach, C., et al. (2010). *Study of teacher preparation in early reading instruction*. NCEE 2010–4036. Washington, DC: National Center for Education Evaluation and Regional Assistance, Institute of Education Sciences, US Department of Education.

Sandak, R., Mencl, W. E., Frost, S. J., & Pugh, K. R. (2004). The neurobiological basis of skilled and impaired reading: Recent findings and new directions. *Scientific Studies of Reading, 8*, 273–292.

Savage-Rumbaugh, S., McDonald, K., Sevcik, R. A., Hopkins, W. D., & Rubert, E. (1986). Spontaneous symbol acquisition and communicative use by pygmy chimpanzees (*Pan paniscus*). *Journal of Experimental Psychology: General, 115*, 211–235.

Scarborough, H. S. (1990). Very early language deficits in dyslexic children. *Child Development, 61*, 1728–1734.

*Schmandt-Besserat, D. (1978). The earliest precursor of writing. *Scientific American, 238*, 50–59.

Schmandt-Besserat, D. (1986). The origins of writing: An archaeologist's perspective. *Written Communication, 3*, 31–45.

Schmandt-Besserat, D. (1992). *From counting to cuneiform*, vol. 1. Austin: University of Texas Press.

Schmeck, R. R. (1988). An introduction to strategies and styles of learning. In R. Schmeck (Ed.), *Learning strategies and learning styles*, 3–19. New York: Plenum Press.

Seghier, M. L., Neufeld, N. H., Zeidman, P., Leff, A. P., Mechelli, A., et al. (2012). Reading without the left ventral occipito-temporal cortex. *Neuropsychologia, 50*, 3621–3635.

*Seidenberg, M. S. (1997). Language acquisition and use: Learning and applying probabilistic constraints. *Science, 275*, 1599–1604.

*Seidenberg, M. S. (2005). Connectionist models of reading. *Current Directions in Psychological Science, 14*, 238–242.

Seidenberg, M. S. (2011). Reading in different writing systems: One architecture, multiple solutions. In P. McCardle, J. Ren & O. Tzeng (Eds.), *Dyslexia across languages: Orthography and the gene-brain-behavior link*. Baltimore: Paul Brooke Publishing.

Seidenberg, M. S. (2012). Computational models of reading: Connectionist and dual-route approaches. In M. Spivey, K. McRae & M. Joanisse (Eds.), *Cambridge handbook of psycholinguistics*. Cambridge: Cambridge University Press.

*Seidenberg, M. S. (2013). The science of reading and its educational implications. *Language Learning and Development, 9*, 331–360.

Seidenberg, M. S., & McClelland, J. L. (1989). A distributed, developmental model of visual word recognition and naming. *Psychological Review, 96*, 523–568.

Seidenberg, M. S., & MacDonald, M. C. (1999). A probabilistic constraints approach to language acquisition and processing. *Cognitive Science, 23*, 569–588.

Seidenberg, M. S., & Petitto, L. A. (1987). Communication, symbolic communication, and language. *Journal of Experimental Psychology: General, 116*, 279–287.

Seidenberg, M. S., & Plaut, D. C. (2014). Quasiregularity and its discontents: The legacy of the past tense debate. *Cognitive Science, 38*, 1190–1228.

Seidenberg, M. S., and Tanenhaus, M. K. (1979). Orthographic effects on rhyme monitoring. *Journal of Experimental Psychology: Human Learning and Memory, 5*, 546–554.

Shanahan, T., & Barr, R. (1995). Reading recovery: An independent evaluation of the effects of an early instructional intervention for at-risk learners. *Reading Research Quarterly, 30*(4), 958–996.

Shankweiler, D., & Liberman, I. Y. (Eds.). (1989). Phonology and reading disability: Solving the reading puzzle. International Academy for Research in Learning Disabilities Monograph Series 6. Ann Arbor: University of Michigan Press.

Siegel, J. (2010) *Second dialect acquisition*. Cambridge: Cambridge University Press.

Sleeter, C. E. (1996). *Multicultural education as social activism*. Albany: State University of New York Press.

Smith, F. (1971). *Understanding reading: A psycholinguistic analysis of reading and learning to read*. New York: Holt, Rinehart & Winston.

Smith, F. (1973). *Psycholinguistics and reading*. New York: Holt, Rinehart & Winston.

Smith, F. (1975). *Comprehension and learning*. New York: Holt, Rinehart and Winston.

Smith, F. (1979). *Reading without nonsense*. New York: Teachers College Press.

Smith, F. (1992). Learning to read: The never-ending debate. *Phi Delta Kappan, 74*, 432–441.

Smith, N. J., & Levy, R. (2013). The effect of word predictability on reading time is logarithmic. *Cognition, 128*(3), 302–319.

Snow, C. (1977). Mothers' speech research: From input to interaction. In C. E. Snow & C. A. Ferguson (Eds.), *Talking to children*, 31–49. Cambridge: Cambridge University Press.

Snow, C. E., Burns, M. S., & Griffin, P. (Eds.). (1998). *Preventing reading difficulties in young children*. Washington, DC: National Academies Press.

*Snowling, M. J. (2006). Language skills and learning to read: The dyslexia spectrum. In M. J. Snowling & M. E. Hayiou-Thomas (2006). The dyslexia spectrum: Continuities between reading, speech, and language impairments. *Topics in Language Disorders, 26*, 110–126.

Snowling M. & Stackhouse, J., Eds. (2006). *Dyslexia, speech and language: A practitioner's handbook*. Hoboken, NJ: John Wiley & Sons.

Snowling, M., & Frith, U. (1986). Comprehension in "hyperlexic" readers. *Journal of Experimental Child Psychology, 42*, 392–415.

Sokal, A. D. (1996). Transgressing the boundaries: Toward a transformative hermeneutics of quantum gravity. *Social Text, 46/47*, 217–252.

Sontag, S. (1978). *Illness as Metaphor*. New York: Farrar Straus & Giroux.

Spencer, L. H., & Hanley, J. R. (2003). Effects of orthographic transparency on reading and phoneme awareness in children learning to read in Wales. *British Journal of Psychology, 94*(1), 1–28.

Sperling, A. J., Lu, Z.-L., Manis, F. R., & Seidenberg, M. S. (2005). Deficits in perceptual noise exclusion in developmental dyslexia. *Nature Neuroscience, 8*, 862–863.

Sperling, A. J., Lu, Z.-L., Manis, F. R., & Seidenberg, M. S. (2006). Motion perception deficits in reading impairment: It's the noise not the motion. *Psychological Science, 17*, 1047–1053.

Sproat, R. W. (2000). *A computational theory of writing systems*. Cambridge: Cambridge University Press.

Stager, C. L., & Werker, J. F. (1997). Infants listen for more phonetic detail in speech perception than in word-learning tasks. *Nature, 388*, 381–382.

Stanovich, K. E. (1980). Toward an interactive-compensatory model of individual differences in the development of reading fluency. *Reading Research Quarterly, 16*, 32–71.

*Stanovich, K. E. (1994). Annotation: Does dyslexia exist? *Journal of Child Psychology and Psychiatry, 35*(4), 579–595.

Stanovich, K. E. (2005). The future of a mistake: Will discrepancy measurement continue to make the learning disabilities field a pseudoscience? *Learning Disability Quarterly, 28*, 103–106.

*Stanovich, P. J., & Stanovich, K. E. (2003). Using research and reason in education: How teachers can use scientifically based research to make curricular and instructional decisions. ERIC. http://files.eric.ed.gov/fulltext/ED482973.pdf.

Starr, M. S., & Rayner, K. (2001). Eye movements during reading: Some current controversies. *Trends in Cognitive Sciences, 5*, 156–163.

*Steiner, D., & Rozen, S. (2004). Preparing tomorrow's teachers: An analysis of syllabi from a sample of America's schools of education. In F. Hess, A. Rotherham, & K. Walsh (Eds.), *A qualified teacher in every classroom? Appraising old answers and new ideas*. Boston: Harvard Education Press.

Stinson, D. W., Bidwell, C. R., & Powell, G. C. (2012). Critical pedagogy and teaching mathematics for social justice. *International Journal of Critical Pedagogy, 4*, 76–94.

Stoeckel, C., Gough, P. M., Watkins, K. E., & Devlin, J. T. (2009). Supramarginal gyrus involvement in visual word recognition. *Cortex, 45*, 1091–1096.

Sunstein, C. (1999). The law of group polarization. John M. Olin Law & Economics Working Paper No. 91 (2nd series). University of Chicago Law School. http://goo.gl/fc5BMb.

Tanaka, H., Black, J. M., Hulme, C., Stanley, L. M., Kesler, S. R., Whitfield-Gabrieli, S., Reiss, A. L., et al. (2011). The brain basis of the phonological deficit in dyslexia is independent of IQ. *Psychological Science, 22*, 1442–1451.

*Taylor, I., & Taylor, M. (2014). *Writing and literacy in Chinese, Korean, and Japanese*. Rev. ed. Philadelphia: John Benjamins Publishing.

Taylor, J., Roehrig, A. D., Hensler, B. S., Connor, C. M., & Schatschneider, C. (2010). Teacher quality moderates the genetic effects on early reading. *Science, 328*, 512–514.

Terry, N. P., Connor, C. M., Petscher, Y., & Conlin, C. R. (2012). Dialect variation and reading: Is change in nonmainstream American English use related to reading achievement in first and second grades? *Journal of Speech, Language, and Hearing Research, 55*, 55–69.

Terry, N. P., Connor, C. M., Thomas-Tate, S., & Love, M. (2010). Examining relationships among dialect variation, literacy skills, and school context in first grade. *Journal of Speech, Language, and Hearing Research, 53*, 126–145.

Thomas, M. S., & McClelland, J. L. (2008). Connectionist models of cognition. In R. Sun (Ed.), *The Cambridge handbook of computational psychology*. Cambridge: Cambridge University Press.

Tobias, E., & Duffy, T. (Eds.). (2009). *Constructivist theory applied to instruction: Success or failure?* Mahwah, NJ: Lawrence Erlbaum.

Tuffiash, M., Roring, R. W., & Ericsson, K. A. (2007). Expert performance in Scrabble: Implications for the study of the structure and acquisition of complex skills. *Journal of Experimental Psychology: Applied, 13*, 124.

*Tunmer, W. E., Chapman, J. W., Greaney, K. T., Prochnow, J. E., & Arrow, A. W. (2013). Why the New Zealand National Literacy Strategy has failed and what can be done about it: Evidence from the Progress in International Reading Literacy Study (PIRLS) 2011 and Reading Recovery monitoring reports. *Australian Journal of Learning Difficulties, 18*, 139–180. http://goo.gl/x6db74.

Turkeltaub, P. E., Gareau, L., Flowers, D. L., Zeffiro, T. A., & Eden, G. F. (2003). Development of neural mechanisms for reading. *Nature Neuroscience, 6*, 767–773.

Tyson, C. A., & Park, S. C. (2006). From theory to practice: Teaching for social justice. *Social Studies and the Young Learner, 19*, 23.

*US Department of Education (DOE). (2010). Thirty-Five Years of Progress in Educating Children with Disabilities Through IDEA. DOE, http://www2.ed.gov/about/offices/list/osers/idea35/history/idea-35-history.pdf.

Van Orden, G. C. (1987). A rows is a rose: Spelling, sound, and reading. *Memory and Cognition, 15*, 181–198.

Van Orden, G. C., Johnston, J. C., & Hale, B. L. (1988). Word identification in reading proceeds from spelling to sound to meaning. *Journal of Experimental Psychology: Learning, Memory, and Cognition, 14*, 371.

Walker, C. B. F. (1990). Cuneiform. In J. T. Hooker (Ed.), *Reading the past: Ancient writing from cuneiform to the alphabet*. Berkeley: University of California Press.

Walker, D., Greenwood, C., Hart, B., & Carta, J. (1994). Prediction of school outcomes based on early language production and socioeconomic factors. *Child Development, 65*(2), 606–621.

*Walsh, K. (2013). 21st-century teacher education. *Education Next, 13*, 18–24.

*Walsh, K., Glaser, D., & Wilcox, D. D. (2006). What education schools aren't teaching about reading and what elementary teachers aren't learning. National Council on Teacher Quality. http://goo.gl/6cD54v.

Washington, J. A., Patton Terry, N., & Seidenberg, M. S. (2013). Language variation and literacy learning: The case of African American English. In C. A. Stone, E. R. Silliman, G. P. Wallach, & B. J. Ehren (Eds.), *Handbook of language and literacy: Development and disorders*. 2nd ed. New York: Guilford Press.

Weaver, C. (1988). *Reading process and practice: From socio-psycholinguistics to whole language*. Portsmouth, NH: Heinemann Educational Books.

Weizman, Z. O., & Snow, C. E. (2001). Lexical output as related to children's vocabulary acquisition: Effects of sophisticated exposure and support for meaning. *Developmental Psychology, 37*(2), 265.

Werker, J. F., & Tees, R. C. (1999). Influences on infant speech processing: Toward a new synthesis. *Annual Review of Psychology, 50*, 509–535.

Wilkinson, C. W. (1980). *Communicating through letters and reports*. 7th ed. Homewood, IL: R. D. Irwin.

Willcutt, E. G., Betjemann, R. S., McGrath, L. M., Chhabildas, N. A., Olson, R. K., et al. (2010a). Etiology and neuropsychology of comorbidity between RD and ADHD: The case for multiple-deficit models. *Cortex, 46*, 1345–1361.

Willcutt, E. G., Pennington, B. F., Duncan, L., Smith, S. D., Keenan, J. M., et al. (2010b). Understanding the complex etiologies of developmental disorders: Behavioral and molecular genetic approaches. *Journal of Developmental and Behavioral Pediatrics, 31*, 533–544.

*Willingham, D. T. (2006). How knowledge helps: It speeds and strengthens reading comprehension, learning—and thinking. *American Educator, 30*, 30–37.

*Willingham, D. T. (2015). *Raising kids who read: What parents and teachers can do*. San Francisco: Jossey-Bass.

Wilson, S. (Ed.). (2009). Teacher quality. Education policy white paper. ERIC. http://eric.ed.gov/?id=ED531145.

Wilson, W. J. (2011). To be poor, black, and American. *American Educator, 35*, 10–23.

*Wolf, M. (2007). *Proust and the squid: The story and science of the reading brain*. New York: Harper.

Wolfram, W. (1974). The relationship of white southern speech to vernacular black English. *Language, 50*(3), 498–527.

Wolfram, W. (2004). The grammar of urban African American vernacular English. In B. Kortmann & E. Schneider, *Handbook of varieties of English*. Berlin: Mouton.

Wood, E., & Barrows, M. (1958). *Reading skills*. New York: Holt, Rinehart & Winston.

Woolfolk, A. (2007). *Educational psychology*. 10th ed. Boston: Allyn & Bacon.

Wray, D. (2004). *Literacy: Major themes in education*, vol. 2. New York: Taylor & Francis.

Yakubovich, I. (2010). Anatolian hieroglyphic writing. In C. Woods, G. Emberling & E. Teeter (Eds.), *Visible language: Inventions of writing in the ancient Middle East and beyond*, 205. Oriental Institute Museum Publications 32. Chicago: Oriental Institute.

Yang, J., McCandliss, B. D., Shu, H., & Zevin, J. D. (2009). Simulating language-specific and language-general effects in a statistical learning model of Chinese reading. *Journal of Memory and Language, 61*, 238–257.

Yeatman, J. D., Dougherty, R. F., Ben-Shachar, M., & Wandell, B. A. (2012). Development of white matter and reading skills. *Proceedings of the National Academy of Sciences, 109*, E3045–E3053.

Zhang, M. (2013). Contrasting automated and human scoring of essays. *ETS R&D Connections, 21*, 1–11.

Zhu, X. (2005). Semi-supervised learning literature survey. Computer Sciences TR 1530, University of Wisconsin–Madison, http://pages.cs.wisc.edu/~jerryzhu/pub/ssl_survey.pdf.

Ziegler, J. C., & Ferrand, L. (1998). Orthography shapes the perception of speech: The consistency effect in auditory word recognition, 5, 683–689.

Ziegler, J. C., & Goswami, U. (2005). Reading acquisition, developmental dyslexia, and skilled reading across languages: A psycholinguistic grain size theory. *Psychological Bulletin, 131*, 3–29.

INDEX

Mark Seidenberg is the Vilas Research Professor and Donald O. Hebb Professor in the department of psychology at the University of Wisconsin-Madison and a cognitive neuroscientist who has studied language, reading, and dyslexia for over three decades. He lives in Madison, Wisconsin.

Photograph by Ethan M. Seidenberg